AF778432

Scannerstuff

NORTHWEST

FREQUENCY DIRECTORY

10th Edition OREGON/WASHINGTON

The Complete Communications Resource for the Pacific Northwest

Dan Rollman, Editor

Scannerstuff has been publishing detailed, comprehensive, locally edited scanner frequency and communications guides for nearly 15 years. Please contact us for additional information on our products.

www.scannerstuff.com info@scannerstuff.com

Cover Photos: Communications towers atop South Mountain in Phoenix, Arizona. Los Angeles County Fire Department "Copter 16", a Bell 205A1, photographed at "Fire Camp 9" while being refilled with water during a brush fire. San Gabriel Fire Department "Engine 51" is a 2002 Pierce Chromo Quantum pumper equipped with a 2000 GPM pump, 500 gallon water tank, and 30 gallon foam tank. Santee Fire Department "Brush 4" is a Type 3 Model 15 engine built by Master Body Works of South Gate during 1997 on a 1996 International 4900 series chassis with a 500 GPM pump, 500 gallon water tank and 20 gallon foam tank. San Bernardino City Fire Department operates the 2000 Emergency-One Titan HPR 4-wheel drive crash truck at San Bernardino International Airport, with a 1500 GPM pump, 1500 gallon water tank, 205 gallons of AFFF foam and a 50 foot Snozzle. Originally delivered and lettered as "Santa Fe 220", Burlington Northern Santa Fe SD75M locomotive 8220 was built at General Motors Electro-Motive Division plant in London, Ontario, Canada during April 1995 and is rated at 4500 horsepower. Cover photos by Steven Isoldi and Dave Greenberg. Photos used with permission. Cover design by Steven Isoldi, Rev5Design.com, and Laura Rollman.

How To Use This Book

This directory will provide you with virtually all the frequencies and information you will need to use a scanner or communications receiver in Oregon and Washington. The following is a brief description of how to read the frequency information contained in this book. In order to save space (pages) and keep the cost of this book down, a "key" to the title of each column is not printed on the top of each page.

The Format for This Directory is as Follows:

[Agency Name]

[Ch#]	[Description of Use]	[PL Tone]	[City]	[Receive]	[TX/System]

Here is an Example:

KOIN Television (Channel 6)

Ch2	Eng. / Traffic Watch	[PL 146.2]	Portland	450.1500	Simplex

Tri County Metropolitan Transportation District (Tri-Met)

A1	Rail Main 1 (East side trains)		Portland	TG: 28208	Portland

Here's How to Interpret It:

1. "KOIN Television (Channel 6)" or "Tri County Metropolitan Transportation District". This is the agency, company, or organization name. Each of the *entries* below (any of the following *indented* lines) falls under this agency.
2. "Ch2" or "A1". The channel number, name or display label in this agency's radios (KOIN Television Channel 2, or A1 in Tri-Met radios).
3. "Eng. / Traffic Watch" or "Rail Main 1 (East side trains)". This area shows a description of the use of this channel, frequency, or talkgroup.
4. "PL 146.2". This is the PL (CTCSS) Tone (if known). See your scanner's instruction manual and specifications for information on whether your radio can use CTCSS tones to reduce interference. Digital PL tones are preceded by "DPL" in this column. **[Dig-P25]** indicates an APCO P25 digital frequency or talkgroup. **[Dig-Enc]** indicates an encrypted frequency or talkgroup that cannot be monitored. **[Analog]** confirms that a particular talkgroup is generally analog.
5. "Portland". This is the city column. This may be either the location of the transmitter of the given frequency, or the base city of the agency or business.
6. "450.1500" This column shows the frequency for this channel. This is the *output* or *receive* frequency. In virtually all circumstances, this is the frequency you will want to listen to. "TG: 28208". The numbers following the abbreviation "TG:" in the frequency column indicate the talkgroup number of a Trunked Radio System. See your TrunkTracker's instruction manual for information on how to program talkgroups.
7. "Simplex". This column shows whether the channel is a simplex or repeated frequency. In some cases, this last column may be blank, indicating either both repeater and simplex operation is often conducted on the frequency, or it is unknown. "Portland". This indicates what trunked radio system the trunked talkgroup is active on. In this case, "Portland" indicates that talkgroup 28208 is active on the City of Portland 800 MHz Trunked Radio System. In some cases, "AM" may be shown in this column to remind listeners that an aircraft band frequency uses "AM" mode.
8. Gray shading indicates new or updated information as compared to the 9th Edition.

CREDITS

Producing the Northwest Frequency Directory takes many people. The people below are just a few of those that have helped produce this and other Scannerstuff books, and who continue to help update our books for the next edition. As the editor, my sincere thanks goes out to each of them. ---Dan Rollman, NWFD Editor

CONTRIBUTORS

Special thanks to those who have been left off this list or requested anonymity

American Fire Service Book
Zander Albertson
Duane Anderson
Don Armstrong
Jeffrey Atwood
A-S NF Crews
Jim Bateman
Brett Bocian
Dave Bohmke
Tom Braeunig
Douglas Brown
Ben Bruegeman
Scott Burke
C. Cheshire
Craig Christensen
Chris Clarke
Tom Colvin
B. Coningham
Cliff Corcoran
D. Coughanour
Wayne Cummings
Richard Curdy
R. Donahoo
Holly Eckel
Brian Eisenhauer
Coconino NF FFs
Coronado NF FFs
Courtney Ellis
Brent Forsyth
John Gesell
Brian Goldtooth
Dave Greenberg
Matt Greenwell
Lane Griswold
John Guetherman
Chris Hale
Michael Hamman
Jim N7MAQ
Tyrone Hardin
Mike Hitt
Robert Homuth
Ron Hutt
ICP Dispatchers
"ICP245"
Steven Isoldi
Jesus Christ
Don Johns
Kaibab NF Crews
I. Kapalczynski
Tom Kelley
F. Kuest
Mike Lanning
Carroll Lantz
Darryl Madison
Carl Maness
Michael Martin
G Martinson
Mike May
Gene McAvoy
Ron McHenry
Danny McLean
Larry Mohler
Gary Moncur
J Moran
J. Robert Moss
F. Barry Mulligan
Jeff Newton
John Nolan
NPS Flag Mon's.
Jason Patterson
C Parris
Brian Potter
PrescottNF E32
David Rice
Barbara Richter
Fred Richter
Mike Rollman
Benjamin Russell
Ron – N8HBF/7
Larney Sager
Dan Seymour
Roy Schahrer
Scott Scheffey
Charles Simmont
Andy Smith
Bob Smith
Jason Smith
Chris Stubblefield
Lyle Tanner
Mike Todd
Tonto NF Crews
Glen Tubbesing
Jon Van Allen
Brian Varine
Duane Vincent
Robert W.
Achim Warth
T Weaver
Brian Whitaker
Jerry Wightman
Kevin Wiley
Dave E. Williams
Everret Wittig
Keith Wyatt
Steven Zielke

CONTENTS

SCANNER USER'S GUIDE

This guide attempts to use very basic terms to answer the most common questions people have about the scanning hobby and scanners themselves. Other more comprehensive books and resources are available from bookstores, electronics stores, and radio shops, as well as on the Internet. It is suggested that you look for one or more of these books if you intend to learn a lot about scanners and/or radios.

- ## What is a Frequency?

A frequency is the term given to the number of cycles a radio wave makes in one second. One cycle (better known as Hertz) means the wave makes one complete cycle in one second. Megahertz (MHz) is one million Hertz. One MHz means that the radio waves makes one million complete cycle in one second. Frequencies in the 800 MHz band would make over 800 million cycles in one second, and so on. Radios, and scanners, are designed to detect the number of cycles made in one second of the signals it receives. In very simple terms, it is able to pick out the signal (or lack thereof) that the user has instructed it to by entering a frequency into memory.

- ## What Types of Stations and Radio Systems Exist?

The most common two types (although there are many more) of radio systems consist of simplex and duplex. A duplex system is most commonly a repeated (or "repeaterized") system. Simplex means that each radio receives and transmits on the same frequency, and that the signal goes directly from the transmitting radio to the receiving radio (provided they are within range of each other). A repeater system means there is an additional "radio" capable of transmitting and receiving at the same time located at a given site (mountain tops generally provide best placement for most repeaters, although many are on towers at ground level or even mobile in a vehicle). Radios in the field are programmed to transmit and receive on two different frequencies. The frequency the radio transmits on is called the "input" for a repeated system, while the frequency the radio monitors is called the "output". When the radio transmits on the input frequency. the repeater picks up the transmission and repeats it on the output frequency. All the other radios in the field hear the transmission from the repeater on the frequency their radios are set to receive, the output frequency. Repeaters allow systems to cover much larger areas, because the range of a repeater is generally much larger than that of a standard radio. Most scanner listeners generally want to monitor the output frequency (the first frequency column or set of numbers in the main portions of this book). See the illustration of two radios using a repeater below.

The most common types of radios or stations include a <u>handheld</u> radio, a mobile radio, and a base radio. A <u>handheld</u> radio is a small somewhat compact radio that can be carried around by an individual. A <u>mobile</u> radio is installed in vehicles and uses an antenna mounted somewhere on top of the vehicle. A <u>base</u> radio is at a fixed location on a desk or in a console. The base radio is often much more powerful than the mobile and portable radio, and uses an outdoor antenna on the building or on a tower. The mobile radio is more powerful than the portable radio, allowing it to transmit farther than the portable radio.

Radios Transmit on: F1
Radios Receive on: F2
Repeater Transmits on: F2
Repeater Receives on: F1

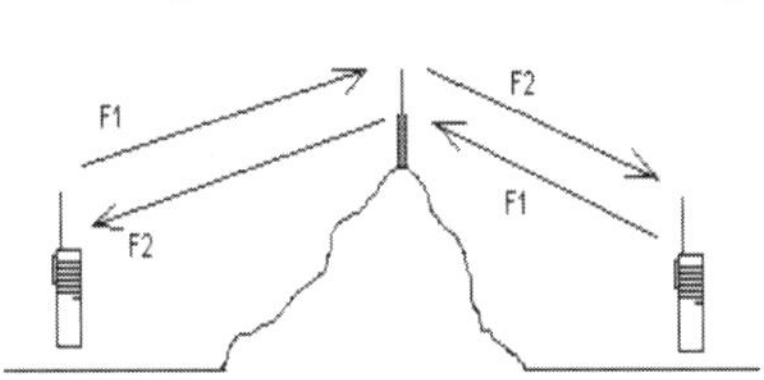

• What is a Trunked Radio System

A Trunked Radio System (TRS) is perhaps the single most complicated type of system a scanner listener often monitors. Trunked radio systems use a number of repeaters tied to a computer, and allow conversations to skip from frequency to frequency without the radio users changing anything. To do this, each radio monitors a "control channel" which, on a scanner, sounds like a buzz or chain saw. The data transmission on this channel tells the system's radios what frequency to switch to next. At various times (after each transmission with an EDACS system, and generally after a certain period of inactivity on a Motorola system), the computer tells all radios on the given channel (or talkgroup) to move to another frequency, using a data transmission on this control channel. TrunkTracker scanners allow reception of the same data instructions that the system's radios receive. This allows the scanner to automatically change frequencies to follow a conversation.

• How Should I Set Up My Scanner's Banks?

In a scanner, banks are simply groupings of channels. Banks allow users to scan only a select set of channels at one time without having to manually lock out unwanted channels located in other banks. How you set up your banks will depend on personal preference and how you use your scanner. But to help, you may want to consider two options. One option is to separate "services" or "categories" by programming, for example, all fire services in one bank, all law enforcement services in another bank, etc. Another option is to dedicate a bank to a particular geographic area, programming fire, police, medical, and other services located in one town or part of town in a bank, and the corresponding services for another town or part of town in a different bank. Remember that each Trunked Radio System you plan to monitor generally must occupy its own bank, although many scanners allow you to also program non-trunked frequencies into a bank containing a single trunked radio system. TrunkTracker scanners have a subset of trunked banks, called "Scan Lists," that permit you to separate categories of trunked talkgroups the same way you might otherwise separate categories of non-trunked frequencies.

• How Can I Improve My Scanners Sound & Reception?

The easiest way to improve reception range and clarity is with an outdoor antenna. Your antenna does not have to be on a large tower or even all that high up in the air (although obviously a tower would work well). A simple rooftop antenna set-up the same way many people set-up television antennas on the roof or attached to the side of the building or around a chimney will work much better than an indoor or handheld antenna. Antennas and mounting kits can be found at most radio and electronics stores, but be sure that the antenna you plan to buy covers the bands (our frequency ranges) you plan to use it for.

• What are the Strange Sounds I Heard? How Do I Get Rid of Them?

Most of these sounds can be attributed to paging systems "bleeding over" on other frequencies or powerful nearby transmitters "overriding" your scanner regardless of whether it's tuned to that frequency. The simplest way to get rid of some of these sounds (if it works) is to adjust the squelch control on the radio. If this does not work, the other option is to get a scanner that has a CTCSS (or PL) decoding option. PL tones are sub-audible tones in most radio transmissions that your ear cannot hear but your radio can. Scanners that have this option will only allow you to hear transmission with the programmed tone, even if other transmissions are active on that frequency. This book has many of these tones listed along with the frequencies for particular entities, but unfortunately they are very hard to find and come by. Read your owner's manual to see if your scanner has this option and for instructions to program these tones into your radio.

MOTOROLA TRUNKING SYSTEMS

Written by: Wiz, City of Seattle, WA
Network Manager, King County Regional Trunked Radio System
Revised: 15 June 2001 Used with Permission

Some History

Two-way radio, or Police radio, are very broad terms that encompass a wide range of different technologies. Lets look at some of the terms you have heard used both correctly and incorrectly. As a subject becomes more technical, it becomes very important that we all speak the same "language."

SIMPLEX - (Portable to Portable Operation):

Simplex is the simplest form of 2-way radio communication. It is when two or more radios transmit and receive on the same frequency. First one radio transmits and the rest of the radios listen. When the transmission is completed, that radio listens on the same frequency for someone else to transmit. Radios transmit directly to each other. The range is limited, because of the low wattage in a portable radio (up to 5 watts), and because these transmissions are line of sight (usually less than 1 mile).

To overcome some of this range limitation, base stations were placed at high locations and wire controlled from a "dispatch" point. This allowed the dispatcher to talk to, and hear from, field units further away but did nothing to improve range between field units. The location of the radio transmitter is called the "radio site" or "site" and is usually given some geographical name that is meaningful to the locals. Because the frequency is normally shared with others, users must wait for a "clear channel" before transmitting much as the old party line telephones (If you are old enough to remember those!).

Scanning Issues: Program in one frequency to listen to the radio users. Can normally hear the base transmissions from a long distance but depending on distance, terrain, and frequency will have very mixed results when the field units transmit. Conversations are easy to follow.

DUPLEX – (Repeater Operation):

To overcome the short range of simplex, we started putting "repeaters" on hills, tall buildings and other high places. Because of the time it would take to receive, record, and then rebroadcast a signal on the same frequency, we went to a two-frequency concept. The field units transmit on one frequency, which the repeater receives. The repeater then instantaneously re-transmits the voice signal on a second frequency, which is received by the field units. Because it is using two frequencies (or a frequency pair), this is called "duplex" operation. The field units are said to be "half duplex" since they use two frequencies, one for transmit and one for receive, but cannot operate on both at the same time. The repeater is said to be "full duplex" since it transmits and receives at the same time. Range for duplex (or repeater) operation is generally 5-10 miles, while operating on 75-100 watts. This range can vary greatly depending on frequency, local terrain, and height of the repeater antenna.

Scanning Issues: Normally program in the repeater output frequency to listen to the radio users. Can normally hear both base and field transmissions from a long distance. If field units go to "talk-around" (to bypass the repeater) they will use the repeater output frequency and scanners will have very mixed results when field units transmit, as if units are simplex (because they are). If field units are in the talk-around mode, the dispatcher will normally not be able to hear them, but a scanner may still pick them up.

SIMULCAST:

To further increase range, a technology called "simulcast" was developed. With this method, two or more transmitters (normally repeaters) on the same frequency are located at multiple high locations (called remote sites). The inputs of each repeater transmitter are tied together so they all transmit the same audio signal at the same time. In this way no matter which repeater the field unit is within range of, it will hear all transmissions. If a field unit (radio) is close to one repeater, it's receiver will be "captured" by the strong signal and the received audio will be clear. If, however, the field unit is approximately equidistant from two or more repeaters, it will receive multiple signals. If these multiple signals are not exact in frequency, timing, "volume", and audio frequency response, the received audio can be garbled to the point it is unintelligible. This equidistant geographical area is called the "overlap" area. For this reason we have to use some rather expensive technical methods to control the transmitter frequency, audio characteristics, and delay of the audio so all transmitters transmit the exact same signal at exactly the same time.

On the repeater receive (or "talk-in") side, we normally feed all the receivers (normally one at each transmitter site and sometimes additional receivers at non-transmitter sites) into a device called a "voter" or "comparator." This unit electronically checks (or compares) each receiver signal for its signal to noise ratio and then selects (or votes) the best signal to be sent to the dispatcher and to be re-broadcast. The location of the simulcast controller (computer) and other equipment (including the voters) that makes simulcast possible is called the "prime site" and may, or may not, be located at one of the remote site locations. Simulcast can be either single frequency or trunked (see next section).

Scanning Issues: Same as for duplex operation. Since coverage area is normally larger than for a single repeater, scanners will also be able to monitor over a larger area. If one of the simulcast transmitters goes out of service, coverage for its area will be reduced. If you are in that area, you may notice a weaker signal or you may stop receiving that system all together. Since the FCC allows only six sites on a license, you may find multiple licenses covering the simulcast system. Note that it is not possible (except under some very special circumstances) for a receiver user to tell which simulcast transmitter site (or combination of sites) is providing the signal being received. Sometimes this may be the farthest site, sometimes the closest, but many times will be a combination of two or more of the remote sites. Thus it is inaccurate and confusing to refer to a received simulcast signal by one remote site name instead of the name for the simulcast system or subsystem.

TRUNKED REPEATER OPERATION:

In the older radio systems, one or more groups of users were assigned to a particular pair of frequencies. If no one in any of those groups were talking, the frequency (or frequency pair) stood idle. At the same time, the groups of users on a different assigned frequency pair may be so busy that each of the groups is constantly waiting for someone to get through talking so they can get a message through. Thus there was a lot of available airtime wasted.

As the number of user groups increased in large metropolitan areas, all available frequencies were assigned to existing user groups (King County being a prime example). As new groups (such as new police departments or additional precincts in an existing police department) were created, no frequencies were available without sharing with an existing user.

A solution, known as "trunking", was borrowed from the telephone industry. With this technology, several frequency pairs are grouped together and put under the control of a computer (called a Trunking Controller). This computer then accepts each "request" to talk, assigns the next available "frequency pair", directs all the radios in the "talk group" to switch to the assigned frequency pair, and sends a "permit to talk" tone to the requesting party. The whole process takes less than half a second. The computer also monitors the activity (not the conversation, but the fact the frequency pair is in use). When the frequency pair becomes idle, it is returned to the "pool" and is available for re-assignment to another user group. By pooling several frequencies, we also "pool" the unused airtime thus allowing more user-groups to work independently. In this manner we can get 3 to 10 times more use from a given set of frequencies than the old non-trunking method of assigning frequencies to users. A trunked repeater system can be either single site or simulcast.

A simple analogy to trunking is the system used to serve in-person customers at a bank. Formerly, customers got in line at individual teller windows to wait for the teller at that window to assist them. All of us have experienced getting in what appears to be the shortest line, when in fact, the transactions in front of us end up taking longer than anticipated, resulting in a longer wait for service than other lines might have offered. Also, you may have experience waiting in an assigned line for your type of transaction, even if another "window" has no waiting customers. Now, many banks have customers form a single queue, and the next available teller waits on the customer at the front of the line.

Rather than being stuck with a specific "teller line" (frequency), the trunking system will accept a request to talk and assign whatever frequency is available to the next request in line. When all frequencies are assigned, the system will put all requests in a waiting queue, and is able to prioritize "customers" in the "line." The next available channel will be assigned to requests with the highest priority. Requests with equal priority will be served on a first-come, first-served basis. Typically, no users should experience any wait except in the event of a major disaster when radio traffic would be unusually high.

Thus, rather than scanning "frequencies," you will be scanning "talk groups."

Scanning Issues: The range that a scanner will work is influenced by the same factors as repeater and simulcast operation, depending on configuration. Frequency, however, is a very different issue. Since a computer is used to assign the next frequency, sometimes as often as every time a field unit keys up, the scanner needs to be able to follow these assignments just as the field units or it will not follow the conversation. To do this, a special scanner that can either be controlled by an external computer or has the internal circuitry to follow the frequency assignments is required. Most of the older trunking scanners also need to "know" all the frequencies in use by the trunking system. Thus, to properly scan, all frequencies used in a trunked system must be programmed into this type of scanner. If you are using one of the computer software products such as Trunker or one of the newest trunking scanners, this is not a requirement, but you must still program the control channel for the system you wish to scan. Since the FCC allows only six sites on a license and the system may use frequencies from more than one "band," you may find multiple licenses covering the trunked or trunked simulcast system. You will need to determine which frequencies on these multiple licenses (probably all the frequencies on all the licenses) are in use for the trunked system you wish to monitor.

When the King County system was first installed (Seattle was the first to implement), Seattle recognized this as an issue and made arrangements with the vendor to sell radios to those wishing to scan at the discounted price Seattle paid. This was an expensive way to scan, but there were no alternatives on the market. Although this offer is still available, Seattle will not program "any" requested talkgroup, will only program Seattle talkgroups (other sub-regions would program their talkgroups), and the radios are several times more expensive than the trunking scanners now on the market.

SmartNet Operation

SmartNet is the Motorola implementation of trunking and is their registered trademark. In a SmartNet trunked repeater system, generally 5 to 28 frequency pairs are grouped together. One of these frequency pairs is called the "control channel" and is used to convey information between the field units and the trunking system. When a mobile or portable unit (also called field, or subscriber, unit) is not listening to a talkgroup conversation, it is monitoring (listening to) the control channel. A SmartNet system can be either single site or simulcast. Note, also, that there is no "requirement" to have the frequencies in any particular order. They can be set up in ascending, descending, or some random sequence and the system will work just as well. It then follows that the frequency order in the scanner does not have a bearing on the scanner operation.

Each subscriber unit is assigned a unique ID number. Each talkgroup is also assigned a unique ID number and turning the "channel" selector knob now selects a "talkgroup" ID number instead of a frequency. The system uses these numbers to track and authorize subscriber use of the system. This tracking starts when the subscriber unit is first turned on and then each time the operator changes the talkgroup.

The SmartNet trunked system is controlled by a Trunked System Controller that is located (in the case of simulcast) at what we call the Prime Site along with any simulcast control equipment. The transmitters are located at what we call Remote Sites (if simulcast) along with the Remote Site Controller. Note that a SmartNet system can be one site only (not simulcast) with everything co-located.

With the introduction of SmartNet, Motorola also introduced a failure mode called "Failsoft." If the trunking system loses its control channel or has certain other failures, it is no longer able to operate in the trunking mode. So instead of going into a condition that stops all communication, the system enters Failsoft. In this state all transmitters (channels) turn on and operate in a "conventional" repeater mode. The subscriber radios are able to recognize this state and switch to a predetermined frequency (one of the trunk system frequencies, but not the control channel frequency) depending on their selected talkgroup. In most systems several talkgroups will share a frequency. Some talkgroups may not be assigned a failsoft frequency and these talkgroups will cease to operate during the failsoft period. If a particular failsoft frequency has also failed, the talkgroups assigned to that frequency will also be off the air during failsoft.

Scanning Issues: Same as Trunked Repeater operation for single site or simulcast, depending on configuration. The location of the prime site has no bearing on the operation of the scanner. The King County system uses Type II trunking. During Failsoft, all transmitters turn on, whether or not there is an on-going conversation. This means a scanner will stop on each frequency and stay there until intervention by the scanner operator.

SmartZone Operation

As much as trunking solved traffic handling ability and simulcast improved wide-area operation, some limitations prevented this from being the panacea we hoped for. Trunking has a limited number of frequencies it can handle (limiting the amount of traffic it can carry). Simulcast has a limited number of sites it can handle (limiting the amount of signal delivered to the coverage area- with larger areas getting less signal for a given number of sites). Simulcast requires a transmitter at every site for each frequency whether or not that site is required for a particular call (increases the system cost), and requires exclusive frequency usage in a very large area (limiting the ability to reuse frequencies- an FCC issue). So we once again turned to the computer to solve these problems and increase the geographical coverage.

Motorola responded to our needs with a system called "SmartZone" (also a Motorola registered trademark) which takes two or more "Smartnet trunking" systems (any combination of simulcast and non-simulcast) and ties them together with a master computer called the Zone Controller. Smart Zone allows us to combine the good points of both single site and simulcast trunking to achieve a better overall system. The job of the Zone Controller is to keep track of every radio in use, what site it is using (from the Zone Controller perspective, each entire simulcast "sub-system" (no matter how many remote sites it has) is called a site and treated as one site), and what talk group that radio is switched to. When a radio is turned on it affiliates with the system, reporting its ID, site it is on, and talkgroup selected. When a user keys a radio, or changes the channel selector knob, the radio talks to the site it is affiliated with, which in turn passes the information (desire to talk or change of talk group) to the Zone Controller. The system thus knows which sites (remember an entire simulcast sub-system is just another site as far as the Zone Controller is concerned) require the audio for a particular talkgroup. In the case of PTT (Press To Talk or keying the radio) activation, the Zone Controller then tells all affected site controllers to make a channel available, and ties the audio of these sites together to complete the call. The individual site controllers then (via the control channel) tell all of the subscriber units with the talkgroup selected which frequency to listen to. All this happens in less than half a second.

Note that, in most cases, if no radio affiliated to a particular site has the talkgroup being talked on selected, the talkgroup audio will not be sent to that site, and anyone scanning that site will not hear the conversation.

Each SmartZone "site" (whether single site or simulcast sub-system) can be programmed for Failsoft operation. Those sites that are programmed for failsoft operation can enter the failsoft mode even while the rest of the SmartZone system is operating normally in the wide-area, trunked, mode. Conversations on a site in failsoft will not be carried on other SmartZone sites.

Scanning Issues: With today's scanners, you can only scan the talkgroups of one SmartZone site at a time. Remember that a simulcast subsystem, no matter how many sites or how may licenses (call signs) is one SmartZone site. This is because all of the remote sites have the same frequencies and any conversation is simultaneously transmitted (simulcast) from all sites. The King County Regional Trunked Radio System uses Type II trunking. Range for each Smart Zone site is same as Trunked Repeater and Simulcast operation, depending on the configuration of the Smart Zone site you are listening to. In addition, if no radio affiliated to the site you are scanning has the talkgroup you wish to listen to selected, you will not hear activity on that talkgroup. Note that programming the frequencies of more than one site into your scanner will not allow you to scan multiple sites simultaneously. This is because the scanner selects only one control channel to listen to and that control channel does not convey any information about frequency assignments for other sites.

In fact, if you program too many frequencies and one or more of the "extra" frequencies is the control channel of another site, the trunking scanner may actually use the wrong control channel. If you have not programmed in all the frequencies for that second site (a very likely event from what I have seen with posted frequency lists) you will probably miss parts or all of many conversations. Again, this is not an issue with computer software products such as Trunker or one of the newest trunking scanners that only require knowledge of the control channel.

Some scanners may claim to be able to scan multiple trunking sites or conventional frequencies as well as trunking systems simultaneously. They do this by sequentially moving from one system or site to the next. While scanning a particular site or frequency, you may miss something on the others. This is because the scanner has only one receiver and in order to know anything about the talkgroup activity on a trunked system, you must listen to the control channel.

So, this begs the question "what is the 'best' site to scan for the talkgroups I want to hear?" The best site to scan for a particular talkgroup is a site that will have at least one subscriber unit affiliated with your desired talkgroup selected most, or all, of the time. And here is the short answer for the King County Regional Trunked Radio System: to scan Seattle talkgroups, program in the Seattle simulcast subsystem; to scan EPSCA talkgroups, program in the EPSCA simulcast subsystem; and to scan King County or Valley Com talkgroups, program in the KC/VC simulcast subsystem. If you can't get a good signal from the "best" simulcast subsystem, you can try another, but be aware that you may not hear any or all of the talkgroup conversations that you are interested in scanning! The same holds true if you want to scan talkgroups that have different "best" sites. Whether or not you can hear a particular talkgroup on a selected SmartZone site fully depends on if an authorized subscriber unit, with that talkgroup selected, is affiliated with the site you are scanning.

OmniLink Operation

OmniLink is a method of tying two or more Smart Zone systems together. OmniLink is also a Motorola registered trademark.

Scanning Issues: Same issues as for Smart Zone.

Other Reading

This paper was meant to give a basic understanding of Motorola trunked two-way radio operation so you might understand some of the weird things that seem to happen as you try to scan. As you become more familiar with your equipment and more comfortable in your understanding of the basic operation of Smart Zone, you may want to read further. Some topics you will probably want to learn more about are:
The format of talkgroup IDs and how to convert from one format to another; Status bits and how they may affect your scanner; The different software scanning products available; Antennas and how they affect your scanner performance; Other manufacturers trunking formats. There are several good papers covering these topics at various technical levels on www.northwestradio.com as well as other scanning related sites.

Glossary: The following are some common terms used in two-way radio and trunking.

Antenna - The device (normally similar to a wire) used to convert the electromagnetic energy in the air to an electrical signal that a receiver can process. The dimensions of an antenna are mathematically related to the frequency it will work best for. Normally, higher frequencies will use smaller antennas.

Base Station - A fixed, or stationary, transmitter-receiver.

Channel - The frequency or talkgroup a radio is operating on.

Comparator - An electronic device to automatically the best of two or more signals. Also called a voter.

Control Channel - One frequency in a trunking system that is used to convey system information such as frequency assigned for a talkgroup call and what talkgroup a radio has selected.

Controller - A computer that is used to control a simulcast or trunking system, or a remote simulcast site.

Dispatch - The act of sending instructions to field personnel in day-to-day operations. Also an abbreviation for Dispatch Center; the place where Dispatchers work.

Duplex - A method of transmitting and receiving via two different “channels” or paths. Full duplex means transmit and receive can occur simultaneously and half duplex means one function needs to wait for the other to be completed.

Failsoft - A trunking system failure mode. Under certain conditions, the trunking system may not be able to continue trunking. Each frequency then operates as a repeater with two or more talkgroups sharing the frequency.

FCC - Federal Communications Commission

Field Unit - A mobile or portable radio. Also called subscriber unit.

Key - To cause a radio to transmit

Mobile - A two-way radio that is designed for installation in a vehicle.

Modulation - The method of putting intelligence on a radio signal.

OmniLink - A wide area trunked radio system consisting of two or more Smart Zone systems.

Portable - A two-way radio that is designed to be carried by a person.

Prime Site - The grouping of electronics that controls a simulcast system. The Zone Controller and its associated equipment can also be called a Prime Site.

PTT - Press To Talk button on a portable or microphone used to cause the transmitter to transmit.

Remote Site - One of two or more simulcast transmitter sites.

Repeater - A radio receiver/transmitter that receives on one frequency and rebroadcasts the signal on a different frequency. Repeaters normally operate in full duplex mode.

RF - Radio Frequency- The 800 MHz signal

Satellite Receiver - One of two or more receivers connected to a voting comparator.

Signal - The radio frequency information received by a receiver that carries voice information.

Simplex - Transmitting and receiving on the same frequency.

Simulcast - Transmitting the same signal from two or more sites simultaneously.

Site - Normally refers to the location of transmitting and receiving equipment. For a Smart Zone system, site also refers to an entire simulcast subsystem.

Site Trunking - This is when a Smart Zone site is "disconnected" from the Zone Controller. The site will continue to process traffic and trunk, but it is acting in a stand-alone manner and cannot share traffic with other Smart Zone sites.

Smartnet - The "brand name" given to a Motorola trunking system. A Smartnet system can be either single-site or simulcast.

Smart Zone - The Motorola "brand name" for a wide-area trunking system made by tying two or more Smartnet systems together. The individual Smartnet systems can be any mix of single-site and simulcast.

Subscriber (Unit) - A mobile or portable radio used by personnel not at fixed locations.

Talkgroup - A specific set of radio users with a common ID that are allowed to talk to each other and are assigned a particular frequency.

Trunk (ed) (ing) - A method of grouping of frequencies so several users can use the grouped resources, thus increasing capacity to carry conversations.

Voter - Another name for "comparator" or "voting comparator." A device that selects the best receive signal from two or more receivers.

Zone Controller - The central computer used to control a Smart Zone trunking system.

AERONAUTICAL (AIRCRAFT) RADIO LISTENERS GUIDE

Written for Scannerstuff by
J. Robert Moss, Airliner World Magazine writer, and Dan Rollman

Listening to commercial or general aviation takes some practice. Like public safety radio, it has its own language. Much of this guide is focused on passenger aviation operations at fairly large airports, though it can also be applied to smaller airports with the understanding that some services (like ramp control) may not be in operation at a smaller airport.

Large Airports

ATIS (Automatic Terminal Info System)

At large airports that serve many airlines, the aircraft follow a fairly standard radio call order. First, the pilots listen to the ATIS (Airport/Aerodrome Automatic Terminal Information System). From the ATIS recording, the pilots learn the wind and weather conditions, and receive current notices about the airport and air traffic system. Controllers update the ATIS recording at least once per hour, more often in changing weather conditions. Each update is assigned a letter (A-Z) that uses the military phonetic alphabet ("Bravo" for B). This letter assignment will be used later to convey to the various ATC (Air Traffic Control) controllers that the pilots have received the most current ATIS information.

Clearance Delivery

After getting the ATIS information, and usually about 20 minutes prior to departure, the flight will contact the Clearance Delivery controller in the tower. This controller's job is to give aircraft their flight plan clearance. The flight plan, which is normally filed about an hour earlier by the airline with ATC, is then reviewed and in some instances modified by ATC. A clearance delivery call often sounds like, "Portland Clearance, Alaska 237, a Boeing MD-80, at Gate 25, with Information Bravo, to Seattle". The controller reads back the clearance containing the instrument departure, initial altitude, and any modifications to the filed flight plan, departure frequency, and transponder code for ATC radar identification. Pilots will usually read back the entire clearance or in some instances just the transponder code. Once complete the controller will say, "read back correct" and sometimes advise which ground control frequency to contact when ready for pushback (push away from the gate) or taxi.

Ramp & Ground Control

Next, depending on the airport, the flight will either contact Ramp Control or Ground Control. Large airports (like Portland and Seattle) have dedicated Ramp Control frequencies, often operated by the largest airline user of the airport, to control the movement of aircraft in and out of gates and in alleys (areas between two parallel concourses). In Phoenix, for example, America West Airlines operates

"Phoenix Ramp" on 131.775, which controls the America West ramp on the North side of Terminal-4 at Sky Harbor Airport. Ramp Control gives aircraft permission to push back from their gate. Once the aircraft has pushed back and is ready to taxi to the runway, the flight will contact Ground Control in the control tower (an FAA controller). At smaller airports where no Ramp Control operation exists, and at larger airports but in areas not covered by a Ramp Control (at Phoenix, Terminal-3, Terminal-2, and south side of Terminal-4), flights contact ground control for both pushback and taxi.

Once on Ground Control, the controller will give the aircraft taxi instructions. As the aircraft approaches the runway, the aircraft will automatically switch to the tower frequency and place the departure frequency in the standby display (a 2nd radio frequency location) on the aircraft radio.

Tower Frequency

Smaller airports generally have only one tower frequency, while larger airports may have two or more, with certain runways controlled on specific tower frequencies. The tower controller generally gives aircraft four types of instructions: hold short, taxi into position, cleared for takeoff, and cleared to land. "Hold short" means do not enter the active runway. "Taxi into position (and hold)" instructs the aircraft to taxi onto the runway and prepare for takeoff, but wait for further takeoff clearance (usually waiting for an aircraft that just landed to clear the runway). "Cleared for takeoff" instructions would then be given. Once the aircraft is airborne, the tower controller will advise the aircraft to "contact departure," and may or may not follow that instruction with a specific frequency on which to contact the departure controller. The aircraft will now switch to the departure frequency given by Clearance Delivery before pushback or by the tower controller.

Departure Frequencies

The departure controller in the TRACON (Terminal Radar Approach Control) will work the aircraft and then possibly pass them on to another sector (different area operated by another departure controller on a different departure frequency). This may sound like "Alaska 237, contact departure on 119.2."

Air Route Traffic Control Centers

Once the aircraft reaches a certain altitude, the aircraft will be handed off to one of 22 Air Route Traffic Control Centers (ARTCC, see map) known on the radio as "Center". Some of the Centers include Los Angeles Center, Albuquerque Center and Seattle Center.

The flight is then passed from Center to Center (if necessary) until it nears its destination and begins its descent. At this point, the order of handoffs from controller to controller is reversed. The flight will be handed off from Center to the TRACON (now called "Approach", but still may go from one approach frequency to the next, before tower) and from the TRACON to the Tower. Once on the ground, the tower will then hand them over to Ground and Ground sometimes to the Ramp for clearance to the gate (at larger airports, as described above).

ACARS

Some airlines use ACARS, which is a data link system (computer terminal located in the cockpit) to receive clearances from clearance delivery, so you may never hear them on Clearance unless Clearance has a modification for them. In this case they are sent a message to the on board computers that says, "contact Clearance Delivery", and they would call in on the clearance delivery frequency.

Small Controlled Airports

Small airports that cater mostly to small general aviation aircraft may only have a tower controller and ground controller.

Smaller Non-Controlled Airports

Even smaller airports, and especially private airparks, may be non-controlled. At these airports, pilots use frequencies called CTAF (Common Traffic Advisory Frequency). Common CTAF frequencies include 122.7, 122.8, 122.975, 123.0, and 123.05. At these fields aircraft "blind call" (announcements without a specific recipient specified or reply required) to each other their position at all times. To activate the runway lights, 3, 5 or 7 clicks of the microphone are used to turn the lights on low, medium or high intensity.

Some of these airports contain Automatic Weather Observing Systems known as AWOS or ASOS which, like ATIS, broadcasts an automated weather synopsis and field conditions every minute on a dedicated frequency. Helicopters taking off or landing at non-controlled heliports (like hospitals) do a similar practice, monitoring a common helicopter frequency (often 123.025).

Some airports may be controlled during daylight hours (or at other times) and thus follow the procedures for controlled airports above, while following the non-controlled airport procedures when the tower is closed. In such a case, the frequency used by the tower controller generally becomes the CTAF frequency when the tower is closed.

Flight Service

For general aviation (private pilots), VFR (visual flight rules) flight plans are activated on the radio by a regional calling Flight Service on frequencies such as 122.1, 122.2, 122.3, 122.4 and 122.6. Arizona's regional Flight Service station, for example, is located in Prescott at Love Field, called "Prescott Radio" on the radio. Flight Service also gives weather advisories and accepts pilot reports. Additionally, aircraft can call Flight Watch to speak to an FAA meteorologist on 122.0.

Practice Areas

Aircraft practicing pilot skills in pre-determined "practice areas" generally use a set of frequencies assigned to the practice areas to communicate with one another.

Company Radio and ARINC

Most major airlines have a sophisticated network of ground-based AM aircraft radio communications transmitters and receivers placed throughout the territory the airline services. These private company-run communications systems offer their respective airlines the ability to communicate with company flight operations personnel and maintenance staff, among others. The frequencies for these airport-based systems are listed in the Aircraft/Airline sections of this book within the county in which the particular airport lies.

A company called ARINC also operates a vast worldwide communications network, called the Air/Ground Domestic Radio Service (and International Radio Service), that airlines can use to communicate with ARINC dispatchers for patches to virtually anyone as part of company and industry operations. The system incorporates more than 100 ARINC-operated VHF radio facilities in the United States and Canada and many more beyond North America, and provides en route coverage above 20,000 feet in the continental United States, Hawaii, and on the coastal regions of western Canada and Alaska, as well as on-ground coverage at most major U.S. airports. The system uses 129.850 MHz (AM) in Western Oregon and Washington state, and 129.400 MHz (AM) from Seattle up north along the coast of Alaska.

The service allows aircrews and ground personnel to immediately communicate about operational control and flight information, Aircraft malfunctions and emergencies, In-flight medical assistance, weather and destination airport information, and aircraft diversions. ARINC's radio network is controlled from the company's San Francisco Communications Center (referred to on air as "San Francisco Radio"), where radio operators monitor VHF frequencies, receive and route calls, and transcribe and deliver messages around the clock.

Brief Aircraft Radio Frequency Terms

Air to Air

Used for aircraft to communicate with each other. General aviation aircraft use 122.75 and 122.85. Helicopters are known to use 123.025

Approach

TRACON (Terminal Radar Approach Control) controller works an "Approach" frequency (sector) primarily handles aircraft inbound to the area or major city.

CTAF

Common Traffic Advisory Frequency. Like airports that are uncontrolled and use a UNICOM frequency, some smaller airports known as Class-D fields, have control towers that are only open certain hours. When the tower closes the primary tower frequency becomes a CTAF. This means that aircraft blind call to each other their position both inbound and outbound from the field. These calls should sound exactly the same as if the aircraft was operating at an uncontrolled field, because the field is uncontrolled when the tower is closed.

Final Approach

Final Approach, also known as “Final Control” are the controllers at major airports responsible for vectoring aircraft into an order that allows a continuous and safe flow in the final stages of flight. Aircraft are usually handed over to Final Control in the final 8000 feet and usually 20 miles from the field. It is this controller’s responsibility to assign the aircraft the appropriate approach Visual or Instrument and then hand them over to tower 10-5 miles from the field.

Flight Service Station

Flight Service is the information branch of the FAA and provides pilots with weather, pre-flight and NOTAM briefings. Flight Service also files the pilot’s flight plan. Most briefings are conducted via telephone however each Flight Service Station maintains an extensive radio system. The most common call on the radio to a FSS is for pilots to activate or close a flight plan.

Helicopter Control

At busy airports such as LAX, one controller in the tower is responsible for the movement of helicopters inbound and outbound from the field. This is to avoid helicopters tying up the tower frequency which could make for an unsafe situation.

Tower

Tower controllers are responsible for aircraft handed from Final Approach/Control that are inbound to a designated runway or runways. These controllers issue land, takeoff, hold short, go around and execute a missed approach clearances.

Unicom

Unicom frequencies such as 122.7, 122.8, 123.0 are used at airports that are classified as “uncontrolled”. This essentially means there is no control tower. At certain airports such as Avalon-Catalina Island off the coast of Los Angeles an airport authority person will announce traffic pattern and wind conditions but does not issue takeoff or landing clearances. Aircraft “blind call” their intentions, See CTAF. Another popular UNICOM frequency is 122.95 This is used by FBO’s for Fuel and parking operations at most airports big and small.

Air Route Traffic Control Center (aka "Center") Map

Seattle Center

Antelope Mountain	124.850, 306.300, 127.050
Beacon Hill	135.525, 120.300, 273.600, 353.900
Cottonwood	120.050, 123.950, 118.550, 282.300, 251.100
Eugene	125.800, 291.700, 132.075
Ferndale	135.150, 124.850, 306.300, 360.700
Fort Lawton	128.500, 306.900
Horton	132.075, 125.800, 121.400, 257.650, 239.000, 291.700
Klamath Falls	134.900, 127.600, 124.850, 327.100, 351.700, 263.050
Lakeside	123.950, 282.300
Lakeview	135.350, 127.600, 335.550, 351.700
Larch Mountain	128.300, 126.600, 269.000, 343.600
Marlin	126.100, 291.600
Medford	135.150, 124.850, 121.400, 239.000, 306.300, 360.700
Mohler	128.450, 307.800
Mullan Pass	128.450, 307.800
Neah Bay	125.100, 319.200
Olympia	128.300, 126.600
Pendleton	135.450, 132.600, 281.400, 321.300, 343.900
Redmond	135.350, 121.350, 128.150, 134.900, 279.600, 263.050, 335.550
Scappoose	124.200, 128.150, 257.750, 317.600
Spokane	119.225, 123.950, 282.300, 335.500
Stampede Pass	134.950, 270.300
The Dalles	135.450, 119.650, 257.600, 281.400
Wenatchee	126.100, 291.600
Whidbey Island	134.950, 125.100, 270.300, 319.200
Yakima	135.525, 132.600, 120.300, 118.550, 273.600, 251.100, 127.050

FIRE DEPARTMENT GUIDE

David Greenberg, Fire Department Coordinator

This fire service operations guide was written as a basic overview of the operations of many fire departments in Arizona, but many concepts are similar in the Pacific Northwest. However, please note that many aspects of this guide, including but not limited to call types, resources dispatched, terminology, and operational methods may vary widely from one fire department to another.

Medical Calls

Medical calls are divided into two categories. BLS (Basic Life Support with EMT's, or emergency medical technicians) and ALS (Advanced Life Support with paramedics). ALS (paramedic) units are able to administer drug intervention which distinguishes them from a BLS (EMT-only) unit. Depending on the fire department, they may have any of the following: ALS and BLS engine companies, ladder companies and rescues or aid units (ambulances). Fire departments that are dispatched by a different neighboring department may have protocols already in place on how they would like their units be dispatched that will differ from others. Additionally, many area's ALS needs are served by a private for-profit ambulance services, such as American Medical Response (AMR), or a not-for-profit entity distinct from the local fire department, such as Medic One organizations in Washington.

CAD & Medical Responses

When a citizen calls 911 for a medical emergency, based on the information given to the call taker, the information is then entered into the CAD (Computer Aided Dispatch) system which decides the type of call and which unit(s) based on resources and proximity are best suited to respond and give the appropriate care. As stated, depending on the type of call, the person in need may receive a fire truck (of any type) and a rescue or aid unit (ambulance) when the likelihood of a transport is significant. If the medical call is not of a serious nature, an ALS transport ambulance will most likely not be dispatched, although they can be special called to the scene by the truck already onscene.

Multi-Unit Medical Calls

In the event of a call with more than one or two patients, additional units (either BLS and/or ALS) may be dispatched or special called to the scene based on additional information gathered from the call taker or by arriving units. In many instances a call may be balanced to a different assignment (upgraded) to fill the requirements for a more serious type of call.

For instance, if two BLS units (an engine and a rescue) are dispatched to a seizure call and they determine the patient has other symptoms that require paramedics, then the call will be balanced to an ALS call requiring that an ALS unit (either an ALS engine or ALS rescue) respond. Additionally, If an engine and rescue are dispatched to an auto accident with one patient and discover on arrival that they have to extricate the victim from the vehicle, the captain onscene may request an extrication-capable unit carrying the Jaws of Life and other specialized tools.

Special Calls and Manpower Requests

The officer in command may at his/her discretion special call any unit that they may need. Sometimes a unit may just be requested to the scene just for use of its manpower and not for its ALS or BLS capabilities. This is called a manpower request.

Fire Incidents and Apparatus Functions

In some parts of the northwest, engine & ladder company (sometimes called truck company) apparatus are staffed by 4 personnel, but other departments may staff fewer firefighters per apparatus as budgets permit.

Engine Companies

An Engine Company is sometimes referred to as a pumper as it carries a tank of water (up to 1,000 gallons) and has a pump capable of extinguishing a small fire within a few minutes, or until a larger water supply can be obtained via a water hydrant or tanker truck. Functions of the engine company include search and rescue, treatment, loss control, and stretching, operating, and pumping hoselines.

Ladder Companies

Ladder Company functions include search and rescue, treatment, ventilation, forcible entry, raising ladders, providing access, checking for fire extension, securing utilities (power/gas), providing lighting, operating aerial ladder pipes, performing overhaul, extrication and loss control. Some departments have ladder trucks that also have a pump and water tank (called Quints) giving them the ability to arrive at a fire scene first and perform some engine company duties. In order to be classified as a Quint, the ladder apparatus must have on board a pump, water tank, aerial device, hoses, and compliment of ground ladders.

Ladder Tenders

The Ladder Tender is a new concept in the United States and use is currently very limited (originally started in Phoenix and Tucson, Arizona and expanded quickly to a handful of Utah departments). The objective of departments operating ladder tenders (LT's) in conjunction with a ladder truck quartered at a fire station is to provide a full level of truck company functions (except aerial ladder), while extending the service life of the larger, more costly ladder truck.

The LT's are purchased at a fraction of the cost of the bigger ladder trucks and generally have a large number of compartments, but no significant pump or water tank, and no large aerial ladder. The functions of the LT's remain largely the same along with the equipment, the exception being that they do not have the larger aerial device and carry fewer ground ladders. The Ladder Captain will generally elect to respond the LT on EMS (medical), special duty, extrication calls, etc. but may have both units respond on a structure fire call, increasing the flexibility of the ladder company crew. If the ladder tender is dispatched to a structure fire call when away from the station, they will notify alarm and have other responding crews notified that they are not in their "big" ladder. Depending on the specifics of the fire call, another "big" ladder may be special called to the scene.

Generally LT's do not have a water pump or tank, but, some departments choose to equip their ladder tenders with small pumps and water tanks, once again allowing them the flexibility to respond to certain types of fire calls such as a car fire.

Rescues / Aid / Medic units

A transport rescue (or aid or medic unit) is an ambulance that is staffed by 2 personnel, who may also be trained firefighters, specifically trained in the transportation of the sick & injured to hospitals, search and rescue, and patient treatment. These units may also carry minimal firefighting tools that include bunker gear and SCBA's (air packs). Some fire departments operate only ambulances (not rescues) that are staffed by either 2 EMT's, 2 paramedics or a combination of the 2 with no firefighting equipment or training.

Water Tanker (Tender)

Many departments have a truck similar to the engine called a water tanker (or tender) that is classified as such because it will carry 1,000 gallons of water or greater. They can look almost identical to an engine or be as large as an 18-wheel semi tractor trailer. Tankers are generally used for two types of calls: Either as a water supply during a structure fire in an area where there are few or no fire hydrants, or at a brush fire acting as an additional water supply. Some will have the capability to move and pump water at the same time (pump & roll) while flowing water through special nozzles located on the truck itself in similar fashion to a construction water tank truck.

Brush Trucks

A brush truck is a smaller version of an engine or pumper about the size of a pickup truck and can be 4-wheel drive equipped. They can be as elaborate as the engine and be custom built or as simple as a flatbed truck with a small pump and tank mounted on a skid that is secured to the back end. These interesting small trucks can be equipped to pump from 60 gallons of water per minute up to 500 GPM. Water tank sizes can range from 200 gallons up to 500 gallons. This type of special unit also operates similar to a water tanker in that they have the ability to pump & roll while effectively controlling brush fires.

Foam Trucks

A foam truck, usually located at a station either on airport property or in very close proximity to an airport is also sometimes referred to as an ARFF (Aircraft Rescue Firefighting) or CFR (Crash Fire Rescue) truck and are specifically designed for aircraft emergencies. These massive rigs are usually driven by only one person that can operate all the functions of the truck from inside the cab but can also be assisted by ground personnel with handlines located on the truck. Not only are they built with large pumps (1500 GPM up 2000 GPM or more) and large water tanks (1000 gallons or greater) but they also carry foam, which is highly effective in extinguishing fires in or around aircraft and aircraft fuel.

Other types of extinguishing agents that may be carried by a Foam Truck are Halon, Dry Chemical and/or Purple K, which is purple in color and similar to dry chemical. Another feature that has been introduced on foam trucks is an aerial device called a Snozzle that is capable of flowing water from its elevated aerial device or through a piercing nozzle that can puncture the "skin" of a fuselage and is able to flow water into the interior of an aircraft. Crash trucks often appear extremely large, with massive tires, in the event of a crash in rough terrain or to easily plow over debris often spread over a large area in a crash to make access to the largest section of fuselage.

Hazardous Materials Teams & Units

The function of the Hazmat team is the protection of life, health, property and environment through control of hazardous materials. Most department hazmat teams are made up of an engine company and its associated hazmat truck that respond together as a "two piece" company. Staffing for the hazmat unit is often one person from the engine company associated with it. The hazmat truck is generally a large panel or "bread" style truck equipped with a variety of equipment such as but not limited to computers, hazardous materials books, disposable hazmat suits and barrels to safely dispose of any equipment that may come in contact with a hazardous material.

Fire departments generally handle HazMat calls, which may include (but are not limited to) large fuel spills, vapor releases, chemical spills or even unmarked containers with an unknown substance. Many fire departments operate HazMat support vehicles that carry all their gear (described above), and generally designate specific engines or ladders to have HazMat trained people on board.

Battalion (and Shift Deputy) Chiefs

Battalion Chiefs often utilize an SUV (Sport Utility Vehicle) type vehicle that is large enough to carry a variety of books and personal firefighting equipment. The function of the "Officer" in Command of such incidents as medical calls involving numerous patients, structure fires, hazmat calls or any other type of call that requires the response of a higher level "supervisor" is to assist in the overall management of an emergency situation.

Utility (Light & Air) Trucks

Several departments have their own light and air truck commonly called a Utility or Air Supply truck. Staffed by only 1 person, this "special unit" has an onboard compressor for refilling firefighter air bottles (SCBA's) and either one or two rotating and extendable light towers for lighting emergency scenes at night. Other equipment or supplies carried might include Gatorade and associated rehab supplies along with a portable misting system and awning.

Scene Operations

Structure Fires

On a working structure fire call, the first arriving unit is referred to as the first due unit (usually an engine) and is responsible for the 3 basic tactical priorities: Rescue, Fire Control and property conservation. They will arrive onscene and announce over their assigned tactical channel their unit designation, a brief description of the situation (building size on a fire, number of vehicles in an accident, etc), and obvious conditions (working fire, hazmat spill, multiple patients, etc.).

The unit should also declare their strategy. In case of a fire, the strategy would be either offensive (interior firefighting operations) or Defensive (exterior firefighting operations). They will also note any obvious safety concerns, assumption and identification of command, and assume and announce an accountability (personnel safety) location.

The second due unit (2nd arriving) usually assists the first unit with their tactical objectives or duties they are assigned by command, while the ladder company will perform their "truck" company duties. The rescue or aid unit will treat any patients or assist other companies as assigned by command.

Once arrived, the Battalion Chief will assume command from the first arriving unit and free them to concentrate on their fireground duties.

Sample structure fire onscene report:

"Engine 1 is onscene of a 2 story house with a working fire on the first floor. Engine 1 is laying a supply line and going in with a hoseline for search & rescue and fire attack in the offensive mode. Engine 1 is assuming Command. Engine 1 has Northside accountability."

Additional Alarms & Staging

In the event the situation escalates, command can call for additional resources by asking for a 2nd alarm or whatever alarm they will need along with any special resources. During a 2nd alarm or greater, a staging sector will usually be setup where incoming units not immediately needed on the scene can park in an area until they are requested and needed. This area will be managed by a Sector officer in communications with Command on a separate radio channel.

Motor Vehicle Accidents

Motor Vehicle Accidents are referred to in a variety of ways by area fire departments. MVA's, traffic collisions, or various radio codes. Patients at these types of calls will be classified in general areas depending on their degree of injuries. Level-1, Priority-1 or Immediate for a patient with life threatening injuries; Level-2, Priority-2 or Delayed for a patient that is not seriously injured; Level-3, Priority-3 or Minor for someone not needing medical attention, having very slight injuries, or refusing treatment. Level-4, Priority-4, DOA, or another code for a fatality.

Units arriving onscene announce the situation almost the same way as a structure fire call. A Treatment sector/area may be set up in a situation that involves multiple patients or in order to treat victims away from a hazardous scene. An LZ (Landing Zone) sector is responsible for securing an area for medical helicopters to land when they are requested.

Brush Fires

A brush fire can be small enough to be handled by either a brush truck, engine or both. If it is in a remote area without a close water supply, units onscene can special call a Water Tender to the scene. In the event of a large brush fire, aerial (aircraft) firefighting equipment may have to be called to assist. These can range from small planes and helicopters to large planes capable of dropping several thousand gallons of water or fire retardant. In the event of the brush fire escalating into a major fire, Command may call for a whatever level of brush response is needed.

SkyWarn Operations

The National Weather Service's SkyWarn Program is a network of spotters, many who are amateur radio operators, that provide NWS meteorologists with information on local storm effects, flooding, rain measurements, lightning strikes, and more. SkyWarn spotters reports are often included in NWS Severe Thunderstorm and other warning releases. Those spotters that are amateur radio operators utilize ham radio repeaters throughout the state to funnel their reports in to "net control" who then pass certain reports on to their weather service office. The SkyWarn nets can be very interesting to listen to during storms. For more information, see the National Weather Service's SkyWarn web page. In Oregon and Washington, the following frequencies are used (note that these are open amateur radio repeaters when storms are not present):

147.260+	[PL 123.0]	Mount Ashland (Medford), OR
146.760-	[PL 100.0]	Rattlesnake Mountain (Richland), WA
146.640-	[PL 100.0]	Johnson Butte (Kennewick), WA
146.860-	[No PL]	Darland Mountain (Yakima), WA
145.270+	[PL 123.0]	Quartz Mountain (Ellensburg), WA
147.080+	[PL 123.0]	Mount Adams (Yakima), WA
146.960-	[PL 100.0]	Lewis Peak (Walla Walla), WA
147.360+	[No PL]	Deschutes County, OR
146.800-	[PL 100.0]	Spout Springs (Umatilla), OR
147.200+	[PL 123.0]	Metro Seattle

SkyWarn Reportable Weather Conditions:

1. Tornado
2. Funnel Cloud
3. Rotation Wall Cloud
4. Wind, >40mph
5. Heavy Rain, > 1/2" in 30 minutes
6. Flooding of any kind
7. Damage / Deaths / Injuries of any kind
8. Hail, greater than pea size
9. Visibility less than 1/4 mile for any reason
10. Snowfall, 1" or more per hour (or unusual)
11. Microburst, visually identified

WEATHER RADIO INFORMATION

National Weather Service – Oregon Transmitter Locations

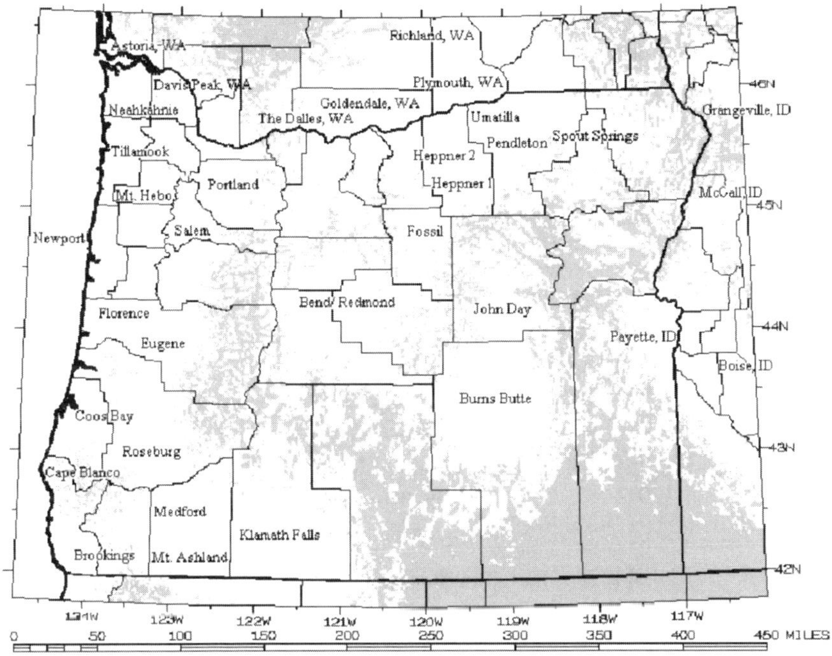

Transmitter Area	Transmitter Site	Frequency	From Office
Bend/Redmond	Powell Butte	162.500 MHz	Pendleton, OR
Brookings	Palmer Butte	162.550 MHz	Eureka, CA
Burns Butte	Harney	162.475 MHz	Boise, ID
Cape Blanco	Port Orford	162.425 MHz	Medford, OR
Coos Bay	Noah Butte	162.400 MHz	Medford, OR
Eugene	Blanton Heights	162.400 MHz	Portland, OR
Florence	Herman Peak	162.500 MHz	Portland, OR
Fossil	Snowboard Ridge	162.550 MHz	Pendleton, OR
Heppner	Robinson Hills	162.425 MHz	Pendleton, OR
Heppner 2	Gleason	162.425 MHz	Pendleton, OR
John Day	Fall Mountain	162.500 MHz	Medford, OR
Klamath Falls	Hogback Mountain	162.550 MHz	Medford, OR
Medford	Blackwell Hill	162.400 MHz	Medford, OR
Mount Ashland	Ashland	162.475 MHz	Medford, OR
Mount Hebo	Mount Hebo	162.525 MHz	Portland, OR
Neahkahnie	Neahkahnie Mtn.	162.425 MHz	Portland, OR
Newport	Otter Rock	162.550 MHz	Portland, OR
Pendleton	Pendleton	162.425 MHz	Pendleton, OR
Portland	Goat Mountain	162.550 MHz	Portland, OR
Roseburg	Dodson Mountain	162.550 MHz	Medford, OR
Salem	Prospect Hill	162.475 MHz	Portland, OR

Spout Springs	Spout Springs	162.400 MHz	Pendleton, OR
Tillamook	Tillamook Hospital	162.475 MHz	Portland, OR
Umatilla	Umatilla Ridge	162.500 MHz	Pendleton, OR

Oregon SAME Codes (see your scanner manual to use)

County	Code	From Transmitters (see above or WA below)
Benton	041003	Eugene, Salem
Clackamas	041005	Portland, Salem
Clatsop	041007	Astoria (WA), Neahkahnie
Columbia	041009	Astoria (WA), Portland
Coos	041011	Coos Bay, Roseburg
Crook	041013	Bend/Redmond
Curry	041015	Brookings
Deschutes	041017	Bend/Redmond
Douglas	041019	Roseburg
Gilliam	041021	Fossil, Goldendale, Heppner, Plymouth
Grant	041023	Johns Day
Harney	041025	Burns Butte
Hood River	041027	The Dalles (WA)
Jefferson	041031	Bend/Redmond
Lane	041039	Eugene, Newport (NW), Florence (W)
Lincoln	041041	Newport, Florence (S)
Linn	041043	Eugene, Salem
Marion	041047	Portland, Salem
Multnomah	041051	Portland
Polk	041053	Salem
Tillamook	041057	Neahkahnie (N), Tillamook, Mount Hebo
Umatilla	041059	Pendleton, Plymouth, Umatilla, Spout Spgs, Heppner 2
Wasco	041065	The Dalles (WA)
Washington	041067	Portland
Yamhill	041071	Salem

National Weather Service – Washington Transmitter Locations

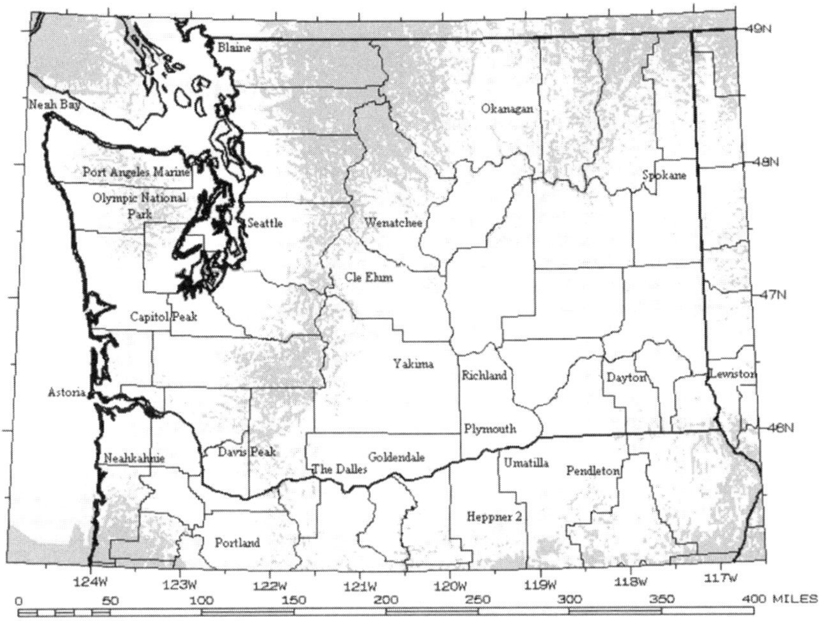

Transmitter Area	Transmitter Site	Frequency	From Office
Astoria	Naselle Ridge	162.400 MHz	Portland, OR
Capitol Peak	Capitol Peak	162.475 MHz	Seattle/Tacoma
Cle Elum	Sky Meadows	162.400 MHz	Pendleton, OR
Davis Peak	Cowlitz County	162.525 MHz	Pendleton, OR
Dayton	Patit	162.525 MHz	Pendleton, OR
Goldendale	Goodnoe Hills	162.525 MHz	Pendleton, OR
Blaine	H Street	162.525 MHz	Seattle/Tacoma
Neah Bay	Bohokus Peak	162.550 MHz	Seattle/Tacoma
Okanagan	Tunk Mountain	162.525 MHz	Spokane, WA
Olympic Nat'l Park	Mount Octopus	162.425 MHz	Seattle/Tacoma
Plymouth	Sillusi Butte	162.425 MHz	Pendleton, OR
Puget Sound Marine	Miller Peak	162.425 MHz	Seattle/Tacoma
Richland	Richland	162.450 MHz	Pendleton, OR
Seattle	Couger Mountain	162.550 MHz	Seattle/Tacoma
Spokane	Spokane Mountain	162.400 MHz	Spokane, WA
The Dalles	The Dalles	162.400 MHz	Pendleton, OR
Wenatchee	Eagle Rock	162.475 MHz	Spokane, WA
Yakima	Ahtanum Ridge	162.550 MHz	Pendleton, OR

Washington SAME Codes (see your scanner manual to use)

County	Code	From Transmitters (see above)
Clallam	053009	Neah Bay, Olympic Nat'l Park, Puget Sound Marine
Clark	053011	Portland, Davis Peak
Cowlitz	053015	Davis Peak, Portland
Grays Harbor	053027	Astoria, Capitol Peak, Olympic Nat'l Park
Island	053029	Puget Sound Marine
Jefferson	053031	Neah Bay, Olympic Nat'l Park, Puget Sound, Seattle
King	053033	Puget Sound Marine, Seattle
Kitsap	053035	Seattle
Kiickitat	053039	Goldendale, Richland, The Dalles (West)
Lewis	053041	Capitol Peak
Mason	053045	Capitol Peak, Seattle
Pacific	053049	Astoria, Capitol Peak, Olympic Nat'l Park
Pierce	053053	Capitol Peak, Seattle
San Juan	053055	Puget Sound Marine
Skagit	053057	Puget Sound Marine
Skamania	053059	Portland, The Dalles (South)
Snohomish	053061	Puget Sound Marine, Seattle
Thurston	053067	Capitol Peak
Wahkiakum	053069	Astoria
Whatcom	053073	Puget Sound Marine, Blaine

Meteorological Service of Canada – Weather Radio

British Columbia Weather Radio Stations

Station	Frequency
Abbotsford	162.550
Alert Bay	162.550
Barry Inlet	162.400
Bowen Island	162.475
Calvert Island	162.400
Castlegar	162.550
Cranbrook	162.400
Cumshewa	162.475
Dundas Island	162.400
Eliza Dome	162.550
Estevan Point	162.475
Kelowna	162.550
Klemtu	162.550
Mount Gil	162.400
Naden Harbour	162.475
Nootka	162.400
Penticton	162.475
Port Alberni	162.400
Port Hardy	162.475
Prince George	162.400
Texada Island	162.550
Van Inlet	162.550
Vancouver-Victoria	162.400
Vernon	162.475
Victoria	162.475

WILDLAND FIRE GUIDE

Mike Lanning, Wildland Fire Coordinator

Also see U.S. Forest Service info in Federal Section.

Large wildland fires (the general term for brush and forest fires) are fought by ground and by air. Wildland firefighting operations are broken down into different stages.

WILDLAND FIREFIGHTING STAGES

Initial Attack (also called a Type 4 incident) is the first attack on a wildfire by federal, state, or municipal firefighting crews. The number and type of resources that respond to the initial attack of a fire varies with the reported size, location (mountainous territory, proximity to accessible roads), fire danger (structures threatened), and fuels (grass, pine trees). Resources vary from a single firefighter (rare) to several pieces of apparatus and crews. The initial attack stage is normally limited to one operational period (day, or part of a day) of firefighting, but mop up may take several days.

Extended Attack (also called a Type 3 incident) is normally characterized by a fire of less than 100 acres in size with firefighting resources varying from several single resources to several task forces. The incident may be divided into several divisions, but does not involve the divisional supervisory complexity that would call for a larger incident management team. An extended attack incident is expected to be contained/controlled in the first operational period (otherwise it will probably transition into a type 2 incident). Some command staff positions including logistics, safety, and liaison may be established. A staging area for incoming equipment and crews may be established and in some cases a small base camp may also be set up. At some point the fire is either contained or transitioned to a Type 2 incident.

Guidelines used by commanders when determining whether to transition to a Type 2 incident include whether the fire will be contained/controlled in the first operational period, a written action plan will be needed for the next operational period, and logistics are becoming complicated (such as a need to establish an incident base and feed, sleep, or supply a large amount of personnel on the fire). The final consideration is whether fire complexity exceeds the capability of extended attack staff and there is a need to fill most or all of the command positions (like logistics, plans, medical, and security).

LARGE FIRES

A Type 2 Incident is the first level at which most or all of the command and general staff positions are activated. The incident commander and Command/General Staff function as a team to command and control the firefighting efforts, including:

a. Supervising a large organization
b. Multiple operational periods
c. Gathering information to develop an action plan
d. Developing an action plan
e. Providing logistical support including the establishment of a operational base and fire camp.

If considerable progress is made, the fire can generally be held at the Type 2 level. Otherwise, the fire may need to be transitioned to a Type 1 incident.

The Type 1 Incident Team is characterized by all command and general staff positions being filled, similar to a Type 2 Incident but on a larger scale. The Type 1 team has access to more resources and logistical tools, and is more equipped to deal with large amounts of personnel (several hundred to several thousand firefighters).

AREA Command. If more than one Type 1 Incident is going on in a generally small area, the incidents may be unified under a larger command structure and referred to as a "Complex". When an Area Command is called in they take overall command of the incidents and help to make sure the Type 1 teams get everything they need. When fire season is in full swing during the summer, several fires may compete for resources nation wide and an incident where an area command is in place will get generally receive priority for resource assignment.

WILDLAND FIREFIGHTING RESOURCES

ENGINES

Type 1 is largest. Apparatus generally become smaller in size as Type number increases, with Type 7 being the smallest.
The following chart shows minimum standards for each Type.

Type	GPM	Tank	Hose			FF	Notes Other Requirements
			2.5"	1.5"	1"		
1	1000	400	1200	400	200	4	Typical structure-firefighting engine. 20ft ladder, 500 GPM monitor/deck gun
2	500	400	1000	500	300	3	20ft ladder.
3	120	320		1000	800	3	Most common forest engine. Many are 4x4.
4	70	750		300	300	3	Not widely used.
5	50	500		300	300	3	Not widely used.
6	50	200		300	300	3	Most common municipal fire brush truck.
7	20	125		200	200	3	Not widely used.

GPM: Gallons Per Minute pumping capacity. **Tank**: Gallons.
Hose: Feet of various size hose. **Other**: Other Equipment
FF's: Minimum number of firefighters.

Prevention Units: Some are equipped as engines with about 50 gallons of water, a small amount of hose and a few hand tools, and 1 or 2 personnel. Generally enough for a large campfire or to attempt to hold a larger fire until help arrives.

HELICOPTERS

The following chart shows minimum standards for each Type.

Type	Seat	Tank	Notes / Examples
1 Heavy Helo	16	700	Erickson Aircrane uses Sikorsky Skycranes called Helitankers which carry 2,500 gallons of water in a mounted tank. Columbia Helicopters uses Boeing's Vertol and Chinook helicopters. Vertol has a 1,100 gal bucket. Chinook has a 3,000 gal bucket. Newer Chinook's also carry 19 FF's.
2	10	300	Medium Helicopter. Bell 205, Bell 212
3	5	100	Light Helicopter. Bell Jet Ranger, A-Star, Bell 206.

AIRTANKERS

Heavy Airtankers (mostly P3 aircraft) carry 1000 to 2000 gallons of retardant.

RADIO PROCEDURES

On the initial attack, the fire traffic can be monitored on the local Forest's Fire Frequency. When the fire progresses into a large fire and a Type 1 or Type 2 team takes over, the team re-assigns the fire to its own set of radio frequencies assigned to their team by the National Interagency Fire Center (NIFC) in Idaho. On a large fire, there can be several frequencies used for different areas of the fire, including one frequency for air tankers, one for helicopters, and several for command.

PROGRAMMING YOUR SCANNER

The following Agencies frequencies (found elsewhere in this directly, largely in the Federal and State sections) should be programmed into your scanner for monitoring Wildfires:

- State Land Department
- Bureau of Indian Affairs Firefighting Crews
- Bureau of Land Management
- Fish & Wildlife Service
- Local Forests, Monuments, and Parks
- National Interagency Incident Management System
- State Fire Mutual Aid

REGIONAL DISPATCH OFFICES

Northwest Dispatch Offices

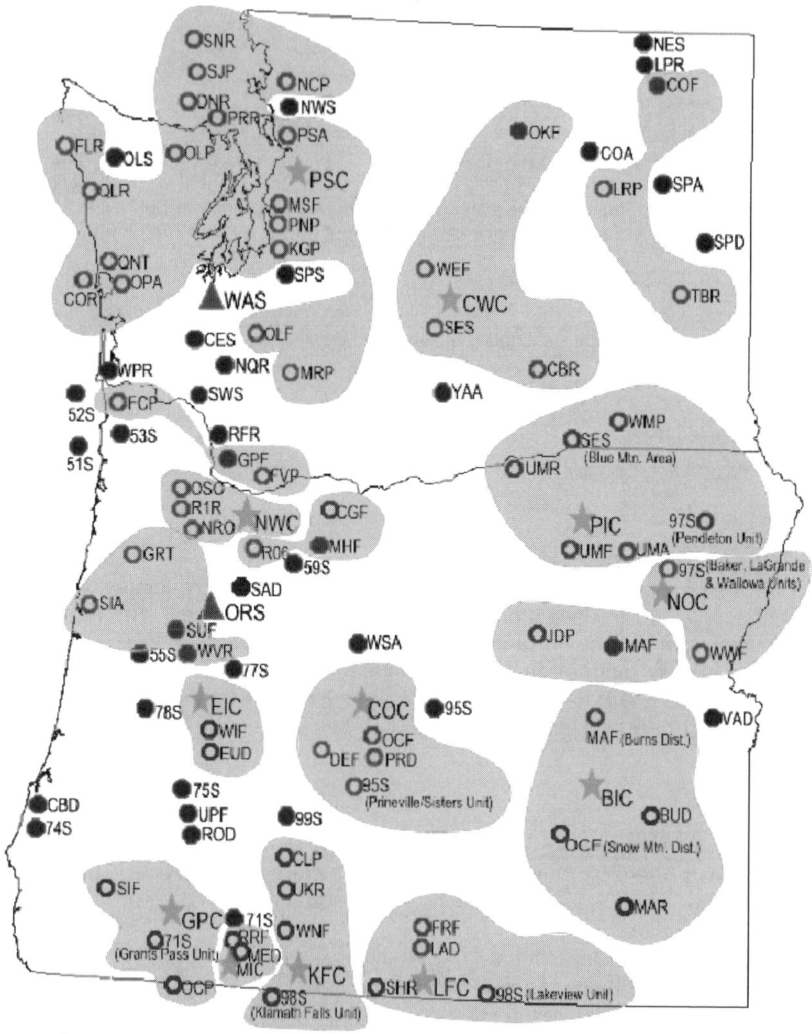

LEGEND

- ★ Interagency Dispatch Office. Serves as dispatch contact for multiple units. May act as a Host Unit for dispatch of Area or National Shared Resources.
- ▲ Oregon/Washington State Headquarters. Dispatch functions for State Resources are generally coordinated through headquarters offices. Notable exception wherein an Interagency center serves as a contact for a state unit or where "neighborhood concept" is used locally.
- ● Federal or State Agency Dispatch Office. Site performs own dispatch functions. May serve as a dispatch contact for other Federal Units and may act as a Host Unit for Area or National Shared Resources.
- ○ Federal or State Agency. Dispatch functions performed by Interagency Dispatch Office.

NWC 04/17/03

INTERAGENCY FIRE CENTERS IDENTIFICATION CODES

BIC	Burns Interagency Communications Center
LFC	Lakeview Interagency Fire Center
COC	Central OR Interagency Dispatch Center
MIC	Medford Interagency Communications Center
CWC	Central WA Interagency Communications Center
NOC	Northeastern OR Interagency Fire Center
EIC	Eugene Interagency Communications Center
NWC	Northwest Interagency Coordination Center
GPC	Grants Pass Interagency Fire Center
PIC	Pendleton Interagency Fire Center
KFC	Klamath Falls Interagency Fire Center
PSC	Puget Sound Interagency Coordination Center

REGION 6 NATIONAL FORESTS (NF) IDENTIFICATION CODES

CGF	Columbia River Gorge Natl. Scenic Area
R06	Regional Office
COF	Colville NF
RRF	Rogue River NF
DEF	Deschutes NF
SIF	Siskiyou NF
FRF	Fremont NF
SUF	Siuslaw NF
GPF	Gifford Pinchot NF
UMF	Umatilla NF
MAF	Malheur NF
UPF	Umpqua NF
MHF	Mt. Hood NF
WWF	Wallowa Whitman NF
MSF	Mt. Baker Snoqualmie NF
WEF	Wenatchee NF
OCF	Ochoco NF
WIF	Willamette NF
OKF	Okanogan NF
WNF	Winema NF
OLF	Olympic NF

BIA AGENCY OFFICES IDENTIFICATION CODES

COA	Colville Agency
SPA	Spokane Agency
OPA	Olympic Peninsula
UMA	Umatilla Agency
PAO	Portland Area Office
WSA	Warm Springs Agency
PSA	Puget Sound Agency
YAA	Yakama Agency
SIA	Siletz Agency

BLM DISTRICTS IDENTIFICATION CODES

BUD	Burns District
PRD	Prineville District
CBD	Coos Bay District
ROD	Roseburg District
EUD	Eugene District
SAD	Salem District
LAD	Lakeview District
SPD	Spokane District
MED	Medford District
VAD	Vale District

NATIONAL PARK SERVICE UNITS

CLP	Crater Lake Nat'l Park
NCP	North Cascades Nat'l Park
FCP	Ft. Clatsop Nat'l Monument
OCP	Oregon Caves Nat'l. Monum.
FVP	Fort Vancouver Nat'l Hist. Site
OYP	Olympic Nat'l Park
JDP	John Day Fossil Beds NM
PNP	Columbia Cascades Support
KGP	Klondike Gold Rush Hist Pk
SJP	San Juan Island Nat'l Historic
LRP	Lake Roosevelt Nat'l Park
WMP	Whitman Mission Nat'l Historic
MRP	Mt. Rainier Nat'l Park

U.S. FISH & WILDLIFE REFUGES (WR) IDENTIFICATION CODES

CBR	Columbia National WR	NQR	Nisqually National WR
SHR	Sheldon-Hart National WR	UMR	Umatilla National WR
LPR	Little Pend Oreille National WR	PAR	Pacific Region
TBR	Turnbull National WR	WOR	Western Oregon National WR
MAR	Malheur National WR	RFR	Ridgefield National WR
UKR	Upper Klamath National. WR	WPR	Willapa National WR

OREGON DEPARTMENT OF FORESTRY (ORS) ID CODES

ORS	State Headquarters	55S	West Oregon District
74S	Coos District	95S	Central Oregon District
51S	Tillamook District	56S	Linn District
75S	Douglas District	97S	Northeast Oregon District
52S	Astoria District	59S	Clackamas-Marion District
77S	Eastern Lane & Linn Districts	98S	Klamath Lake District
53S	Forest Grove District	71S	Southwest Oregon District
78S	Western Lane District	99S	Walker Range District

WASHINGTON DEPT. OF NATURAL RESOURCES (WAS) ID CODES

CES	Central Region	NES	Northeast Region
OLS	Olympic Region	SPS	South Puget Region
WAS	State Headquarters	NWS	Northwest Region
SES	Southeast Region	SWS	Southwest Region

TERMINOLOGY

AFMO — Assistant Fire Management Officer

Air Attack — Spotter plane. When not on a fire, its call sign is its tail number. When on a fire its call sign is ____ Air Attack, where ___ is filled in with the forest or area the aircraft is operating in (i.e. "Tonto Air Attack"). Manned by a pilot who picks up a forest representative at a nearby airport who will coordinate firefighting between ground and air units from the Air Attack plane.

Black Line — Cut a fire break and burn back from it.

Bladder Bag — Backpack pump ("piss pump"). Backpack holds about 5 gallons of water with hand operated nozzle used during mop up. Firefighters have other slang names for back pack pumps that are not supposed to be used on the radio but sometimes are.

Complex — Two or more fires in same general area, managed by the same team. The Rodeo and Chediski fires of 2002 were a complex until they burned together.

Contain — Fire line all the way around the fire. Forward rate of spread has stopped. When a fire is 25% contained, the fire line is 1/4 way around fire. Usually, a declaration of 50% containment means that means fire line is cut around the back of fire but the active head of the fire is still spreading.

Control — All active flames extinguished, mop up in operation.

Cubey Container: Square plastic jugs (about 10 gallons) in a box used for holding water for drinking etc.

Fire Brand — Sparks, smoldering bits of paper, or grass debris that become airborne and help spread the fire.

Fire Line — A clear cut area around the fire to stop spread.

FMO — Fire Management Officer (Some forests call them Division Chief or Battalion Chief). The equivalent of a Battalion chief in a city fire department.

Fussee — Road Flare for starting back fires

Heavy — Usually a heavy airtanker, but also large stumps

Helitack — Firefighters who respond to the initial attack on a fire by helicopter and repel out to begin cutting fireline. They also work on large fires usually with firefighter shuttles in the helicopter and bucket work. Once a fire is a large fire and no longer an initial attack, a helitack pulls off and turns their line-cutting operation over to other crews.

Helitanker Heavy Lift Helicopter (Sikorsky Skycrane) with a fitted tank carrying 2500 gallons of water.

Helo Helicopter

Hotshot crew: 20 person crew composed of "elite" of wildland firefighters. "Hotshots" handcut fire line using hand tools such as a shovel, Pulaski, chain saw, etc. and carry enough equipment on their trucks to travel anywhere in the US and be self contained for the first 24 hours of a fire.

Lead Plane Small twin engine Beachcraft Baron that leads airtankers, showing them where to drop retardant.

LEO Law Enforcement Officer. US Forest Service, Bureau of Land Management, US Fish & Wildlife, Bureau of Indian Affairs officers are federal with same law enforcement powers as agents with other federal agencies.

Lookout Lookout tower or person assigned to a high spot on a fire to keep an eye on conditions for crew safety.

Tower Tower on top of a mountain to help spot fires.

MAFFS Modular Airborne Fire Fighting System. A modular system that slips into the back of National Guard C-130's and uses compressed air to spray retardant out the back through pipes, usually called out only after all civilian air tankers are committed to other assignments.

McCloud Double-headed tool with wide hoe and a steel tooth rake.

Mop Up Process of working in towards the center of a fire from the fire line to extinguish hot spots. On large fires, mop up is only done a few yards into the fire along the perimeter and the middle is allowed to burn itself out.

Out Fire is out and not manned any longer

Patrol Fire Patrol / Prevention unit. Carries limited water.

Prevention Fire Patrol / Prevention unit. No water.

Pulaski Double-headed tool with an axe on one side and a pick/hoe combination on the other.

Red Bag Back pack holding personal gear (socks, clothes, etc.)

Retardant Thick firefighting mixture (not a suppression agent) made of water, clay, grass, seed, fertilizer, and dye, dropped from an aircraft.

S.E.A.T. Single Engine Air Tanker (800 gallon crop duster) can operate off dirt roads close to a fire. Pronounced "seat"

Severity Patrols: Local fire departments are contracted by the federal government to assist in patrolling the forests with their brush units during time of extreme fire danger. The idea is more units in the forest means less likelihood that a fire will go unnoticed before it becomes large.

Skidder Wheeled tractor with a dozer blade on the front and a towing arm on the back. Used for pulling large logs, etc.

Smoke Jumpers: Highly elite team of individuals crazy enough to jump from airplanes into forest fires to cut fire lines.

Spot A place where fire has crossed outside of the main fire boundary line. On large fires, fire brands can travel miles and start a spot fire where they lands.

Strike Team Five engines and a strike team leader (usually in a separate truck) assigned as a group to a specific task (including protecting structures, holding a road, etc.). Perhaps the wildland equivalent of a municipal fire department's "alarm" assignment.

Tactical Tender: Large Engine / Water Tender Combo. Full Staffing.

Tanker Air Tanker, not a water truck. A water truck is a "water tender" or "tender".

Task Force Several strike teams together on same assignment.

Type 2 Crew: 20 person crew, not as highly trained as a hotshot crew.

Water Tender: Truck carrying a large amount of water. (Contrast with the old name, tanker, now solely used to refer to air tankers to avoid confusion.)

Yellow Shirt: Nomex (fire resistant) shirt worn by fire crews.

TERMS & ABBREVIATIONS

The following terms and abbreviations are used in this book, or may be used over the radio. This list can be used not only to see what is meant by abbreviations in this book, but also as a guide to interpreting things that may be heard over the radio.

A & B Assault & Battery
A Aid or Ambulance
Act Activity
Actor Suspect Or Subject
ADW Assault With a Deadly Weapon
AG Attorney General
Aggr. Aggravated
Air Helicopter Or Aircraft
AKA Also Known As
Alc. Alcohol
ALI Automatic Line Identification (911 Address, pronounced "Alley")
ALOC Altered Level Conscious
ALS Advanced Life Support
Amb Ambulance
APB All Points Bulletin
ARJIS Automated Regional Justice Information System
Arr Arrival
AS Air Supply Unit
ASAP As Soon As Possible
Asg. Assignment
Assoc Association
ATL Attempt To Locate
Bail Running From Car
BC Battalion Chief
Bird Helicopter / Aircraft
BLS Basic Life Support
Boost Shoplifting
BR Brush Truck
BreakOff End The Pursuit
BP Border Patrol
CA County Attorney
CAD Computer Aided Dispatch
Cap Off Shoot
Ch Channel
Cite (N) Citation
Cite (V) Issue Citation
Cnty. County
Co Company
Code 2 No Lights or Siren (or Code 1)
Code 3 With Lights and Siren
Code No Pulse or Breathing
Code Arrest: No Pulse/Breath
Com'l Commercial
Comm Communication
Coop Cooperative
Coord Coordination
CP Command Post
CRT Crisis Response Team
Ctr. Center
CV Command Van
D & D Drunk And Disorderly
DA District Attorney
Defib Defibrillate(or)
Delayed Non-Life-Threatening
Dep Departure
Desk 911 Operator
Detec Detective
Dispo Disposition. End result.
Dist District
DL Driver's License
DOA Dead On Arrival
DOB Date Of Birth
DPS Dept. of Public Safety
DRE Drug Recognition Expert
Drive-By Victim ID's Suspect
DUI Driving under Influence
DWI Driving While Intoxicated
E Engine
EKG Electrocardiogram
Emerg Emergency
EMS Emergency Medical Service
EMT Emergency Medical Technician
EOC Emergency Operations Center
Eyeball Visual On Subject
F Foam Unit
FBO Fixed Base Operator
FD Fire Department
FI Field Interview
Fld. Field
FLIR Forward Looking Infrared
Ft Fort
FTA Failure To Appear (Court)
FYI For Your Information
GOA Gone On Arrival
Gov't Government
GTA Grand Theft Auto
Gore Point: Triangle area between Freeway Lanes and on/off-ramps
HazMat Hazardous Materials
HazTech: HazMat Technician
Heli Helicopter (+Helio and Helo)
Hit Stolen / Wanted
HM Hazardous Materials
Hook Tow Truck
Hook-Up Handcuff Or Tow
HQ Headquarters
HR Heavy Rescue
IC Incident Commander
ID Identification
Immediate: Life Threatening Injuries

Immig. Immigration
Inc Incorporated
Inj Injury
Instr Instrument
Interagency: One agency assisting another
Intox Intoxicating / Intoxilizer
Inv Involved
IPS Intensive Probation Service
J/ Just
Jctn. Junction
Juv Juvenile
KDT Mobile Data Transponder
L Ladder
Landline Telephone
LC Loss Control
Level Patient Status
Level-1 Life Threatening
Level-2 Non Life Threatening
Level-3 Minor Injuries
Lic License
Line Fire Hose
Liq Liquor
Lk Lake
M Medic Unit (ALS Ambulance)
Maint. Maintenance
MDT Mobile Data Terminal
Mechanism: Injuries should be worse
MI Mandatory Insurance
Mic Microphone
Minor Minor Injuries
Misd. Misdemeanor
Misrep Misrepresent
Mngt Management
Msg. Message
Mtn Mountain
Mutual Aid: Agency assisting other
MVA Motor Vehicle Accident
NCIC National Crime Info Computer
NIP Not In Possession (License)
OD Overdose
Ofc. Office / Officer
Ofcr. Officer
Ofnc. Offence
Oper Operating
Opp. Opposite
OPS Operations
OT Overtime (Work)
P.R. Payroll Number
Patch Radio Link
PD Police Department
Perimeter Area Around Scene
PIO Public Info Officer
Pk Peak
PM Paramedic Unit (ALS Ambulance)
Point (control): Traffic Control
Prev. Prevention
Priority-1 Life Threatening
Priority-2 Non Life Threatening
Priority-3 Minor Injuries
Priv. Private
Prog. Progress
Prop. Property
R Rescue
R.O. Registered Owner
R.P. Reporting/Responsible Party
RAJ Run-Away Juvenile
RE Reserve Engine
Regist. Registration
Regs Regulations
Reh. Rehabilitation
Resp Response
RIC Rapid Intervention Crew (Safety)
RIV Rapid Intervention Vehicle
Rptr. Repeater
Rsv. Reserve
Running Red: Emergency Lights
Rwy. Runway
S & R Search & Rescue
S Support Truck
SD Sheriff's Department
Slim Jim Device to open locked car door
SO Sheriff's Office
Sort Special Oper'ns Response Team
SQ Squad
Sta. Station
Subj. Subject
Subst. Substation
Supv. Supervisor
Surv. Surveillance
Susp. Suspicious
Svcs. Services
SWAT Special Weapons/Tactics
T Water Tender
Tac Tactical
TC Traffic Collision
TOU Tactical Operations Unit
Trk Truck, or Ladder Truck
TraumaCode:Code Arrest Due to Trauma
TRS Trunked Radio System
TRT Technical Rescue Team
TX'er Transmitter
UHF Ultra-High Frequency
Unk Unknown
UTL Unable To Locate
Veh Vehicle
VHF Very High Frequency
Vict Victim
VIN Vehicle ID Number
Vio. Violence / Violent
Vlg. Village
Vly Valley
Vol. Volunteer
WT Water Tender

FEDERAL GOVERNMENT

Note: Federal law enforcement agencies utilize encryption or other devices for sensitive communications. Only "routine" communications can be monitored using the information contained in this book.

General Federal Agency Encryption Tip

Many federal government agencies use various types of digital encryption to securely transmit sensitive information. Most scanner listeners will find encrypted communications annoying to listen to because when using only a scanner, the transmission will sound much like open squelch. When you hear a "squelch" like sound in the federal bands, and adjust your radios squelch knob with no relief, it is a good bet that what you are hearing is encrypted transmissions. A good way to keep from being annoyed by encryption transmissions you can't understand anyway, while still hearing transmissions that are "in the clear" or transmitted in conventional, analog mode, is to program the PL (CTCSS) tone (if known, and if your radio is CTCSS capable) with the frequency. Because digital transmissions do not transmit a standard PL (CTCSS) tone, your radio will be silenced when transmissions are encrypted, but you will still hear all communications in conventional analog mode (which transmit the programmed PL).

Alcohol Tobacco & Firearms, United States Bureau of (BATF)

Ch1	Primary Direct	[PL 100.0]	State-Wide	165.2875	Simplex
Ch2	Tactical	[PL 100.0]	State-Wide	166.5375	Simplex
Ch3	Primary Repeater	[PL 100.0]	State-Wide	165.2875	166.5375
Ch4	Treasury Department	[PL 100.0]	State-Wide	166.4625	Simplex
Ch5	Tactical		State-Wide	165.9125	Simplex
Ch6	Arson Detail Direct		State-Wide	173.8875	Simplex
Ch7	Tactical / Metro Seattle	[Dig-P25]	State-Wide	168.0000	Simplex
Ch8	Arson Detail Repeater		State-Wide	173.8875	166.5375
	Tactical Uses		State-Wide	166.5875	Simplex
	Portland Area Operations		Portland	TG: 15632	Portland

Aviation Administration, Federal (FAA) - Facilities Maintenance

Ch1	State-Wide	172.9250
Ch2	State-Wide	172.9500
Ch3	State-Wide	172.9750
Ch4	State-Wide	172.8500
Ch5	State-Wide	172.8750
Ch6	State-Wide	172.9000
Ch7	State-Wide	172.8250
Ch8	State-Wide	172.1250
Ch9	State-Wide	172.1500
Ch10	State-Wide	172.1750
Ch11	State-Wide	166.1750

Boise Interagency Fire Cache (BIFC)

* See National Interagency Fire Center

Bonneville Power Administration

State-Wide 172.5250 Repeater

Communications Commission, United States Federal (FCC)

State-Wide Operations	State-Wide	167.0500	Simplex

Court Houses, Federal

See Marshall Service, United States

Customs and Border Protection (CBP) – Homeland Security

163.6250	163.6750	163.7250	163.7750
165.8500	165.9750	168.8625	162.8250

Radio Codes

Code	Meaning	Code	Meaning	Code	Meaning
10-01	Not Receiving Well	10-16	Pick-Up Illegals	10-33	Dangerous Subject
10-02	Receiving Well	10-17	Inspection Party	10-34	Check local file
10-03	Disregard	10-18	Any traffic ?	10-35	False claim check
10-04	Acknowledgment	10-19	No traffic	10-36	Smuggler index
10-05	Be On the Lookout	10-20	Location	10-38	Meet at
10-06	Busy	10-21	Telephone (call)	10-40	No record found
10-07	Unavailable	10-22	Report to...	10-42	Unable to talk now
10-08	Available	10-23	Restricted radio	10-43	Switch to Channel
10-09	Repeat message	10-24	Normal radio traffic	10-46	Drug Activity
10-10	Urgent Assist!	10-26	Single man traffic	10-50	Immigration check
10-11	Investigate a subject	10-27	Info from civilian	10-51	Warrant
		10-28	Registration Check	10-86	Change unit
10-12	Visitors	10-29	Wanted Check	10-87	Change assign.
10-13	Enroute	10-30	English speaker	10-96	Escorting female
10-14	ETA	10-31	Vehicle Pursuit		
10-15	In Custody	10-32	Dual man stop		

Drug Enforcement Administration, United States (DEA)

Ch	Description	Tone	Area	Frequency	Input/Mode
Ch1	Operations Repeater	[PL 156.7]	State-Wide	418.6250	416.0500
Ch2	Operations Repeater	[PL 156.7]	State-Wide	418.9000	416.3250
Ch3	Tactical / Alternate	[PL 156.7]	State-Wide	418.7500	Simplex
Ch4	Metro Seattle	[PL 156.7]	State-Wide	418.6750	Simplex
Ch5	Operations Repeater	[PL 156.7]	State-Wide	418.8250	415.6000
Ch6	Operations Repeater	[PL 156.7]	State-Wide	418.9500	416.2000
Ch7	Operations Repeater	[PL 156.7]	State-Wide	418.9750	417.0250
Ch8	Tactical / Alternate	[PL 156.7]	State-Wide	418.1750	Simplex
Ch9	Also possibly FBI use		State-Wide	419.2500	Simplex
Ch10	Interagency Coordination		State-Wide	418.0500	Simplex
Ch11	Also possibly FBI use		State-Wide	419.3000	Simplex
Ch12	Also possibly FBI use		State-Wide	419.3750	Simplex
	Air Surveillance		State-Wide	139.7000	Simplex

120.7750 120.3750 282.4250 260.8000 303.8250
336.6000 353.9000 254.2000 243.6000 381.8000
283.4000 361.8000 281.4000 355.9000 132.9500

Ch	Description	Tone	Area	Frequency	Input/Mode
	Air Surveillance (FM Mode Use)		State-Wide	418.5000	Simplex
U	Air Support (U-Uniform / Trackers)		State-Wide	350.4500	AM
	Portland Area Operations		Portland	TG: 16432	Portland

Radio Codes

Code	Meaning	Code	Meaning	Code	Meaning
10-1	Can't Copy	10-9	Repeat Message	10-22	Disregard
10-2	Received Message	10-10	Prisoner Present	10-25	Meet with __ at __
10-4	Acknowledgment	10-15	Residence	10-28	Registration
10-5	Relay Message	10-16	Change to Ch_	10-30	Subscriber Info
10-6	Stand-By	10-19	Return	10-33	Emergency
10-7	Out of Service	10-20	Location	10-99	Need Help!!!
10-8	In Service	10-21	Telephone	10-100	Tactical Ops

Emergency Management Agency, Federal (FEMA)

Ch	Description	Area	Frequency	Mode
	Nation-Wide Operations Use	State-Wide	138.2250	Repeater
	Nation-Wide Operations Use	State-Wide	141.7250	
Ch1	Nation-Wide Operations Use	State-Wide	165.6625	
Ch2	Nation-Wide Operations Use	State-Wide	164.8625	
Ch1	National Emergency Warning System	State-Wide	167.9750	Repeater
Ch2	National Emergency Warning System	State-Wide	167.9750	
Ch3	National Emergency Warning System	State-Wide	169.8750	
Ch4	National Emergency Warning System	State-Wide	167.9250	
Ch1	Task Force Operations Radio	State-Wide	418.5750	Simplex
Ch2	Task Force Operations Radio	State-Wide	418.0750	Simplex
	Search & Rescue Operations Simplex	State-Wide		

408.5125 409.4875 410.4875 410.5125 413.2125
416.0375 416.8125 416.9375 417.5875 417.6625

	Unknown Use Simplex Operations		State-Wide		

138.2250 139.8250 142.9750 142.3500 142.3750
142.4250 143.0000 166.2250 168.3500 418.0500

Forest Service (USFS) - Interagency Management System

	Interagency Common		State-Wide	168.5500	Simplex
	Interagency Common		State-Wide	168.3500	Simplex
	Nationwide Air Guard	[PL 110.9]	State-Wide	168.6250	Simplex

General Services Administration (GSA)

	Operations Repeater		State-Wide	417.2000	419.1750

Gifford Pinchot National Forest

	South Repeater – Mt. St. Helens		Areawide	172.2250	Repeater
	North Repeater		Areawide	171.4250	Repeater
	East Repeater		Areawide	172.3250	Repeater

Highway Traffic Safety Administration, National (NHTSA)

Ch1	Operations Repeater		State-Wide	040.9700	040.3900
Ch2	Operations Direct		State-Wide	040.9700	
Ch3	Alternate / Tactical		State-Wide	040.2600	
Ch4	Alternate / Tactical		State-Wide	040.3900	

Immigration & Customs Enforcement (ICE) – Homeland Security

A-1	Primary Direct	[PL 100.0]	State-Wide	165.2375	Simplex
A-2	Primary Repeater	[PL 100.0]	State-Wide	165.2375	166.4375
A-3	Treasury Department	[PL 100.0]	State-Wide	166.4625	Simplex
A-4	Local Operations		State-Wide	166.5875	
Ch5	Secondary Direct		State-Wide	163.1250	Simplex
Ch6	Secondary Repeater		State-Wide	163.1250	164.3250
Ch7	Alternate Direct		State-Wide	165.4375	Simplex
Ch8	Alternate Repeater		State-Wide	165.4375	166.3000
U	Air Support (U-Uniform)		State-Wide	282.4250	AM
U	Air Support (U-Uniform / Trackers)		State-Wide	350.4500	AM
V	Air Support (V-Victor)		State-Wide	132.9500	AM
X	Air Support (X-Ray)	[PL 100.0]	State-Wide	165.7375	
	Aircraft Operations		State-Wide	132.0500	AM
	Aircraft Operations		State-Wide	120.3250	AM
	Area Operations		State-Wide	168.7000	164.4000

Terminology

Alpha####	Enforcement Agent	Papa#	Patrol Vehicle
Cothen#	Comm Technicians	Sierra#	Internal Affairs
India#	Customs Inspector	Tango#	Project Alliance Task Force
Lima#	Customs Air Branch	##Whiskey	Air Interdiction Support
Omaha#	Customs Aircraft		

Interagency Fire Center, National (NIFC)

Air Tactics - 1 (SW OR – Medford)		166.6750	Simplex
Air Tactics - 1 (C WA – Wenatchee/Ellensburg/Ritzville)		166.6750	Simplex
Air Tactics - 2 (S OR – Lakeview/Klamath Falls)		169.1500	Simplex
Air Tactics - 2 (E OR – La Grande)		169.1500	Simplex
Air Tactics - 3 (NE WA – Spokane)		169.2000	Simplex
Air Tactics - 3 (SE OR – Burns)		169.2000	Simplex
Air Tactics - 4 (C OR – Prineville)		170.0000	Simplex
Air Tactics - 5 (E/C OR – John Day/Pendleton/Walla)		167.9500	Simplex
Air Tactics - # (W WA – Seattle/Olympia)		166.6875	Simplex
Command — 1	State-Wide	168.7000	170.9750
Command — 2	State-Wide	168.1000	170.4500
Command — 3	State-Wide	168.0750	170.4250
Command — 4	State-Wide	166.6125	168.4000
Command — 5	State-Wide	167.1000	169.7500
Command — 6	State-Wide	168.4750	173.8125
Command — 7	State-Wide	162.9625	171.7875

	Command — 8	State-Wide	169.5375	164.7125
	Command — 9	State-Wide	170.0125	165.2500
	Command — 10	State-Wide	170.4125	165.9625
	Command — 11	State-Wide	170.6875	166.5750
	Command — 12	State-Wide	173.0375	167.3250
	Forest Service Tactical - 1	State-Wide	168.0500	Simplex
	Forest Service Tactical - 2	State-Wide	168.2000	Simplex
	Forest Service Tactical - 3	State-Wide	168.6000	Simplex
	Interior Dep't. Tactical - 1	State-Wide	166.7250	Simplex
	Interior Dep't. Tactical - 2	State-Wide	166.7750	Simplex
	Interior Dep't. Tactical - 3	State-Wide	168.6000	Simplex
	Logistics - 1	State-Wide	414.6500	410.7750
	Logistics - 2	State-Wide	415.4000	411.4000
	Logistics - 3	State-Wide	415.5000	411.5000
	Logistics - 4	State-Wide	417.3000	411.7500
	Logistics - 5	State-Wide	417.3500	411.9250
	Logistics - 6	State-Wide	417.5000	412.1500
	Logistics - 7	State-Wide	417.8000	412.2000
	Air Guard	State-Wide	168.6250	Simplex
	Air to Air, Air to Ground, Helo & Plane	State-Wide	122.9250	AM
	Air to Air, Air to Ground: Helo	State-Wide	122.8500	AM
	Air to Air, Air to Ground: Helo	State-Wide	123.0250	AM
	Air to Air, Air to Ground: Helo	State-Wide	123.0750	AM
	Air to Air, Air to Ground: Plane	State-Wide	122.9750	AM
	Air Tanker Base / Initial Call-Up	State-Wide	123.9750	AM
	Air to Ground: Helicopters	State-Wide	123.0500	AM
	Air to Air (SW OR – Medford)		124.0750	AM
	Air to Air (C WA – Wenatchee/Ellensburg/Ritzville)		132.5500	AM
	Air to Air (S OR – Lakeview/Klamath Falls)		132.4500	AM
	Air to Air (E OR – La Grande)		133.2250	AM
	Air to Air (NE WA – Spokane)		132.4250	AM
	Air to Air (SE OR – Burns)		125.0750	AM
	Air to Air (C OR – Prineville)		132.3750	AM
	Air to Air (E/C OR – John Day/Pendleton/Walla)		132.4750	AM
	Air to Air (W WA – Seattle/Olympia)		132.1250	AM
	Federal Aviation Administration	State-Wide	169.3500	AM
	Flight Following	State-Wide	168.6500	Simplex
	Forest Service Law Enforcement	State-Wide	168.0250	Simplex
	Government All-Call	State-Wide	163.1000	Simplex
	Government All-Call	State-Wide	168.3500	Simplex
	Incident CMD Call-Up (+Smokejumpers)	State-Wide	168.5500	Simplex

Internal Revenue Service (IRS), United States

Ch1	Operations Direct	[PL 77.0]	State-Wide	165.9500	Simplex
Ch2	Repeater Reverse (Input Simplex)		State-Wide	167.0000	Simplex
Ch3	Operations Repeater	[PL 77.0]	State-Wide	165.9500	167.0000
Ch4	Treasury Dept. Common	[PL 100.0]	State-Wide	166.4625	Simplex
Ch5	Alternate Direct		State-Wide	415.0000	Simplex
Ch6	Alternate Repeater		State-Wide	415.0000	418.2500
Ch7	Tactical		State-Wide	418.2500	Simplex
Ch1	Primary Direct		State-Wide	418.2250	Simplex
Ch2	Primary Repeater		State-Wide	418.2250	414.7000
Ch3	Secondary Direct		State-Wide	418.1750	Simplex
Ch4	Secondary Repeater		State-Wide	418.1750	415.7250

Investigation, United States Federal Bureau of (FBI)

Ch1	Tactical Reverse	State-Wide	167.4875	Simplex
Ch2	Operations Simplex	State-Wide	167.3375	Simplex
Ch3	Operations Simplex	State-Wide	167.2375	Simplex
Ch4	Nationwide Use	State-Wide	167.5625	Simplex

Ch5	Tactical Repeater	State-Wide	163.9375	167.4875
Ch6	Tactical Simplex	State-Wide	167.7125	Simplex
Ch7	Tactical Simplex	State-Wide	167.2625	Simplex

Land Management Bureau, United States

Baker Operations	Baker	166.3750	166.9750
Burns/Medford Operations	Burns/Medford	166.3500	166.9500
Coos Bay Operations	Coos Bay	166.2000	166.8000
Lakeview Operations	Lakeview	166.3250	166.9250
Prineville Operations	Prineville	163.1500	167.0250
Roseberg Operations	Roseberg	166.3000	166.9000
Salem Operations	Salem	171.6750	166.8500
Tactical Operations	Countywide	166.1500	Simplex

Land Management, U.S. Bureau of - Interagency Management

* See National Interagency Fire Center

Marshal Service, United States (USMS)

Ch1	Nationwide Repeaters - Prisoner Trans.	State-Wide	163.2000	163.8125
Ch2	Nationwide Direct - Prisoner Trans.	State-Wide	163.2000	
Ch3	Local Repeater	State-Wide	164.6000	163.8125
Ch4	Local Direct	State-Wide	164.6000	Simplex
Ch5	Local Repeater (and Ch7)	State-Wide	162.7875	170.8500
Ch6	Local Direct (and Ch8)	State-Wide	162.7875	Simplex
Ch9	Federal Prisons (External Operations)	State-Wide	170.8750	Simplex
Ch10	Federal Prisons (Internal Operations)	State-Wide	170.9250	Simplex
Ch11	Nationwide Repeaters (+ 12, 13, 14)	State-Wide	163.2000	163.8125
Ch15	Federal Courthouse Security / OPS	State-Wide	170.7500	Simplex
Ch16	Federal Courthouse Security / OPS	State-Wide	170.8500	Simplex

Mount Hood National Forest

Ch1	Bear Springs / Barlow / Hood River		Countywide	169.9250	
Ch2	Clackamas, Estacada		Countywide	170.5250	
Ch3	Zig-Zag, Columbia Gorge		Countywide	169.9500	
Ch4	Hood River Tanker Base		Countywide	168.6250	Simplex
	Tactical Operations		Countywide	168.2000	
	OR Dept. Forestry - RED Net		Countywide	151.3400	
	ICS - Calling		Countywide	168.5500	Simplex
	Air Operations (AirTac-4)	[PL 110.9]	Countywide	170.0000	Simplex
	Air Guard	[PL 110.9]	Countywide	168.6250	Simplex
	TRS Link		Countywide	TG: 18704	Portland

Mount Rainier National Park (Ashford, WA)

Ch1	Operations Repeater	Countywide	171.7750	172.6250
Ch4	Law Enforcement	Countywide	172.4500	Simplex

North Cascades National Park (Skagit County, WA) 166.7500 165.9750

Olympic National Forest

Ch1	East End	[PL 123.0]	164.8000
Ch2	West End	[PL 123.0]	164.8250
Ch3		[PL 123.0]	164.1250
Ch4		[PL 123.0]	164.9625

Olympic National Park 168.5250 169.5500

Maintenance	168.3500	

Prisons Bureau (Marshal Service), United States

Ch1	Federal Prison - External OPS	State-Wide	170.8750	Simplex
Ch2	Federal Prison - Internal OPS	State-Wide	170.9250	Simplex
Ch3	Federal Prison - Emergency	State-Wide	170.6500	Simplex
Ch4	Federal Prison - Emergency	State-Wide	170.8250	Simplex
Ch5	Federal Prison - Emergency	State-Wide	170.9000	Simplex

Secret Service, United States (USSS)

Operations Repeater	State-Wide	165.7875	Repeater

	Operations Repeater		State-Wide	165.3750	Repeater
	Protection Detail		State-Wide	165.7875	
	Protection Detail		State-Wide	166.7000	
	Tactical Operations		State-Wide	164.6500	
	Tactical Operations		State-Wide	165.2125	
	Tactical Operations		State-Wide	166.5125	
	Treasury Dept. Common	[PL 100.0]	State-Wide	166.4625	Simplex
	Paging Services		State-Wide	167.0250	
	Air Force 1 & 2 - Ground to Air		State-Wide	407.8500	
	Air Force 1 & 2 - Air to Ground		State-Wide	415.7000	

Suislaw National Forest

Ch1	Operations Direct	[PL 103.5]	Portland	162.6125	Simplex
Ch2	Operations Repeater	[PL 103.5]	Portland	162.6125	162.0250
Ch3	Alternate Simplex	[PL 103.5]	Portland	163.3750	Simplex
Ch4	Alternate Simplex	[PL 103.5]	Portland	164.1500	Simplex
Ch5	Alternate Simplex	[PL 103.5]	Portland	164.9625	Simplex
Ch6	Air Operations (AirTac-4)	[PL 103.5]	Portland	170.0000	Simplex
	Air Guard	[PL 103.5]	Portland	168.6250	Simplex

Transportation Safety Board, National (NTSB)

	Operations Repeaters		State-Wide	165.7625	166.1750
	Operations Simplex		State-Wide	166.0250	
	Operations Simplex		State-Wide	418.2000	

Transportation Security Administration (TSA)

S1	Air Screeners – Primary	[Dig-P25]	State-Wide	172.1500	Simplex
S2	Air Screeners – Secondary	[Dig-P25]	State-Wide	172.9000	Simplex
S3	Air Screeners – Secondary	[Dig-P25]	State-Wide	169.3000	Simplex
R1	Air Screeners – Repeater	[Dig-P25]	State-Wide	172.9000	169.3000

Weather Service, National (NWS)

Nation-Wide Use - Weather Information

Ch1	Weather Radio Broadcasts	State-Wide	162.5500	Simplex
Ch2	Weather Radio Broadcasts	State-Wide	162.4000	Simplex
Ch3	Weather Radio Broadcasts	State-Wide	162.4750	Simplex
Ch4	Weather Radio Broadcasts	State-Wide	162.4250	Simplex
Ch5	Weather Radio Broadcasts	State-Wide	162.4500	Simplex
Ch6	Weather Radio Broadcasts	State-Wide	162.5000	Simplex
Ch7	Weather Radio Broadcasts	State-Wide	162.5250	Simplex

* See appropriate County section for local info

Maintenance Service	State-Wide	162.1500
Link To Radio & TV Stations	State-Wide	162.0500

Input To Remote Sites

410.1000 410.5750 415.9000 416.3750

OREGON

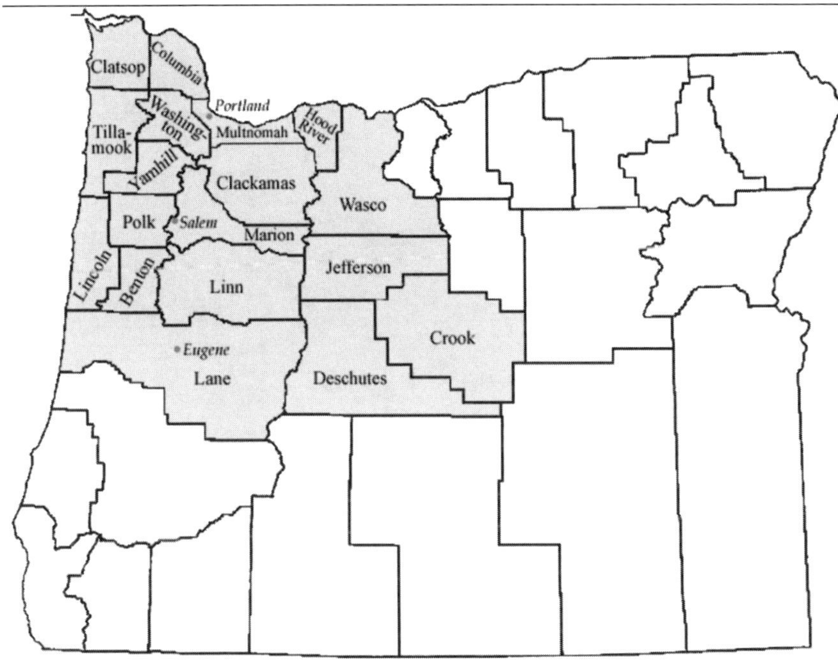

Conventional Mutual Aid & Coordination

Statewide Interagency – Oregon Ops 1 – Fire / EMS 1	866.3250
Statewide Interagency – Oregon Ops 2 – Law Tac 1	866.3875
Statewide Interagency – Oregon Ops 3 – Local Gov't	866.7500
Statewide Interagency – Oregon Ops 4 – Fire / EMS 2	866.7750
Statewide Interagency – Oregon Ops 5 – Law Tac 2	866.8000
All Oregon Search & Rescue Groups	155.8050

STATE GOVERNMENT

Oregon Department of Forestry / Fish / Wildlife

Ch#	County	Repeater Site / Description	PL Tone	Receive	Transmit
Ch1B	Washington	Buxton	[PL 146.2]	151.1450	159.2850
Ch1C	Josephine	Manzanita	[PL 179.9]	151.1450	159.2850
Ch1D	Tillamook	Bald Mountain West	[PL 146.2]	151.1450	159.2850
Ch1D	Josephine	Fielder Mountain	[PL 179.9]	151.1450	159.2850
Ch1F	Josephine	Sexton	[PL 179.9]	151.1450	159.2850
Ch2A	Klamath	Walker Mountain	[PL 131.8]	151.1525	159.3150
Ch2B	Klamath	Bald Mountain East	[PL 131.8]	151.1525	159.3150
Ch3A	Clatsop	Nicolai	[PL 131.8]	151.1600	159.2925
Ch4B	Douglas	Simulcast	[PL 146.2]	151.1600	159.4275
Ch4E	Douglas	Canyon Mountain	[PL 146.2]	151.1600	159.4275
Ch5A	Union	Mount Emily Pendleton	[PL 131.8]	151.1600	159.4275
Ch5C	Wallowa	Courtney	[PL 131.8]	151.1600	159.4275
Ch5F	Union	Spout Springs	[PL 131.8]	151.1600	159.4275
Ch5G	Morrow	Black Mountain	[PL 131.8]	151.1600	159.4425
Ch6A	Lincoln	Euchre	[PL 167.9]	151.1675	159.3225
Ch6B	Tillamook	Hebo	[PL 167.2]	151.1675	159.3225
Ch6E	Benton	Mary's Peak	[PL 167.9]	151.1675	159.3225
Ch6F	Tillamook	Hebo	[PL 146.2]	151.1675	159.3225
Ch7A	Jackson	Tallowbox	[PL 179.9]	151.1750	159.2925
Ch7B	Josephine	King Mountain	[PL 179.9]	151.1750	159.2925
Ch7C	Jackson	Soda Mountain	[PL 179.9]	151.1750	159.2925
Ch7D	Crook	Grizzly East	[PL 162.2]	151.1750	159.2925
Ch7E	Deschutes	Sugarpine	[PL 162.2]	151.1750	159.2925
Ch7F	Jackson	Roxy Ann	[PL 179.9]	151.1750	159.2925
Ch8A	Lane	Noti	[PL 131.8]	151.1825	159.3075
Ch8D	Douglas	Roman Nose	[PL 131.8]	151.1825	159.3075
Ch9D	Washington	South Saddle	[PL 146.2]	151.1900	159.4425
Ch9E	Tillamook	Rector Ridge	[PL 146.2]	151.1900	159.4425
Ch10A	Grant	Dixie	[PL 162.2]	151.2050	159.3750
Ch10A	Klamath	Hamaker	[PL 131.8]	151.2050	159.3750
Ch10B	Coos	Signal Tree	[PL 151.4]	151.2050	159.3750
Ch10C	Morrow	Madison	[PL 162.2]	151.2050	159.3750
Ch10C	Curry	Edson	[PL 151.4]	151.2050	159.3750
Ch10D	Grant	Aldrich	[PL 162.2]	151.2050	159.3750
Ch10D	Curry	Bosley Butte	[PL 151.4]	151.2050	159.3750
Ch10D	Curry	Grizzly Mountain West	[PL 151.4]	151.2050	159.3750
Ch10F	Wheeler	Snowboard	[PL 162.2]	151.2050	159.3750
Ch10F	Douglas	Dean Mountain	[PL 151.4]	151.2050	159.3750
Ch10C	Morrow	Madison	[PL 162.2]	151.2050	159.3750
Ch10G	Klamath	Yainax Butte Klamath	[PL 131.8]	151.2050	159.3750
Ch10H	Klamath	Welch Butte	[PL 131.8]	151.2050	159.3750
Ch11C	Marion	Mount Horeb	[PL 179.9]	151.2050	159.4375
Ch11E	Skamania (wa)	Skamania	[PL 179.9]	151.2050	159.4350
Ch11F	Yamhill	High Heaven	[PL 179.9]	151.2050	159.4300
Ch12D	Lane	Mount Nebo	[PL 162.2]	151.2200	159.3975
Ch12F	Lane	Farview	[PL 162.2]	151.2200	159.3975
Ch13A	Union	Mount Emily LaGrande	[PL 131.8]	151.2425	159.4050
Ch13D	Baker	Beaver Ridge	[PL 131.8]	151.2425	159.4050
Ch13E	Linn	Scott Mountain	[PL 167.9]	151.2425	159.4050
Ch13G	Linn	Coffin Mountain	[PL 167.9]	151.2425	159.4050
Ch14C	Klickitat (WA)		[PL 162.2]	151.4375	159.3975
Ch15A	Lake	Black Cap	[PL 131.8]	154.1150	159.2550

Ch15C	Klamath	Yainax Butte Lakeview	[PL 131.8]	154.1150	159.2550
Ch15F	Lake	Round Pass	[PL 131.8]	154.1150	159.2550
Ch16	State-wide	Tactical – Red Net		151.3400	Simplex
Ch17	State-wide	Tactical – White Net		151.3100	Simplex
Ch18	State-wide	Tactical – Blue Net		159.2625	Simplex
	State-wide	Tactical – Green Net		172.2250	Simplex
Ch19	State-wide	NICS – Callup		168.5500	Simplex
Ch20	State-wide	NICS – ODF		159.2400	Simplex
Ch21	State-wide	NICS – USFS		168.2000	Simplex
Ch22	State-wide	NICS – BLM		173.6750	Simplex
Ch23	State-wide	NICS – Fire Mutual Aid		154.2800	Simplex
Ch24	State-wide	NICS – W DNR		151.4150	Simplex
Ch25	State-wide	Fire Cache – TAC 1		151.2350	Simplex
Ch26	State-wide	Fire Cache – TAC 2		151.2500	Simplex
Ch27	State-wide	Fire Cache – TAC 3		151.3550	Simplex
Ch28	State-wide	Fire Cache – TAC 4		151.3850	Simplex
Ch29	State-wide	Fire Cache – TAC 5		151.4150	Simplex
Ch30	State-wide	Fire Cache – TAC 6		151.4600	Simplex
Ch31	State-wide	Fire Cache – TAC 7		151.4750	Simplex
Ch32	State-wide	Fire Cache – TAC 8		151.4900	Simplex
Ch33	State-wide	Fire Cache – NET 1 Repeater	[PL 131.8]	151.4000	159.4650
Ch34	State-wide	Fire Cache – NET 1 Back-up	[PL 131.8]	151.4300	159.4500
Ch35	State-wide	Fire Cache – NET 2 Repeater	[PL 146.2]	151.4450	158.8950
Ch36	State-wide	Fire Cache – NET 2 Back-up	[PL 146.2]	151.4300	159.4500
Ch37	State-wide	Fire Cache – NET 3 Repeater	[PL 162.2]	151.2800	159.2700
Ch38	State-wide	Fire Cache – NET 3 Back-up	[PL 162.2]	151.4300	159.4500
Ch39	State-wide	Interagency – V CALL	[PL 156.7]	155.7525	Simplex
Ch40	State-wide	Interagency – V TAC 1	[PL 156.7]	151.1375	Simplex
Ch41	State-wide	Interagency – V TAC 2	[PL 156.7]	154.4525	Simplex
Ch42	State-wide	Interagency – V TAC 3	[PL 156.7]	158.7375	Simplex
Ch43	State-wide	Interagency – V TAC 4	[PL 156.7]	159.4725	Simplex
	State-wide	Air Operations		122.8500	AM
	State-wide	Air Operations		122.9500	AM
	State-wide	Air Operations		125.2750	AM

Oregon State Government	Portland	453.8000	458.8000
Emergency Medical Services	Portland	150.7750	
Emergency Medical Services	Portland	150.7900	
Forestry Department	Portland	453.2250	458.2250
Forestry Department - Low Power	Portland	453.5625	
Oregon State University	Portland	453.5250	458.5250
University Electronic Services	Portland	453.1250	
Miscellaneous Frequencies	Portland	151.3550	

154.1600	154.1900	155.5500	453.2125	453.4125
453.6125	453.6625	453.7125	453.8125	453.9625
458.2125	458.4125	458.5250	458.5625	458.6125
458.6625	458.6875	458.7125	458.8125	458.9625

Oregon Department of Transportation

ODOT Region Map

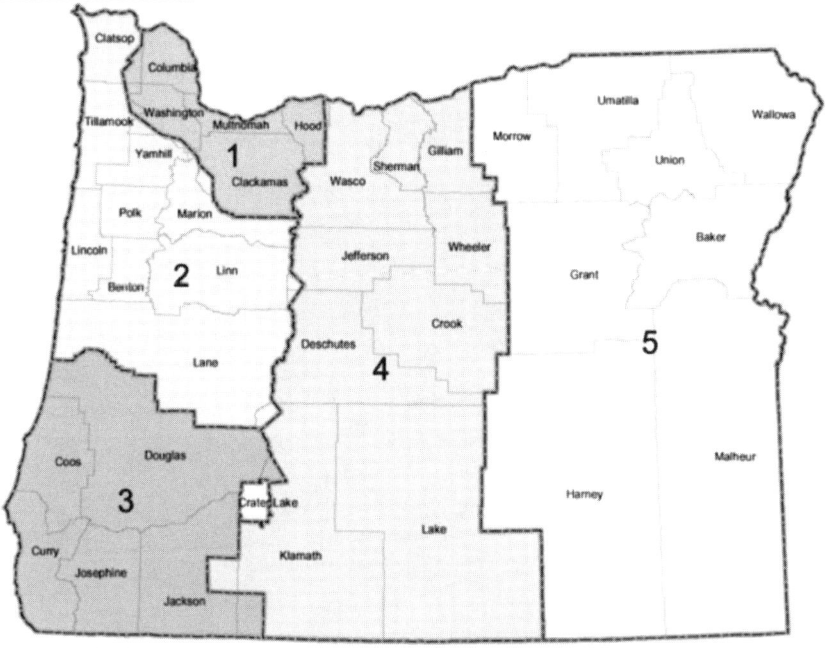

ODOT Link to 154.1000	Portland	TG: 17008	PDX
COMET Dispatch/Operations	Portland	TG: 17168	PDX
COMET Tac (Tactical/Car to Car)	Portland	TG: 17072	PDX
District 1: Clatsop, Tillamook Cnty.	State-Wide	159.0000	Repeater
District 11: Lake County	State-Wide	151.0700	Repeater
Region 1: (2A/2C) - Outside PDX	State-Wide	047.1800	
Region 1: (Dist.2B) - Metro Portland	State-Wide	047.1200	
Region 2: Northwest Oregon	State-Wide	047.0600	
Region 3: Southwest Oregon	State-Wide	047.0400	
Region 4: Central Oregon	State-Wide	047.1600	
Unknown Use	State-Wide	154.1000	Repeater
Unknown Use	State-Wide	155.7450	
Unknown Use	State-Wide	155.8200	Repeater

Oregon State Police

Repeater Locations by Channel Number

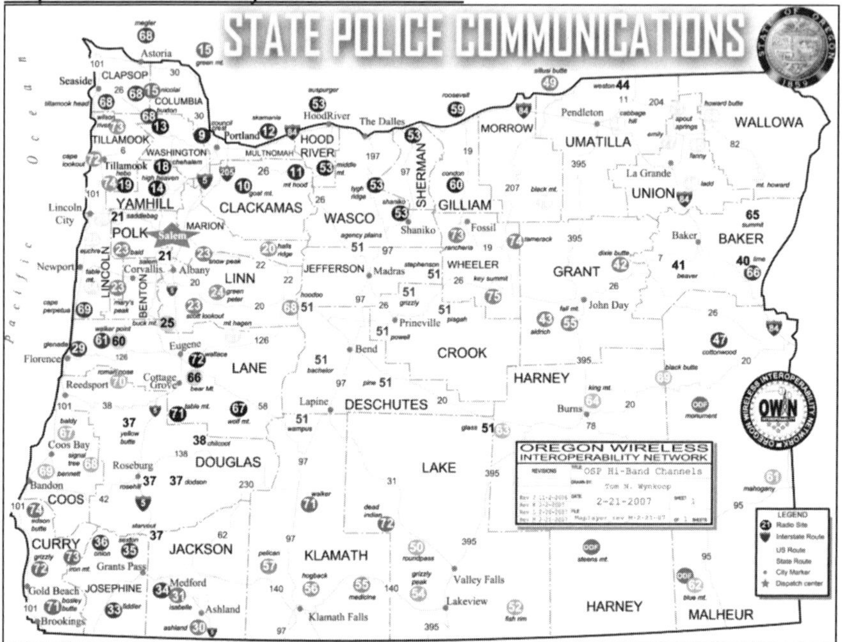

Instructions & Notes: Numbers that are circled and shaded are radio channel numbers. Identify radio channels in your area to monitor. Numbers not circled or shaded are highway numbers. Listen to the Receive frequency for wide area monitoring. Listen to the Transmit frequency to hear only mobile radios which are within several miles of your location and handheld. Multiple channels share the same Receive PL (listed below), Receive Frequency (listed below) and Transmit Frequency (listed below), but each has a different Transmit PL (not shown below) to activate different area repeaters.

Station (#)	Channel(s)	Receive PL	Receive	Transmit
Portland (10)	9 10 11 12	[PL 179.9]	154.9350	159.0900
Astoria (13)	68	[PL 179.9]	156.1500	155.4450
McMinnville (14)	13 14 18 19	[PL 179.9]	156.2250	159.1050
Saint Helens (15)	15	[PL 179.9]	154.9050	155.4600
Tillamook (17)	72 73 74	[PL 179.9]	154.8600	158.9100
Salem (21)	21	[PL 156.7]	155.9100	159.1500
Mill City (21A)	20 24 68	[PL 156.7]	154.8600	155.4450
Albany (23)	23	[PL 156.7]	154.6650	159.0300
Springfield (25)	25 60 66	[PL 156.7]	154.8450	155.5200
Cottage Grove (25A)	71 72 67	[PL 156.7]	154.9050	155.4600
Blue River (25B)	27	[PL 156.7]	154.7700	155.3100
Newport (26)	26 28 64	[PL 156.7]	154.9350	155.5800
Florence (29)	29 61 69	[PL 156.7]	154.7850	159.2100
Medford (31)	30 31	[PL 131.8]	154.6800	159.0300
Grants Pass (31A)	33 34 35 36	[PL 131.8]	154.7850	159.2100
Coos Bay (33)	67 68 69 70	[PL 131.8]	154.8600	155.4450
Gold Beach (34)	71 72 73 74	[PL 131.8]	155.9100	159.1500
Klamath Falls (36)	55 56 57	[PL 131.8]	154.6950	155.5800
Roseburg (37) (Most Towers)	37	[PL 131.8]	154.8150	155.5050
Roseburg (37) (Chikoot Tower)	38	[PL 131.8]	154.9200	155.5050
Lakeview (38)	50 52 54	[PL 131.8]	155.4600	154.8450
Baker (41)	40 41 65	[PL 151.4]	155.9100	159.0300

John Day (41A)	42 43 55	[PL 167.9]	154.9050	155.4600
LaGrande (46)	45 46	[PL 131.8]	154.9050	155.4600
Enterprise (46A)	67 68	[PL 123.0]	154.9050	155.4600
Enterprise (46A)	71	[PL 151.4]	154.9050	155.4600
Ontario (47)	47 66	[PL 151.4]	154.9350	155.5050
Burns (47A)	61 62 63 64 69	[PL 151.4]	154.8600	155.4450
Burns (47A Infill areas use ODF area Repeater)		[PL 151.4]	151.1900	151.4600
Pendleton (48)	48 70	[PL 151.4]	154.9050	155.4600
Umatilla (48A)	49	[PL 167.9]	154.6500	159.0300
Milton-Freewater (48B)	44	[PL 192.8]	154.6500	159.1500
Bend (51)	51	[PL 192.8]	154.7850	159.0300
Gilcrest (51A)	71 72	[PL 127.3]	154.9350	159.0900
The Dalles (53)	53	[PL 192.8]	155.9100	159.2100
Fossil (53A)	73 74 75	[PL 192.8]	154.9350	159.1500
Arlington (53B)	59 60	[PL 192.8]	155.9100	159.2100

Oregon Station Police Channels other than Patrol Station Primary Operations

22	Executive Security	[PL 179.9]	Salem	154.7850	155.5050
100	Tactical – Portable Rptr.	[PL 156.7]	State-Wide	156.0300	158.8950
75	Fish & Game Enforcement	[No PL]	State-Wide	158.8950	Simplex
76	Search & Rescue	[No PL]	State-Wide	158.9050	Simplex
77	OR Police Emerg. Net.	[No PL]	State-Wide	155.4750	Simplex
78	FireNet Mutual Aid	[No PL]	State-Wide	154.2800	Simplex
79	HEAR Medical Net	[No PL]	State-Wide	155.3400	Simplex
	Tactical/Car to Car		Portland	TG: 17424	Portland
	TRS Link to 154.935		Portland	TG: 17392	Portland

Radio Codes

12-01	In Service
12-02	Out of Service
12-03	Return
12-04	Contact by Phone
12-05	Repeat Last
12-06	Contact Location
12-07	Registration Check
12-08	Owner Check
12-09	Status Check
12-10	License Check
12-10A	License Invalid
12-11	License Info
12-12	Unable to Copy
12-13	Prepare to Copy
12-14	Relay to
12-15	Emergency Msg.
12-16	Collision
12-16A	Fatal Collision
12-16V	Injury Collision
12-17	Injury Collision
12-18	Request Ambul.
12-19	Request Tow
12-20	Wanted Check
12-20A	Confidential Info
12-21	No Wants
12-22	Prior Record - M
12-23	Prior Record - F
12-24A	Felony Warrant
12-24B	Misdem. Warrant
12-25	Wanted Similarity
12-26	Out of Service
12-27	Call on Arrival
12-28	Suspicious Subject
12-29	Disturbance
12-30	Reckless Driver
12-31	DUI
12-32	Intoxicated Subject
12-33	Emergency Traffic
12-34	Resume Normal
12-35	Abandoned Car
12-36	Illegal Shooting
12-37	Advise Conditions
12-38	Switch to _
12-39	Attention
12-40	Stand-By
12-41	State Traffic
12-42	No Traffic
12-43	Disregard
12-44	HazMat Spill
12-45	Burglary Alarm
12-46	Phone Number
12-47	Records Unavail.
12-48	Records Avail.
12-49	Death
12-49A	Homicide
12-50	Phone Message
12-51	Radio Maint.
12-52	Radio Tech
12-53	Power Outage
12-54	Radio Check
12-55	Radio Test
12-56	No Back-Up
12-57	Motorist Assist
12-58	Narcotics
12-59	Late Return
12-65	Roll Call
12-88	Off-Duty
12-94	Clear
12-96	Vehicel Stop
12-97	Radio Check
12-98	Need Help
12-99	Need Help NOW

BENTON COUNTY

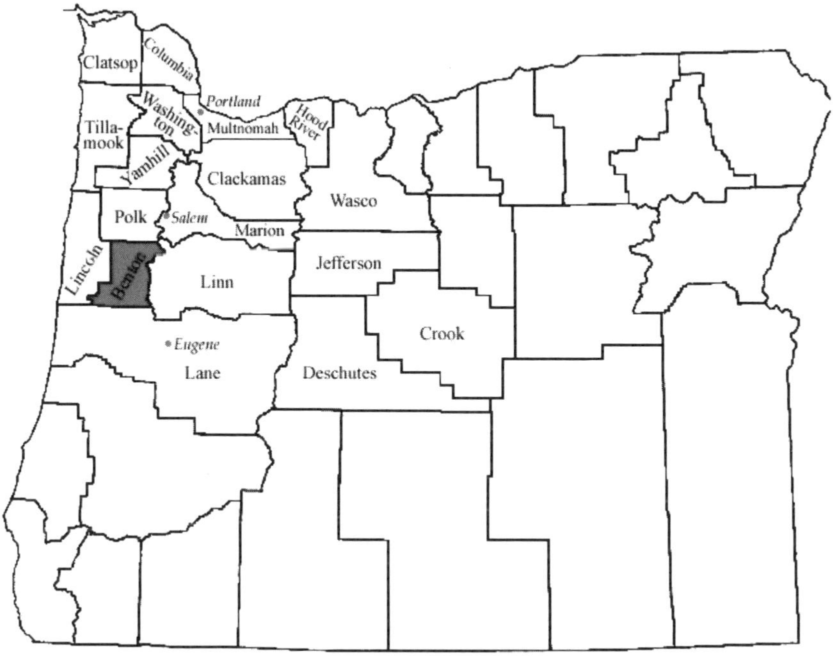

Government & Public Safety

Benton County Sheriff's Office			Countywide	156.0900	Repeater?
Ch1	Operations	[PL 131.8]	Countywide	155.4900	156.0900
	Operations	[PL 167.9]	Countywide	155.4900	155.6250
	Tactical		Countywide	155.0700	
Tac-4	Tri County Net (w/ Lincoln & Linn)		Countywide	155.0100	Simplex
	Search & Rescue		Countywide	155.8050	Simplex
	Corrections / Jail		Countywide	155.8800	
	Tactical		Countywide	153.9125	Simplex

Benton County Law Enforcement Radio Designators

4##	Benton County Sheriff's Office
8##	Corvalis Police Department
11##	Philomath Police Department
41##	Benton County Corrections

Benton County Fire / EMS

Ch1	Dispatch	[PL 167.9]	Countywide	154.3400	156.1950
Ch2	Tactical		Countywide	154.1900	
Ch3	Linn County FD Dispatch		Countywide	154.3100	
Ch4	Statewide Fire Mutual Aid		Countywide	154.2800	
	Tactical		Countywide	150.7750	Simplex

153.8300 154.4300 158.8200 150.8050

Blodgett Summit Rural Fire Protection District

See Benton County Fire / EMS

Agency	Location	Frequency	Input
Corvallis Fire Department			
See Benton County Fire / EMS			
Corvallis Police Department			
See Benton County Sheriff's Office			
Corvallis Public Works	Corvallis	153.8000	Simplex
	Corvallis	156.2400	158.9550
	Corvallis	159.0975	155.1975
	Corvallis	453.6750	Simplex
	Corvallis	453.3125	458.3125
	Corvallis	452.4250	457.4250
Hoskins Kings Valley Rural Fire Protection District			
See Benton County Fire / EMS			
Monroe Rural Fire Protection District		155.8275	159.0525
Also See Benton County Fire / EMS			
Oregon Dept. Forestry [PL 167.9]	Mary's Peak	151.1675	159.3225
Oregon State Police [PL 156.7]	Albany	154.6650	159.0300
Oregon State University Police	Corvallis	155.7450	158.8650
Philomath Rural Fire Protection District			
See Benton County Fire / EMS			
Philomath Police Department	Philomath	154.0850	Simplex

Miscellaneous

Agency	Location	Frequency	Input
Benton County Dial A Bus	Corvallis	151.6550	Simplex
Operations Repeater	Corvallis	152.2850	157.5450
Consumers Power	Countywide	047.8600	048.4200
	Countywide	048.0400	048.5400
Corvallis Municipal Airport			
CTAF / UNICOM	Corvallis	123.0000	AM
Cascade Approach / Departure	Corvallis	127.5000	AM
Seattle Center Approach / Dep	Corvallis	125.8000	291.7000
Corvallis Country Club	Corvallis	467.3125	Simplex
Evanite Fiber Manufacturing Plant	Corvallis	464.8750	Simplex
Operations Repeater	Corvallis	464.4750	469.4750
Georgia Pacific Corporation	Philomath	463.7125	Simplex
Lumber Production	Philomath	463.8250	Simplex
Lumber Production	Philomath	464.2125	Simplex
Good Samaritan Hospital	Corvallis	152.2700	157.5300
463.8125 464.2125	Corvallis	152.3450	157.6050
464.9875 467.1625	Corvallis	152.4050	157.6650
Oregon Department of Corrections			
Philomath Facility	Philomath	453.2000	458.2000
453.6875 453.7625 453.9125	Philomath	458.9375	Simplex
Pacificorp Electric Utility	Corvallis	153.5600	159.6450
Spring Hill Country Club	Albany	452.1875	Simplex
	Albany	457.1875	Simplex
Trysting Tree Golf Club		159.9975	Simplex
Union Pacific Railroad	Mary's Peak	160.9500	
Wiley Creek Community Hospital	Lebanon	457.5375	Simplex

Willamette & Pacific Railroad	Mary's Peak	160.2150	
Yard / Maintenance of Way	Countywide	160.3650	
Road – 1	Mary's Peak	160.7700	
Road – 2	Countywide	160.8450	
Road – 3	Countywide	160.8750	

Schools

Corvallis School District	Corvallis	152.4200	Simplex
First Student Bus Transportation	Philomath	151.9850	159.5250
Laidlaw Transit	Corvallis	461.9250	466.9250
Oregon State University	Corvallis	453.3375	Simplex
151.9250 152.8700 151.7000 151.7600 464.3250			
464.5750 461.3875 461.4375 461.5625 462.0875			
463.2125 463.6375 463.9375 463.9875 464.2375			
464.4375 466.3875 466.4375 466.5625 467.0875			
Housing Maintenance	Corvallis	464.2000	469.2000
Repeaters	Corvallis	453.5250	458.5250
453.7000 453.8250 453.8750	Corvallis	460.3250	465.3250
College of Forestry	Corvallis	151.6250	Simplex
College of Forestry – Research	Corvallis	151.9250	Simplex
College of Forestry – Research	Corvallis	152.8700	Simplex
Alumni Association	Corvallis	463.2125	Simplex
463.5375 463.8125 463.9125 465.8125 468.9125			
Philomath School District			
Bus – First Student Inc Contract	Philomath	151.9850	159.5250

Weather Information

National Weather Service – NOAA Weather Radio		
Conditions & Forecasts – Eugene	162.4000	Simplex
Conditions & Forecasts – Salem	162.4750	Simplex
Benton County SAME Alert Code: 041003		

Clackamas County

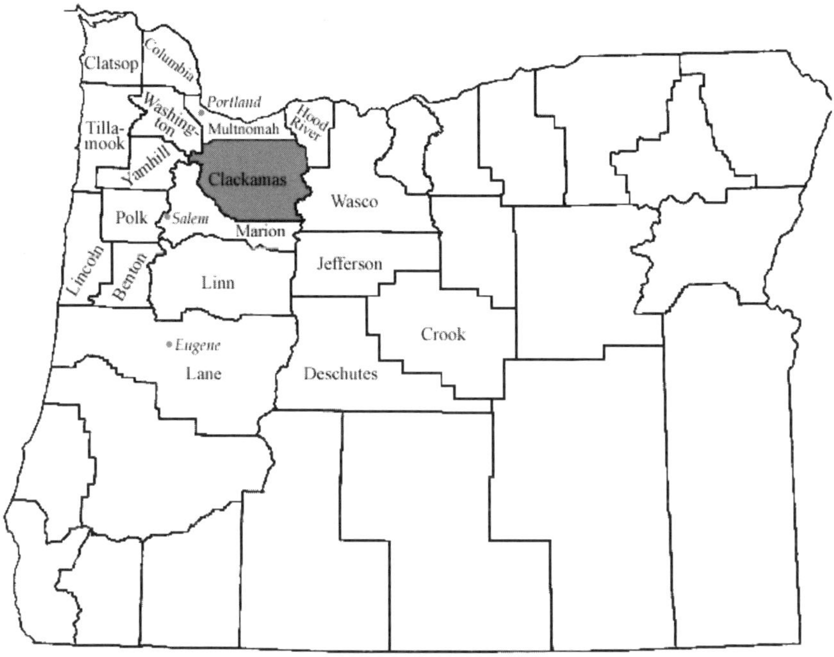

Government & Public Safety

Clackamas County Trunked Radio System ("Clackamas" talkgroups)
Clackamas County Communications Agency Trunked Radio System
TRUNKED RADIO SYSTEM (Motorola SmartZone)
TrunkTracker Setting: "Type II 800"
Clackamas County Simulcast Zone: Oregon City
866.2375 866.4875 867.1625 867.1875 867.4375*
867.8625* 868.2625* 868.7625* 866.3000 866.8500
Goat Mountain Intellirepeater Estacada
867.7250 867.8375 867.9875 868.5125* 866.9375

Clackamas County Public Safety Dispatching
Clackamas County Communications (CCOM) is the dispatching entity for most Clackamas County law enforcement and fire/life-safety agencies. The agencies not dispatched by C-COM are dispatched by Lake Oswego (LOCOM) and include Lake Oswego, Milwaukie, and West Linn. Whether dispatched by CCOM dispatchers or LOCOM dispatchers, virtually all public safety agencies in Clackamas County operate on talkgroups on the Clackamas County Trunked Radio System, which is operated as a subsystem of the Washington County Consolidated Communications Agency trunked radio system.

Clackamas County Law Enforcement Callsigns

Unit Designators Example: # X # #

First Digit (numeric)	Second Digit (alpha)	Third Digit(s) (number)	
		(Varies by Agency)	
		Sheriff Numbers:	
1: 2100-0700 Shift	D: Patrol Deputy	1-7:	Westside Cars
2: 0700-1700 Shift	F: Marine Patrol	10-34:	99E/McLoughlin
3: 1100-2100 Shift	G: Estacada	40-59:	82nd Avenue / Clackamas Cars
4: 1700-0300 Shift	K: Civil Division	60-69:	Boring / Mtn.
5: K-9 Units	L: Lake Oswego P.D.	70-73:	East Cars
6: Special (DARE, PIO)	N: Molalla P.D.	76-79:	Metro Cars
7: Reserve Deputies	P: Oregon City P.D.	80-85:	South Cars
8: Traffic / Special	R: Sandy P.D.	90-93:	Central Sgt's.
9: SWAT Team Members	S: West Linn P.D.	94-96:	Outlying Sgts.
10: Search & Rescue	T: Wilsonville	97-99:	Roving Sgt's.
	V: CART (Gangs)		
	Z: Detectives		

American Medical Response — Countywide 155.2350 Simplex

AMR Northwest is a private ambulance company with a contract to provide Advanced Life Support and Basic Life Support medical transportation in Clackamas County.

Beavercreek Rural Fire District

See Clackamas County Fire Agencies

Boring Fire Protection District

See Clackamas County Fire Agencies

Canby Rural Fire District #62

Ch1	Canby Fire 1	Canby	TG: 55536	Clackamas
Ch2	Canby Fire 2	Canby	TG: 55504	Clackamas

Canby Police Department — Canby 155.5950

See Clackamas County Law Enforcement Agencies, Plus:

Desk Sergeant	Canby	TG: 58384	Clackamas
Administration	Canby	TG: 58416	Clackamas
Detectives	Canby	TG: 58448	Clackamas
OLD UHF Repeater	Canby	453.1000	458.1000
OLD UHF Repeater	Canby	453.5500	458.5500

Canby City Services — Canby 154.0250

Public Works Department	Canby	TG: 58352	Clackamas

Clackamas County Fire Agencies

Clackamas County TRS Talkgroups

A1	C-COM Dispatch Countywide	Countywide	TG: 51408	Clackamas
A2	C TAC 2 – LOCOM Minor Incidents	Countywide	TG: 51344	Clackamas
A3	C TAC 3 – LOCOM Major Incidents	Countywide	TG: 51312	Clackamas
A4	C TAC 4 – C-COM Minor Incidents	Countywide	TG: 51280	Clackamas
A5	C TAC 5 – C-COM Major Incidents	Countywide	TG: 51248	Clackamas
A6	C TAC 6	Countywide	TG: 51216	Clackamas
A7	C TAC 7	Countywide	TG: 51184	Clackamas
A8	C TAC 8	Countywide	TG: 51152	Clackamas
A9	PUB SAFE 1 – Public Safety	Countywide	TG: 64656	Clackamas
A10	PUB SAFE 2 – Public Safety	Countywide	TG: 64624	Clackamas
A11	TRAIN 1 – Training	Countywide	TG: 51120	Clackamas
A12	TRAIN 2 – Training	Countywide	TG: 51088	Clackamas
A13	ADMIN – Administration	Countywide	TG: 51440	Clackamas
A14	Simplex 1	Countywide	868.4375	Simplex
A15	Simplex 2	Countywide	866.0375	Simplex
A16	MAYDAY Channel	Countywide	868.3875	Simplex
	VHF F-2 Patch (Molalla)	Countywide	TG: 55600	Clackamas

VHF F-4 Patch (Polivka)	Countywide	TG: 55568	Clackamas
Talkgroup Patches to Old VHF Frequencies			
C-Ch2 Operations - South County & Metro	Countywide	TG: 55600	Clackamas
C-Ch4 Operations - East County	Countywide	TG: 55568	Clackamas
Old VHF Frequencies			
V-Ch1 Dispatch Countywide	Countywide	154.1900	
V-Ch2 Operations - South County & Metro	Countywide	153.8900	
V-Ch3 Operations - North County & Metro	Countywide	154.2200	
V-Ch4 Operations - East County	Countywide	154.3250	
V-Ch5 Operations - Sandy Area & Training	Countywide	153.8300	
V-Ch6 Training	Countywide	154.2050	
V-Ch7 Mayday	Countywide	154.3400	
V-Ch8 State Fire Marshall (Mutual Aid)	Statewide	154.2800	Simplex
VHF Patches to Portland TRS			
P-Ch1 Dispatch Countywide (Link to V-Ch1)	Countywide	TG: 15568	Portland
P-Ch3 Operations - North Cnty (Link to V-Ch3)	Countywide	TG: 16048	Portland
P-Ch4 Operations - East Cnty (Link to V-Ch4)	Countywide	TG: 16080	Portland

Clackamas County Law Enforcement Agencies (CCOM)

A1	CCOM LAW 1 – Primary Dispatch	Countywide	TG: 64912	Clackamas
A2	CCOM LAW 2 – Secondary / Service	Countywide	TG: 64880	Clackamas
A3	CCOM LAW 3 – Tertiary / Car to Car	Countywide	TG: 64848	Clackamas
A4	CCOM SERV	Countywide	TG: 64816	Clackamas
A5	CCOM TAC 1	Countywide	TG: 64784	Clackamas
A6	CCOM TAC 2	Countywide	TG: 64752	Clackamas
A8	Common Law – Interagency	Countywide	TG: 65008	Clackamas
A9	CC PS1 – Public Safety	Countywide	TG: 64656	Clackamas
A10	CC PS2 – Public Safety	Countywide	TG: 64624	Clackamas
	Training	Countywide	TG: 64720	Clackamas
	Search & Rescue	Countywide	TG: 64336	Clackamas
	Marine Detail	Countywide	TG: 64688	Clackamas
	Detectives	Countywide	TG: 64688	Clackamas
	CCSO Sheriff Desk Sergeant	Countywide	TG: 64432	Clackamas
Old VHF Sheriff Frequencies		Countywide	159.0300	
Ch1	Primary / Dispatch	Countywide	155.4300	Repeater
Ch2	Secondary / Service NET	Countywide	155.1900	Repeater
Ch3	Car to Car	Countywide	155.7600	Repeater
Ch4	Tactical Simplex	Countywide	155.6100	
Ch6	Interagency (O.P.E.N.)	Countywide	155.4750	

Clackamas County Government

Animal Control	Countywide	045.4400	
Conventional Repeater	Countywide	851.6875	806.6875
Conventional Repeater	Countywide	856.2125	811.2125
Conventional Repeater	Countywide	856.7125	811.7125
School District 12J	Countywide	853.4875	808.4875
Emergency Management	Countywide	046.5800	
Emergency Medical	Countywide	155.2800	
Public Safety Department	Countywide	155.0550	
Road Maintenance	Countywide	156.1650	
Road Maintenance	Countywide	159.0600	Repeater
Road Maintenance	Countywide	159.1650	Repeater
Search & Rescue	Countywide	155.8050	
	Countywide	453.4000	458.4000
	Countywide	453.7000	458.7000
	Countywide	154.8450	Repeater
Miscellaneous Use Simplex	Countywide	155.5200	Simplex
Miscellaneous Use Simplex	Countywide	156.1050	Simplex

158.7600 158.8500 158.9700

Clarkes Rural Fire Protection
See Clackamas County Fire Agencies

Colton Rural Fire Protection District
See Clackamas County Fire Agencies

Estacada Rural Fire District #6
See Clackamas County Fire Agencies

Estacada City Public Works
	Estacada City Public Works	Estacada	045.4000	Simplex

Estacada Police Department
See Clackamas County Law Enforcement Agencies

Gladstone Fire Department
See Clackamas County Fire Agencies

	Gladstone FD Talk	Gladstone	TG: 55824	Clackamas

Gladstone City Services
	Gladstone City Services	Gladstone	155.1450	Simplex
	Public Works Department	Gladstone	TG: 55824	Clackamas

Gladstone Police Department
See Clackamas County Law Enforcement Agencies

	Gladstone Police Tactical	Gladstone	TG: 55856	Clackamas
	OLD VHF Frequency	Gladstone	155.7000	Simplex
	OLD VHF Frequency	Gladstone	159.0300	Simplex

Lake Oswego Fire Department
A1	LOCOM Dispatch	Countywide	TG: 51056	Clackamas
A2	C TAC 2 – LOCOM Minor Incidents	Countywide	TG: 51344	Clackamas
A3	C TAC 3 – LOCOM Major Incidents	Countywide	TG: 51312	Clackamas
A4	C TAC 4 – C-COM Minor Incidents	Countywide	TG: 51280	Clackamas
A5	C TAC 5 – C-COM Major Incidents	Countywide	TG: 51248	Clackamas
A6	C TAC 6	Countywide	TG: 51216	Clackamas
A7	C TAC 7	Countywide	TG: 51184	Clackamas
A8	C TAC 8	Countywide	TG: 51152	Clackamas
A9	PUB SAFE 1 – Public Safety	Countywide	TG: 64656	Clackamas
A10	PUB SAFE 2 – Public Safety	Countywide	TG: 64624	Clackamas
A11	TRAIN 1 – Training	Countywide	TG: 51120	Clackamas
A12	TRAIN 2 – Training	Countywide	TG: 51088	Clackamas
A13	ADMIN – Administration	Countywide	TG: 51440	Clackamas
A14	Simplex 1	Countywide	868.4375	Simplex
A15	Simplex 2	Countywide	866.0375	Simplex
A16	MAYDAY Channel	Countywide	868.3875	Simplex

Lake Oswego City Government
	Lake Oswego City Government	Lk. Oswego	045.2800	Simplex
	Public Works 1	Lk. Oswego	TG: 57488	Clackamas
	Public Works 2	Lk. Oswego	TG: 57456	Clackamas
	City Emergency Operations Center	Lk. Oswego	TG: 50992	Clackamas
	OLD Operations Simplex	Lk. Oswego	045.5200	Simplex

159.0300 453.9375 458.9375

Lake Oswego Law Enforcement Communications (LOCOM)
A1	LOCOM LAW 1 – Primary Dispatch	Countywide	TG: 57776	Clackamas
A2	LOCOM LAW 2 – Secondary / Service	Countywide	TG: 57744	Clackamas
	LOCOM TAC 1 – Tactical	Countywide	TG: 57712	Clackamas
	LOCOM TAC 2 – Tactical	Countywide	TG: 57680	Clackamas
	LOCOM Public Safety 1	Countywide	TG: 57552	Clackamas
	LOCOM Public Safety 2	Countywide	TG: 57520	Clackamas

Lake Oswego Police Department
See Lake Oswego Law Enforcement Communications (LOCOM), Plus:

	Training	Lk. Oswego	TG: 57616	Clackamas
	Administration	Lk. Oswego	TG: 57584	Clackamas
U-Ch1	Old Primary - Dispatch Direct	Lk. Oswego	460.4500	465.4500
U-Ch2	Old Secondary - Car to Car	Lk. Oswego	460.2250	465.2250
U-Ch3	Old Primary - Direct	Lk. Oswego	460.4500	

Milwaukie City Government | Milwaukie | 453.8750 | 458.8750
156.0000 159.0300 453.3000

Public Works - Low Power	Milwaukie	453.9875	Simplex
Public Works - Low Power	Milwaukie	458.9875	Simplex
Public Works – 1	Milwaukie	TG: 3184	Portland
Public Works – 2	Milwaukie	TG: 3216	Portland

Milwaukie Fire Department

See Clackamas County Fire Agencies

Milwaukie Police Department

Dispatch by LOCOM Dispatchers on Portland TRS Talkgroups Below:

P1	Dispatch / Operations	Milwaukie	TG: 3120	Portland
P2	Tactical/Car to Car	Milwaukie	TG: 3152	Portland
P6	Tactical/Car to Car	Milwaukie	TG: 3056	Portland
P8	Tactical/Car to Car	Milwaukie	TG: 3088	Portland
T1	MilTac – Tactical/Car to Car	Milwaukie	TG: 3024	Portland

Molalla Rural Fire Protection

See Clackamas County Fire Agencies

Molalla City Government | Molalla | 046.5200 | Simplex

Public Works	Molalla	TG: 58224	Molalla

Molalla Police Department

See Clackamas County Law Enforcement Agencies, Plus:

Desk Sergeant	Canby	TG: 58256	Clackamas
Administration	Canby	TG: 58288	Clackamas
Detectives	Canby	TG: 58320	Clackamas

Oregon City City Government | Oregon City | 045.0800 | Simplex

Public Works Department	Oregon City	TG: 58096	Clackamas
City Emergency Operations Center	Oregon City	TG: 58064	Clackamas

Oregon City Fire Department

See Clackamas County Fire Agencies

Oregon City Police Department

See Clackamas County Law Enforcement Agencies, Plus:

Desk Sergeant	Oregon City	TG: 58128	Clackamas
Administration	Oregon City	TG: 58160	Clackamas
Detectives	Oregon City	TG: 58192	Clackamas

Oregon State Police

North County (Sta 10)	[PL 179.9]	Portland	154.9350	159.0900
South County (Sta 21A)	[PL 156.7]	Mill City	154.8600	155.4450

Sandy Fire District

See Clackamas County Fire Agencies

Sandy City Government | Sandy | 046.5400 | Simplex

	Public Works Department	Sandy	TG: 56976	Clackamas
Tac-7	Tactical Simplex	Sandy	868.0500	Simplex
	Public Safety 2	Sandy	868.7375	Simplex

Sandy Police Department

See Clackamas County Law Enforcement Agencies, Plus:

Desk Sergeant	Sandy	TG: 57008	Clackamas
Administration	Sandy	TG: 57040	Clackamas
Detectives	Sandy	TG: 57072	Clackamas

West Linn City Government | West Linn | 154.0550

Public Works Department	West Linn	TG: 56848	Clackamas
City Emergency Operations Center	West Linn	TG: 56912	Clackamas

West Linn Police Department | West Linn | 158.7600

See Lake Oswego Law Enforcement Communication (LOCOM), Plus:

Administration	West Linn	TG: 56880	Clackamas
Detectives	West Linn	TG: 56944	Clackamas

Wilsonville City Government	Wilsonville	154.9800	
Transit F1	Wilsonville	TG: 61200	Clackamas
Transit F2	Wilsonville	TG: 61168	Clackamas
Public Works 1	Wilsonville	TG: 61264	Clackamas
Public Works 2	Wilsonville	TG: 61232	Clackamas
Sewer 1	Wilsonville	TG: 61328	Clackamas
Sewer 2	Wilsonville	TG: 61296	Clackamas

Wilsonville Police Department

See Clackamas County Law Enforcement Agencies

Clackamas Fire District & Apparatus Numbering Plan

Nmbr.	Department	Location	Apparatus
1000	Gladstone FD	Main Station	E101 E103 R101
1100	Estacada FD	Main Station	E111 E114 R111 SQ111 WT115
		George Station	E112 WT118
1200	Colton FD	Main Station	E125 E129 WT124 R121
		Elwood Station	E123 WT126
1300	Clarkes FD	Main Station	E132 WT134 R130
		Upper Hill	E131 WT133
1400	Boring FD	Main Station	E140 WT140 WT141 R140
		Eagle Cr. Sta.	E148 SQ148
		Damascus Sta.	E149 R149 Brush149
1500	Beavercreek FD	Main Station	E152 E153 WT158 R155
WL	West Linn FD	Willamette Sta.	E172
		Bolton Station	E182
		Sunset Station	E192 R192
		Robinwood Sta.	E232 R231
		Rosemont Sta.	E242 E244 WT245
2000	Lake Oswego FD	Westlake Sta.	E210 E213
		Jean Road Sta.	E212 BR212
		South Shore	E211 T211
		Main Station	E214 M214 DiveRescue214 BC1
2500	Hoodland FD	Main Station	E251 E254 WT251 R251 BR251
		Brightwood	E252 WT252
		Gov't Camp	E253 R253
3100		Oak Lodge	E10 TR10 R10
3200	County Fire	11300 S Fuller Road	E1 TR1 BC2
		Milwaukie	E2 E22
		Oak Lodge	E3 TR3 SQ4 E23 TR23
		Lake Road	E4
		Mount Scott	E5
		Happy Valley	E6 BR6 E26
		SE 172nd	E7
		SE 130th Avenue	E8 BT8 E28
		Holcomb	E9 BC3
		Beavercreek	E10 E20 BR10 WT10 RH10
		Redland	E11 E31 BR11 WT11
		Logan	E12 BR12 WT12
		Clarkes	E13 R13 BR13 WT13
		Clarkes	E14 WT14
		Oregon City – Main	T15
		Oregon City – Hilltop	E16
		Oregon City – Southend	E17
3500	Milwaukie FD	Main Station	E54
		40th Station	E55 R51
3600	Canby FD	Main Station	E61 E65 E67 WT66 WT68 R64
		Hwy170 Sta.	E62

3700	Sandy FD	Main Station	E71 E72 E74 WT71 R77 BR72
		Dover Station	E75 WT75
		Roslyn Lake Sta.	E76
3700	Molalla FD	Main Station	E82 E84 WT82 R83 Amb81 Amb82
		Mulino Station	E81 E83 WT81
		Sawtell Station	E85 WT85

Please see Multnomah County section for Portland area non-public safety information:

Airports, Amateur Radio, Schools, Industries, Recreation, Media, etc.

Weather Information

National Weather Service – NOAA Weather Radio

Conditions & Forecasts – Portland	162.5500	Simplex
Conditions & Forecasts – Salem	162.4750	Simplex

Clackamas County SAME Alert Code: 041005

Clatsop County

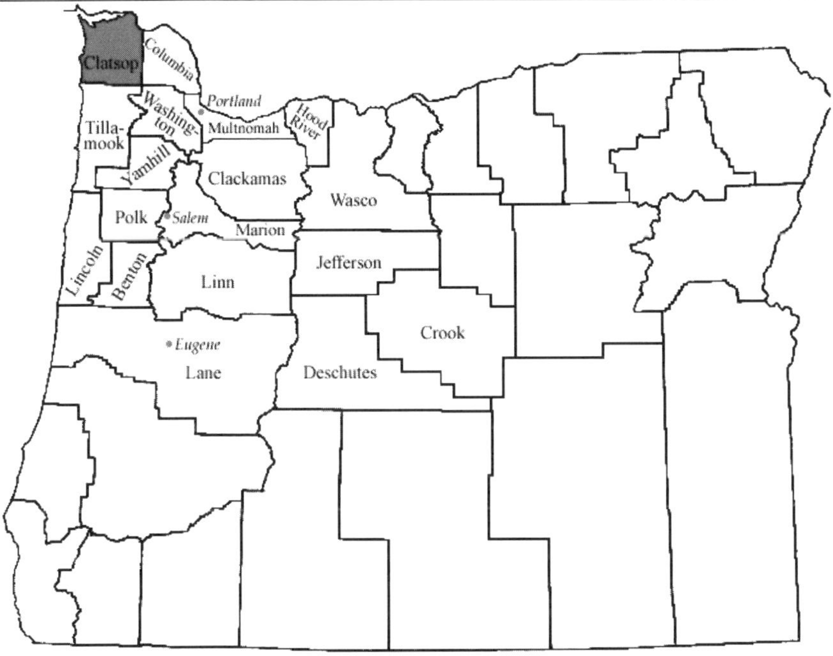

Government & Public Safety

Agency	Location	Frequency	Output
Astoria City Government	Astoria	460.5500	Simplex
Port of Astoria Operations	Astoria	155.9250	Simplex
Public Works	Astoria	158.8200	Simplex
Astoria Fire/Rescue Department: See North Clatsop County Fire Departments			
Astoria Police Department [PL 77.0]	Astoria	155.2500	155.6850
Cannon Beach Rural Fire District	Cannon Bch.	155.6400	Simplex
Cannon Beach Police	Cannon Bch.	158.7600	Simplex
	Cannon Bch.	158.8050	Simplex
Clatsop County Government			
Public Works	Countywide	151.0400	Simplex
Public Works	Countywide	156.0450	Simplex
Clatsop County Sheriff's Department	Countywide	155.5500	Simplex
Operations Repeater [PL 107.2]	Countywide	155.7900	155.1300
[PL 123.0]	Countywide	154.8000	Simplex
	Countywide	155.8050	Simplex
	Countywide	158.9550	Simplex
Low Power	Countywide	453.6375	Simplex

Clatsop County Sheriff's Department – Radio Codes

12-Codes

12-01	In Service
12-02	Out of service
12-03	Return to office
12-04	Call office by phone
12-05	Repeat message
12-06	Contact address or complainant
12-07	Motor vehicle registration
12-08	Motor vehicle reg & owner
12-09	Escort
12-10	Check driver's license
12-11	Check description & D.L.
12-12	Unable to copy, change loc.
12-13	Prepare to copy message
12-14	Relay to station
12-15	Locate for emerg. Message
12-16	Motor vehicle accident
12-16A	Motor vehicle accident (fatal)
12-16B	Motor vehicle accident (Injury)
12-17	Motor vehicle accident (Ambul)
12-18	Dispatched ambulance
12-19	Distatched tow car
12-20	Check for wants
12-20A	Can subject hear radio
12-21	No record or reports
12-22	Prior misd. record/not wanted
12-23	Felony record / not wanted
12-24	Subject wanted - A or B
12-25	Similar subject record
12-26	Base station out of service
12-27	Call by radio on arrival
12-28	Suspicious person
12-29	Disturbance/fight
12-30	Reckless driver
12-31	DUI (intoxicated driver)
12-32	Intoxicated person
12-33	Emergency - clear channel
12-34	Resume normal traffic
12-35	Abandoned motor vehicle
12-36	Illegal hunting
12-37	Advise road & weather contitions
12-38	Swith Radio freq. (F1-F2-F3-F4)
12-39	Attention all stations
12-40	Standby, busy
12-41	Your traffic
12-42	No traffic
12-43	Disregard previous traffic
12-46	Telephone number of your station
12-47	Computer down
12-48	Computer is now up
12-49	Death investigation
12-49A	Possible homicide
12-55	Equip test
12-50	Message not for radio traffic,
12-51	Radio repairs requested
12-52	Radio technician enroute
12-53	Regular power out on emergency
12-54	Testing station voice quality
12-56	No help available
12-57	Bomb threat
12-58	Found explosive
12-59	Explosion reported
12-60	Fire reported at _____
12-61	What is your location
12-62	Officer in pursuit
12-64	Loud & clear
12-68	Correct time
12-71	Burglary alarm
12-72	Burglary in progress
12-73	Robbery in progress
12-99	Officer needs assistance

Numeric Codes

Code 1	Routine Response
Code 2	Urgent Response
Code 3	Emergency Response
Code 4	No Further Assistance Needed
Code 5	Surveillance
Code 6	Traffic Stop
Code 7	Meal Break
Code 8	Prisoner in Custody
Code 11	Off Duty
Code 20	Pick-up
Code 99	Urgent Assistance Needed

Agency / Channel		Location	Frequency	Input
Elsie Vine Maple Fire District		Astoria	154.4300	Simplex
Medix Ambulance Service				
Operations Repeater		Countywide	155.8950	150.9950
Alternate Simplex		Countywide	155.2200	Simplex
North Clatsop County Fire Departments				
Ch1 Orange - Dispatch		Countywide	158.9550	Simplex
Ch2 Blue - Tactical / Fireground		Countywide	154.2650	Simplex
Ch3 Black - Tactical / Fireground		Countywide	154.3250	Simplex
Ch4 Elsie Vine Maple Fire District		Countywide	154.4300	Simplex
Oregon Dept. of Forestry	[PL 131.8]	Nicolai	151.1600	159.2925
Oregon State Police (Sta 13)	[PL 179.9]	Countywide	156.1500	155.4450
Seaside Fire & Rescue Department		Seaside	155.8650	Simplex
Dispatch & Operations Repeater		Seaside	154.3850	153.8900

Seaside Police	[PL 173.8]	Seaside	154.6500	155.6100
Warrenton Fire Department		Warrenton	158.8200	Simplex
		Warrenton	158.9550	Simplex

Miscellaneous

Astoria Golf & Country Club	Warrenton	451.3375	Simplex
	Warrenton	456.3375	Simplex
	Warrenton	154.6000	Simplex
Astoria Regional Airport	Astoria	122.8000	AM
Seattle Center – Approach / Departure	Astoria	124.2000	317.6000
McMinnville Radio – Astoria RCO	Astoria	122.3000	AM
Brusco Tug & Barge	Astoria	851.3625	806.3625
Clatskanie People's Utility District	Westport	153.6200	Simplex
	Westport	173.3750	Simplex
Columbia Memorial Hospital	Astoria	035.6800	Simplex
Ambulance to Hospital Channel	Astoria	155.2800	Simplex
Ambulance to Hospital Channel	Astoria	155.3400	Simplex
Hospital Operations – Repeater	Astoria	451.8625	456.8625
Columbia River Bar Pilots	Astoria	151.6850	Simplex
Columbia River Maritime Museum	Astoria	151.7150	Simplex
Nike Factory Outlet	Seaside	466.0250	Simplex
	Seaside	467.2000	Simplex
	Seaside	467.3750	Simplex
Northwest Natural Gas	Astoria	153.4400	Simplex
Providence Seaside Hospital	Seaside	451.9125	456.9125
Ambulance to Hospital Channel	Seaside	155.2800	Simplex
Ambulance to Hospital Channel	Seaside	155.3400	Simplex
Qwest Telephone Company	Astoria	151.9850	Repeater
Seaside Municipal Airport	Albany	122.9000	AM
Tolovana Inn Vacation Villages	Astoria	461.0625	Simplex
West Oregon Electric Cooperative	Westport	048.5200	047.9200

Weather Information

National Weather Service – NOAA Weather Radio

Conditions & Forecasts – Astoria (WA)	162.4000	Simplex
Conditions & Forecasts – Neahkahnie Mountain	162.4250	Simplex

Clatsop County SAME Alert Code: 041007

COLUMBIA COUNTY

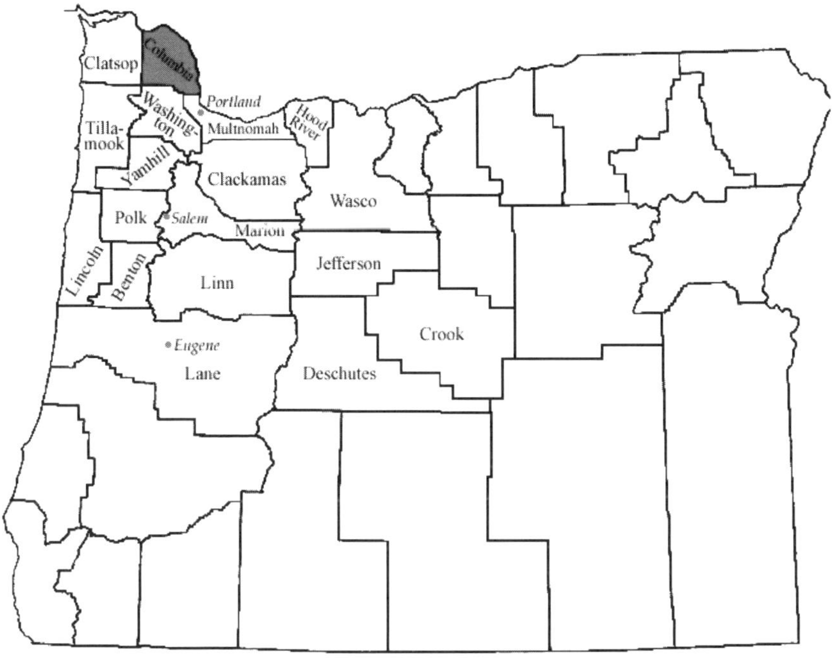

Government & Public Safety

Note: All Columbia County Public Safety Repeater channels are simulcast and broadcast countywide from the Dispatch Center, Goose Hill (Woodland), Columbia Heights (Longview), Benson Point (Clatskanie), and Corry Hill (Vernonia).

Clatskanie City Government			Clatskanie	153.8450	Simplex
Columbia County Fire Districts					
Dispatches Clatskanie, Mist-Birkenfeld, Rainier, Saint Helens, Scappoose, Vernonia					
Ch1	DISPATCH	[PL 186.2]	Countywide	154.1300	Repeater
Ch2	MEDNET – Med Incidents	[PL 186.2]	Countywide	153.8750	Repeater
Ch3	TAC 3 – Tactical		Countywide	154.7400	Repeater
Ch4	TAC 4 – Tactical		Countywide	154.8750	Repeater
Ch5	COMMON – Public Safety Mutual Aid		Countywide	154.3550	Repeater
Ch6	FIRENET – Fire Mutual Aid		Countywide	154.2800	Simplex
Ch7	ODF – State Forestry Firefighting		Countywide	151.1450	Simplex
Ch8	ODF RED – State Forestry Red Net		Countywide	151.3400	Simplex
Ch9	MOB 1 – Mobilization 1		Countywide	154.2950	Simplex
Ch10	MOB 2 – Mobilization 2		Countywide	154.4000	Simplex
	Scappoose Simplex Operations		Scappoose	155.2200	Simplex
Columbia County Law Enforcement			Countywide	155.6325	Simplex
Ch1	OPS – Operations	[PL 186.2]	Countywide	154.7850	Repeater
Ch2	SUP – Support (Records)	[PL 186.2]	Countywide	154.7100	Repeater
Ch3	CHAT – Countywide Car to Car		Countywide	154.8300	Repeater

Ch4	AREA – Local area Simplex	Countywide	159.4725	Simplex
Ch5	COMMON – Public Safety Mutual Aid	Countywide	154.3550	Repeater
Ch6	OPEN – Law Enforcement Interagency	Countywide	155.4750	Simplex
Ch7	OSP15 – State Police Columbia County	Countywide	154.9050	
Ch8	SAR – Search & Rescue	Countywide	155.8050	Simplex
Ch9	LERN – Law Enforcement Radio Net	Countywide	155.3700	Simplex
Ch10	NAT CALL – National Intersystem	Countywide	155.7525	Simplex
Ch11	Tactical Simplex	Countywide	158.8125	Simplex
Ch12	Tactical Simplex	Countywide	155.6325	Simplex
	Narrowband Repeater	Countywide	155.1375	158.7675
Columbia County Government		Countywide	155.7150	154.9800
Maritime Fire & Safety Association		Countywide	150.9800	154.5850
	Operations Repeater	Countywide	159.4800	158.4450
Oregon State Police (Sta 15) [PL 179.9]		Saint Helens	154.9050	155.4600
Saint Helens City Government		Saint Helens	159.1350	Simplex

Miscellaneous

Armstrong World Industries Plant			153.1250	Simplex
Associated Oregon Loggers		Countywide	451.7000	456.7000
		Countywide	452.2750	457.2750
	Primary Operations	Countywide	451.6250	456.6250
Boise White Paper Plant		St. Helens	452.0125	457.0125
Cascade Natural Gas		Rainier Hill	048.5000	049.1200
Clatskanie Public Utility District		Clatskanie	153.3050	159.6300
Columbia River Public Utilities Dist		Columbia City	152.4200	Simplex
	159.6150 159.7350 159.8250 160.6950 161.5350			
Longview School District		Rainier	153.0350	158.3700
Portland General Electric	[PL 203.5]	Countywide	160.0500	
		Countywide	158.3100	153.0650
Rainier School District		Saint Helens	461.3500	466.3500
Saint Helens School		Saint Helens	451.5375	Simplex
	451.2375 451.3125 451.4125	Saint Helens	456.5375	Simplex
Scappoose Industrial Airpark		Scappoose	122.8000	AM
	Portland Approach / Departure	Scappoose	124.3500	AM
	Clearance Delivery	Scappoose	121.6500	AM
US Gypsum Plant		Rainier	464.7000	469.7000
	463.5500 463.8500 464.5750	Rainier	452.1750	457.1750
	461.7250 452.0500 452.1000	Rainier	452.1250	Simplex
	463.4750 461.1750 461.6250	Rainier	461.6500	Simplex
Vernonia Airfield		Vernonia	122.9000	AM

Weather Information

National Weather Service – NOAA Weather Radio

Conditions & Forecasts – Astoria (WA)	162.4000	Simplex
Conditions & Forecasts – Portland	162.5500	Simplex

Columbia County SAME Alert Code: 041009

CROOK COUNTY

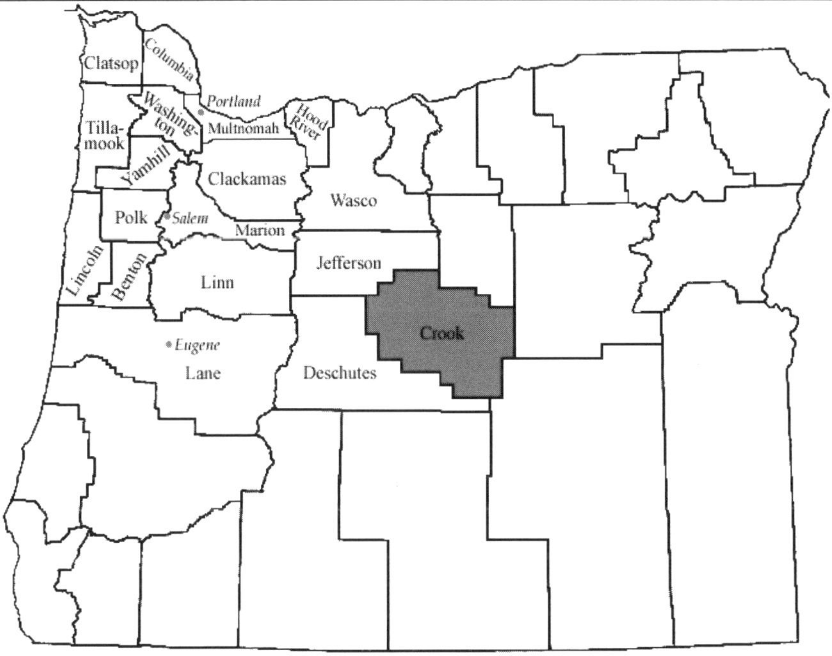

Government & Public Safety

Crook County Sheriff's Office		Countywide	158.9550	Simplex
Dispatch		Countywide	154.8900	155.9700
Tactical		Countywide	155.1900	Simplex
Tactical		Countywide	155.9700	Simplex
Tri-County / Mutual Aid		Countywide	155.2500	Simplex
Crook County Fire & Rescue		Countywide	154.9650	155.5950
Tri-County Shared Fire/EMS Tactical System (Deschutes, Jefferson and Crook Counties)				
Tac-5 Tactical		Countywide	154.3850	Simplex
Tac-6 Tactical		Countywide	155.5200	Simplex
Tac-7 Tactical		Countywide	155.5500	Simplex
Tac-8 Tactical		Countywide	153.8300	Simplex
Tac-9 Tactical		Countywide	159.2400	Simplex
Crook County Government				
Parks Department		Countywide	453.9625	Simplex
Parks Department		Countywide	458.9625	Simplex
Road Maintenance		Prineville	153.9200	155.9250
Ochoco Irrigation District		Prineville	158.9400	Simplex
Operations Repeater		Prineville	153.9200	155.9250
		Prineville	153.4250	Simplex
Oregon Dept. of Forestry	[PL 162.2]	Grizzly East	151.1750	159.2925
Oregon State Police (Sta 51)	[PL 192.8]	Bend Station	154.7850	159.0300
Oregon Youth Authority		Prineville	866.6125	

Agency	Channel	Tone	Location	Frequency	Mode/Input
Prineville City Government			Prineville	154.0400	Simplex
	Water Department		Prineville	159.9975	Simplex
Prineville Fire Department			Prineville	154.9650	155.5950
Prineville Police Department			Prineville	154.1150	Simplex

Miscellaneous

Note: Federal agencies use encryption or other devices for sensitive communications. Only "routine" communications can be monitored using the info in this book.

Agency	Channel	Tone	Location	Frequency	Mode/Input
Administrative School District #1			Prineville	461.2250	466.2250
Bureau of Land Management					
	Prineville Operations		Prineville	163.1500	Repeater
	Prineville Repeater Input		Prineville	167.0250	
	Calling (Smoke Jumpers channel)		Prineville	168.5500	
	Tactical		Prineville	166.1500	
	Tactical		Prineville	166.2250	
Central Electric Cooperative			Prineville	153.7100	Repeater
Crook County Schools			Prineville	461.4375	Simplex
Federal Bureau Of Investigation					
B1	Operations Repeater		Prineville	167.2625	Repeater
B2	Operations Simplex		Prineville	167.2625	
Ochoco National Forest					
F1	Operations Repeater	[PL 131.8]	Prineville	169.1750	Repeater
F2		[PL 131.8]	Prineville	169.9750	
F3		[PL 131.8]	Prineville	169.1250	Simplex
F4		[PL 146.2]	Prineville	169.1250	Simplex
F5	Projects Channel		Prineville	170.5000	
	Fire Tac 1		Area	168.2000	
	Fire Tac 2		Area	168.0500	
	Fire Tac 3		Area	168.6000	
Pacificorp			Prineville	153.5900	159.7500
Prineville Airport			Prineville	122.8000	AM
	Seattle Center – Approach / Departure		Prineville	128.1500	257.7500
Western Forestry Communications			Prineville	158.3350	153.2450

Weather Information

National Weather Service – NOAA Weather Radio

Conditions & Forecasts – Bend/Redmond	162.5000	Simplex
Crook County SAME Alert Code: 041013		

DESCHUTES COUNTY

Central Oregon Coordinator -- Duane Anderson

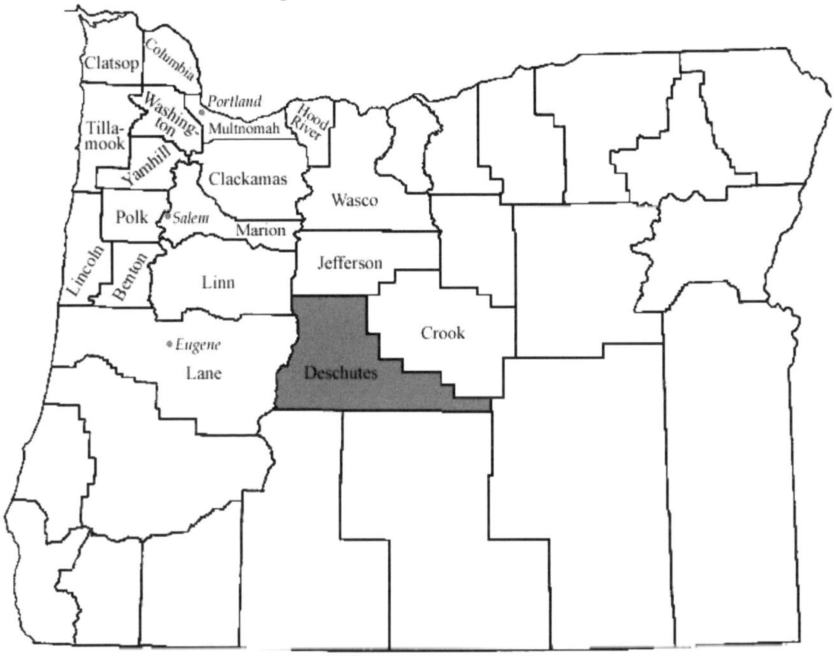

Government & Public Safety

Air Life Helicopter / Air Ambulance	Bend	463.0125	468.0125
Dispatch	Bend	155.2350	Simplex
Bend, City of	Bend	155.8200	154.9800
Dial-A-Ride – Dispatch	Countywide	TG: 63408	Deschutes
Dial-A-Ride – Alternate	Countywide	TG: 63392	Deschutes
Public Works – Main	Countywide	TG: 63504	Deschutes
Public Works – Streets	Countywide	TG: 63472	Deschutes
Public Works – Projects	Countywide	TG: 63440	Deschutes
Bend Fire Department			
Dispatch	Bend	154.1450	153.7700
Fire Tac 2	Bend	155.2200	
Fire Tac 3	Bend	153.8300	
Fire Tac 5	Bend	154.3850	
Fire Tac 6	Countywide	155.5200	
Fire Tac 7	Countywide	155.5500	
Fire Tac 8	Countywide	153.8300	
Fire Tac 9 – ODF Forestry	Countywide	159.2400	

Bend Fire Department – Station Locations & Apparatus

301	1212 Simpson Ave.	BC311, E321, L351, M371, M370
302	19850 4th Street	BR342
303	61800 Country Club Road	E323, WT333, BR343, M373
304	64420 Hamby Road	E324, WT334, BR344, M374
305	63377 Jamison Street	E325, WT335, BR345, R365, M375

Bend Police Department

Bend PD Dispatch 1	Countywide	TG: 64144	Deschutes
Bend PD Dispatch 2	Countywide	TG: 64112	Deschutes
Bend PD Dispatch 3 Car to Car	Countywide	TG: 64080	Deschutes
Bend PD Events Tac #12	Countywide	TG: 64048	Deschutes
Bend PD Volunteer Channel	Countywide	TG: 64016	Deschutes

Black Butte Ranch Fire District

Black Butte 153.8300 Simplex

See Northwest Fire Department

Black Butte Ranch Police Department

Sunriver / Black Butte Ranch PD Disp.	Countywide	TG: 64224	Deschutes
Sunriver / Black Butte Ranch PD Tac.	Countywide	TG: 64192	Deschutes

Camp Sherman Fire Services

See Northwest Fire Department

Cloverdale Fire Services

See Northwest Fire Department

Deschutes County Fire Services

Tri-County Shared Fire/EMS Tactical System (Deschutes, Jefferson and Crook Counties)

Tac-5 Tactical	Countywide	154.3850	Simplex
Tac-6 Tactical	Countywide	155.5200	Simplex
Tac-7 Tactical	Countywide	155.5500	Simplex
Tac-8 Tactical	Countywide	153.8300	Simplex
Tac-9 Tactical	Countywide	159.2400	Simplex

Numbering System

First Digit	Second Digit	Third Digit
1 LaPine Fire	1 Chief	Station Number
2 Sunriver Fire	2 Engine	
3 Bend Fire	3 Water Tender	Example:
4 Redmond	4 Brush Truck	121 LaPine FD Engine 1
5 Crooked River Ranch	5 Ladder Truck	
6 Cloverdale	6 Rescue	
7 Sisters-Camp Sherman	7 Medic / Ambulance	
8 Black Butte Ranch	8 Investigator / Admin	
	9 Utility Vehicle	

Deschutes County Government

Road Maintenance	Sunriver	156.1800	Simplex
Solid Waste / Sanitation	Countywide	453.3500	Simplex

Deschutes County Sheriff's Office

TRUNKED RADIO SYSTEM (Motorola Simulcast): TrunkTracker Setting: Type II 800

856.2125 857.2125 858.2125 859.2125 860.2125
856.9375 857.9375* 858.9375* 859.9375* 860.9375*

Ch1	Sheriff Dispatch 1	Countywide	TG: 64784	Deschutes
Ch2	Sheriff Dispatch 2	Countywide	TG: 64752	Deschutes
Ch3	Sheriff Dispatch 3	Countywide	TG: 64720	Deschutes
Ch4	Sheriff TAC 1 (Encrypted)	Countywide	TG: 64688	Deschutes
Ch5	Special Services (S&R/Mntn/Marine)	Countywide	TG: 64656	Deschutes
Ch6	Alternate	Countywide	TG: 64624	Deschutes
Ch7	Alternate	Countywide	TG: 64592	Deschutes
Ch8	Car to Car	Countywide	TG: 64560	Deschutes
Ch9	Public Safety Campus Maintenance	Countywide	TG: 64528	Deschutes
Ch10	Alternate	Countywide	TG: 64496	Deschutes
Ch11	Alternate	Countywide	TG: 64464	Deschutes
Ch12	Conflagration	Countywide	TG: 64432	Deschutes
Ch13	Regional	Countywide	TG: 64400	Deschutes
Ch14	Adult Corrections	Countywide	855.9875	
	VHF Repeater (Remote Areas)	Countywide	155.1300	
	Juvenile Jail	Countywide	155.3700	153.7400
	Search & Rescue – F16	Countywide	155.8950	

Search & Rescue – F17	Countywide	155.8050	

Deschutes County Sheriff Substations

Headquarters	63333 Highway 20, Bend
Alfalfa Station	26185 Willard Road, bend
LaPine Station (South)	51590 Huntington Road, LaPine
Terrebonne Station (North)	8222 N. Hwy 97, Terrebonne
Sisters Station	703 N. Larch, Sisters

La Pine Fire Department La Pine 154.1750 158.9850

101 Main Station
102 Cascade Lakes Station
201 Sun River Station

Northwest Fire Departments

Dispatch	Sisters/BBR	154.4150	150.8050

Northwest Fire Department – Station Locations & Apparatus

601	Main Station	C610, C612, BR641, E620, E621
602	Cloverdale Substation	E622, BR642, WT632
701	Main Station	C710, C711, E721, WT701, BR740, BR749, R760, M770, M771, M772
702	Tollgate Station	E722, WT732
703	Squaw Creek Station	E723, WT733, BR743
704	Camp Sherman Station	E724, BR744, WT734, M774
801	Black Butte Ranch	C812, E821, BR841, WT831, L851, M871, M872

Oregon Department of Corrections Bend 460.0250 465.0250

Operations Repeater	Bend	460.6000	465.6000

Oregon Dept. of Forestry [PL 162.2] Sugarpine 151.1750 159.2925

Oregon State Police (Sta 51) [PL 192.8] Bend Station 154.7850 159.0300

Redmond Police Department

TRUNKED RADIO SYSTEM (Motorola): TrunkTracker Setting: Type II 800
856.7625 857.7625* 858.7625* 859.7625* 860.7625*

Dispatch	Redmond	TG: 16	Redmond
Car to Car	Redmond	TG: 48	Redmond
Events Tac	Redmond	TG: 80	Redmond

Redmond Fire Department

Dispatch	Redmond	154.0700	158.8950

401 NW Dogwood Ave
402 C Avenue
403 Airport
404 NW 67th Avenue

Sisters City Government Sisters 155.0250 Simplex

Sisters Fire Services

See Northwest Fire Department

Sunriver Police Department

Sunriver / Black Butte Ranch PD Disp.	Countywide	TG: 62224	Deschutes
Sunriver / Black Butte Ranch PD Tac.	Countywide	TG: 62192	Deschutes

Warm Springs Police Department

Operations	Warm Sprgs.	155.4900

Airports

Aeronautical Radio Incorporated	Redmond	129.0250	AM
Airline Company Operations	Redmond	130.4000	AM
Airline Company Operations	Redmond	130.5250	AM
Airline Company Operations	Redmond	132.0000	AM
Roberts Field - Redmond Airport (RDM)			
Seattle Center Approach / Departure	Redmond	128.1500	257.7500
ATIS (Airport Info) & ASOS (Weather)	Redmond	119.0250	AM
Emergency	Redmond	121.5000	243.0000
Ground Control	Redmond	121.8000	AM
High Altitude	Redmond	135.4500	281.4000
Redmond City Airport Operations	Redmond	155.0850	FM
Tower	Redmond	124.5000	256.8000
UNICOM	Redmond	122.9500	AM
USFS Airtanker Base	Redmond	123.9750	AM
Airport Operations	Redmond	155.0850	Simplex
Bend Municipal Airport	Bend	123.0000	AM
Seattle Center Approach / Departure	Bend	128.1500	257.7500
Federal Express Corp			
Enroute Aircraft Radio	Redmond	122.8750	AM
Sisters Eagle Air Airport	Sisters	122.9000	AM
Sunriver Airport	Sunriver	122.8000	AM
Seattle Center Approach / Departure	Sunriver	128.1500	257.7500
McMinnville Radio – Sunriver RCO	Sunriver	122.3000	AM
Skywest Airlines			
Ground Operations	Redmond	460.6500	465.6500

Federal Government

Note: Federal law enforcement agencies utilize encryption or other devices for sensitive communications. Only "routine" communications can be monitored using the information contained in this book.

Bureau of Indian Affairs

Warm Sprgs. 166.6125

168.4000 171.7750 172.4250

Ch	Channel	PL	Location	Frequency	Mode
	Deschutes National Forest				
	Air Tactical		Area	169.1500	
	Air to Ground – Primary		Area	170.0000	Simplex
	Air to Ground – Secondary		Area	166.6875	Simplex
	Air to Air Tactical		Area	125.8750	AM
	Air to Ground		Area	127.9000	AM
Ch1	Newbery & East Divisions	[PL 123.0]	Bend	169.8750	Repeater
Ch2	Cascade Division	[PL 123.0]	Bend	170.4750	Repeater
Ch5	Projects		Bend	170.5000	Simplex
	Fire Tac 1		Area	168.2000	Simplex
	Fire Tac 2		Area	168.0500	Simplex
	Fire Tac 3		Area	168.6000	Simplex
	Redmond Tanker Base		Redmond	119.8500	AM
	Redmond Tanker Base		Redmond	123.9750	AM
	Drug Enforcement Administration				
Ch3	Bend Simplex Operations		Bend	418.7500	Simplex
Ch5	Bend Repeater		Bend	418.8250	Repeater

Federal Bureau Of Investigation				
A1	Operations Repeater	Bend	167.5875	Repeater
A2	Operations Simplex	Bend	167.5875	Simplex
	Tactical	Bend	167.7375	Simplex
Land Management Bureau (BLM)		Prineville	163.1500	Repeater

Hotels & Country Clubs

Bend Golf and Country Club	Bend	151.8950	Simplex
	Redmond	461.0750	Simplex
Operations Repeater	Bend	451.6625	456.6625
Black Butte Ranch	BBR	461.6000	466.6000
Broken Top Golf Club	Bend	451.5000	Simplex
452.1875 457.1875 461.9500 464.6000 466.9500			
Operations Repeater	Bend	461.9500	466.9500
Eagle Crest Resort	Redmond	452.0000	Simplex
461.0750 463.5375 464.7125			
Greens at Redmond	Redmond	154.5700	Simplex
Inn of the Seventh Mountain	Bend	451.4000	
Juniper Golf Club	Redmond	451.2375	Simplex
451.3375 457.7125 457.8125 461.4875			
Riverhouse Motor Inn	Bend	464.3750	
River's Edge Golf Club	Bend	451.2375	Simplex
	Bend	456.2375	Simplex
Rock Springs Guest Ranch	Bend	464.2500	
Sunriver Golf Resort	Sunriver	462.2125	Simplex
462.2875 462.3625 467.8125	Sunriver	467.1875	Simplex
462.4625 467.7875 467.4375	Sunriver	467.2375	Simplex
Tetherow Golf Club	Bend	452.4875	457.4875
Widgi Creek Golf Club	Bend	150.9500	Simplex
151.6250 151.9550			

Hospitals

St. Charles Med Center Redmond	Redmond	155.3400	Simplex
Operations Simplex	Redmond	155.4000	Simplex
St. Charles Medical Center	Bend	155.3400	Simplex
	Bend	155.4000	Simplex
Hospital Paging System	Bend	453.0750	Simplex
Pine Mountain Repeater	Brothers	452.9750	457.9750

Recreation

Bend Factory Outlets	Bend	461.6000	Simplex
High Desert Museum	Bend	461.0125	Simplex
461.0375 461.0625 461.0875			
HooDoo Ski Bowl	Santiam Pass	157.6200	152.3600
Ski Patrol	Santiam Pass	155.1600	Simplex
Ski Patrol	Santiam Pass	155.2950	Simplex

Mt. Bachelor Ski Resort			
Operations on Back Side of Mountain	Mt Bachelor	855.3125	810.3125
TRUNKED RADIO SYSTEM (Motorola Type II) 855.8375 856.8625 857.8625			
Security and Shuttle Busses	Mt Bachelor	TG: 48	Mt. Bachelor
Main Operations	Mt Bachelor	TG: 80	Mt. Bachelor
Maintenance	Mt Bachelor	TG: 112	Mt. Bachelor
Grooming	Mt Bachelor	TG: 144	Mt. Bachelor
Instructors	Mt Bachelor	TG: 176	Mt. Bachelor
Second Nature Cascades Camp	Bend	451.4000	456.4000

Schools

Bend – LaPine School District	Bend	461.2750	466.2750
Bend High School	Bend	151.7450	Simplex
Central Oregon Community College	Bend	151.5500	159.5400
Crook County School District	Prineville	464.4500	469.4500
Deschutes County Schools			
Campus Maintenance	Countywide	TG: 64528	Deschutes
Oregon Youth Challenge Program	Bend	153.1250	158.2200
Redmond School District	Redmond	152.3000	Simplex
Sagewalk Outdoor School	Bend	461.4125	466.4125
Sisters School District	Sisters	155.1750	Simplex

Utilities

Avion Water Company	Bend	461.1000	
Cascade Natural Gas	Bend	048.5000	048.3000
Central Electric Cooperative	Redmond	158.1450	Simplex
Central Oregon Irrigation District	Redmond	153.5300	Simplex
Deschutes Valley Water District	Culver	452.1000	Simplex
Midstate Electric Cooperative	Sunriver	451.0750	456.0750
Operations Repeater	Sunriver	451.2500	456.2500
North Unit Irrigation District	Madras	462.5250	467.5250
PG&E Gas Transmission	Area	462.4750	467.4750
Pacific Power	Area	153.5900	159.7500
153.1500 158.1300			
Sunriver Water LLC	Sunriver	153.4100	Simplex
Operations Repeater	Sunriver	452.3625	457.3625

Weather Information

National Weather Service – NOAA Weather Radio

Conditions & Forecasts – Bend/Redmond	162.5000	Simplex

Deschutes County SAME Alert Code: 041017

HOOD RIVER COUNTY

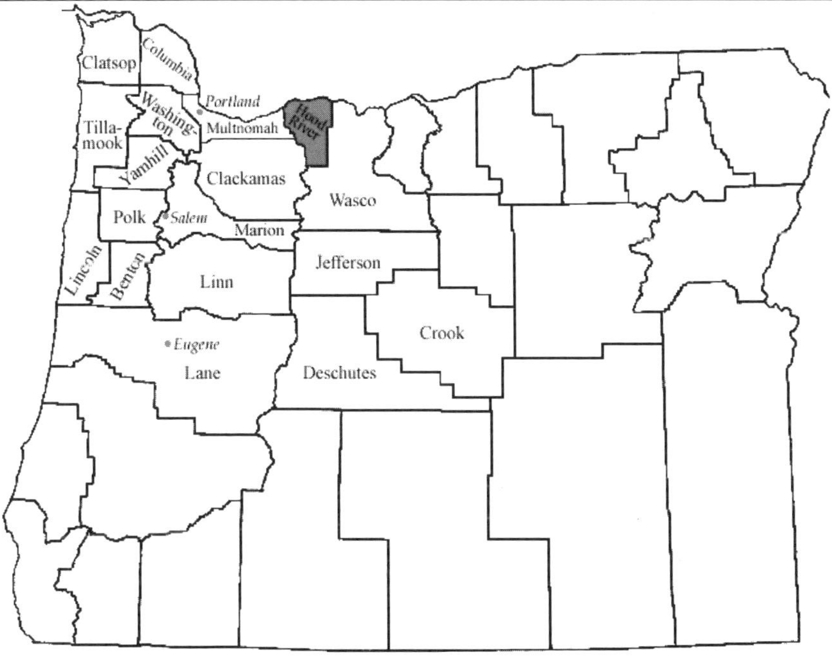

Government & Public Safety

Cascade Locks Fire Rescue – See Hood River County Fire Services

			Location	Frequency	Frequency
Columbia River Intertribal Fish Commission				156.2250	151.1300
			Hood River	155.5425	153.7925

Dee Rural Fire Protection District – See Hood River County Fire Services

Hood River County Fire Services

Ch	Description	Tone	Location	Frequency	Frequency
Ch1	Countywide Dispatch	[PL 114.8]	Countywide	158.8050	153.9950
Ch2	TAC 2 – Tactical		Countywide	154.1750	Simplex
Ch3	TAC 3 – Tactical	[PL 114.8]	Countywide	154.2350	Simplex
Ch4	TAC 4 – Cascade Locks Dispatch		Countywide	154.3700	Simplex
Ch5	TAC 5 – Tactical		Countywide	153.9500	Simplex
Ch6	TAC 6 – Tactical		Countywide	154.2650	Simplex

			Location	Frequency	Frequency
Hood River County Sheriff's Office			Hood River	154.0550	154.9500
Ch1	Dispatch	[PL 192.8]	Hood River	156.0900	155.6400
Ch2	Tactical	[PL 114.8]	Hood River	155.0100	Simplex
			Hood River	453.4500	458.4500
Hood River County Transporation Department				453.5500	458.5500
Hood River City Services			Hood River	153.8150	Simplex
			Hood River	158.7600	Simplex

Hood River Fire Department – See Hood River County Fire Services

Hood River Police Department – See Hood River County Sheriff Radio

	Location	Frequency	Frequency
Odell Fire District	Odell	154.1750	Simplex
	Odell	154.3700	Simplex

Oregon State Police (Sta 53) [PL 192.8]	The Dalles	155.9100	159.2100
West Side Fire District	Hood River	154.1750	Simplex
	Hood River	154.3700	Simplex

Miscellaneous

Cascade Locks State Airport	Cascade Lks	122.9000	AM
Duckwall Pooley Fruit Processing	Hood River	151.7150	159.6750
Hood River Electric Cooperative	Odell	153.5300	Simplex
Ken Jernstedt Airfield / Port of Hood River		453.0375	Simplex
CTAF – Common Traffic Advisory	Hood River	122.8000	Simplex
Mount Hood Ski Patrol	Parkdale	154.0850	158.9550
Providence Hood River Memorial Hospital		462.2750	467.7750
Paging System	Hood River	467.8500	Simplex

Weather Information

National Weather Service – NOAA Weather Radio

Conditions & Forecasts – The Dalles	162.4000	Simplex

Hood River County SAME Alert Code: 041027

JEFFERSON COUNTY

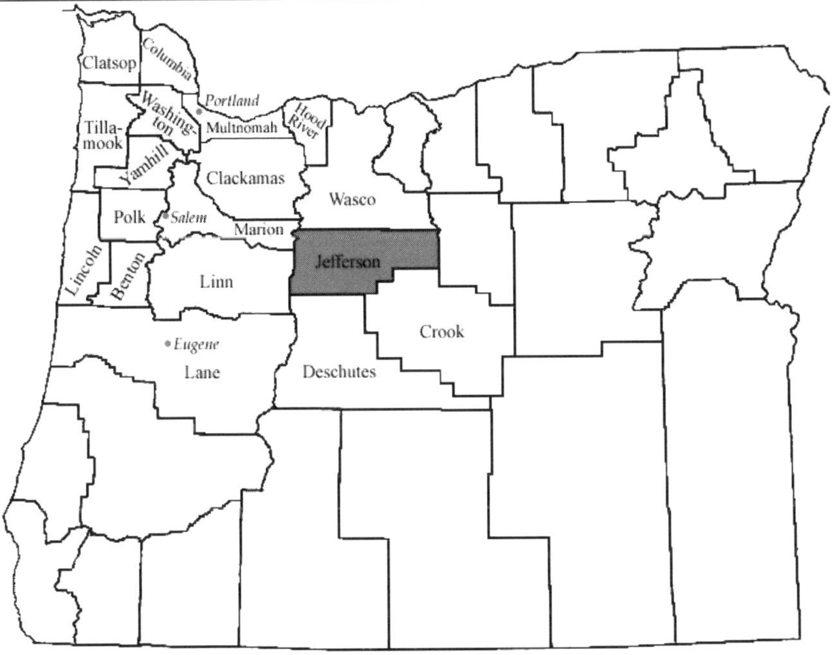

Government & Public Safety

Agency		Location	Frequency	Input
Confederated Tribes of the Warm Springs Reservation				
Police	[PL 162.2]	Warm Springs	155.4900	155.0100
		Warm Springs	153.9800	158.9400
Crooked River Ranch Rural Fire		Crooked River	154.3925	Simplex
Jefferson County EMS	[PL 131.8]	Countywide	155.2800	154.2050
Prineville Repeater		Prineville	155.8650	153.8150
Jefferson County Fire District				
Dispatch		Countywide	154.2500	150.7750
Tactical		Countywide	154.4450	Simplex
Jefferson County Government		Countywide	154.1000	Simplex
Jefferson County Sheriff's Office		Countywide	155.7000	154.7750
Dispatch		Countywide	153.8600	155.4150
Tactical #15		Countywide	155.4750	Simplex
Jail		Countywide	859.4875	
Madras Police Department		Madras	153.8600	
		Madras	155.9400	Simplex
Oregon State Police (Sta 51)	[PL 192.8]	Bend Station	154.7850	159.0300
Three Rivers Fire Department		Culver	154.2050	158.8650

Miscellaneous

Cascade Natural Gas	Countywide	048.5000	048.3000
City-County Airport	Madras	122.8000	AM
Seattle Center Approach / Departure	Madras	128.1500	257.7500
House on the Metolius Ranch Resort	Camp Sherm.	159.7650	Simplex
	Camp Sherm.	159.8850	Simplex
Lake Billy Chinook State Airport	Culver	122.9000	AM
Mountain View Hospital	Madras	451.8875	456.8875
National Weather Service – NOAA Weather Radio			
Conditions & Forecasts – Bend/Redmond		162.5000	Simplex
Jefferson County SAME Alert Code: 041031			
Redmond School District	Terrebonne	152.3150	157.5750

LANE COUNTY

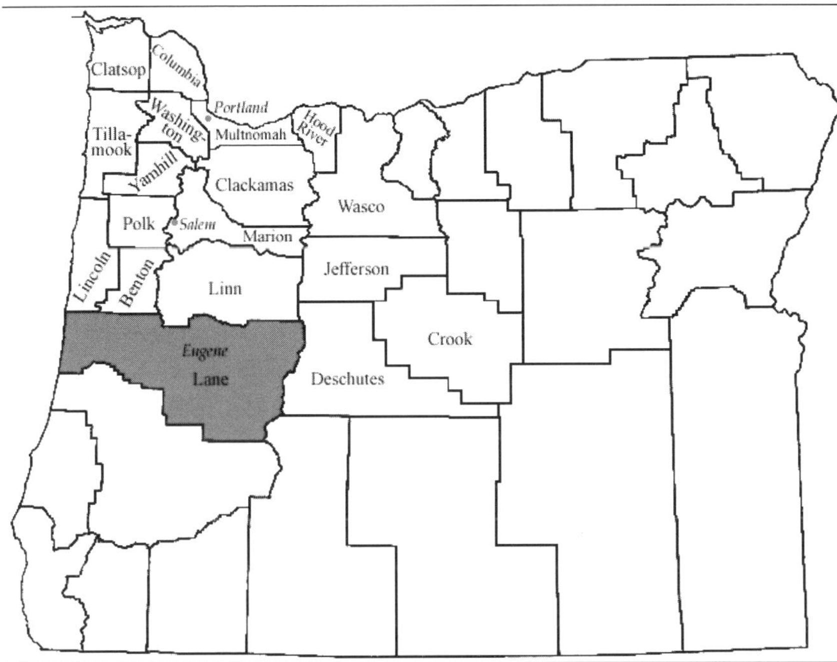

Government & Public Safety

HEARS Net: 155.1600 155.3400 155.4000

Coburg City Government	Coburg	155.0850	Simplex
Coburg Police Department	Coburg	153.7550	Simplex
Cottage Grove Fire Department	Cottage Grv	154.4000	Simplex
Cottage Grove Police Department	Cottage Grv	460.3250	465.3250
Cottage Grove City Government	Cottage Grv	155.1600	Simplex
Public Works Department	Cottage Grv	155.8800	Simplex
Public Works Department	Cottage Grv	155.8350	Simplex
Creswell Fire Department	Creswell	154.1450	Simplex

Eugene FireComm System

Ch1	City Dispatch	Eugene	154.3550	
Ch2	City Fireground I	Eugene	154.2650	
Ch3	County Dispatch	Eugene	154.4300	
Ch4	County Fireground I	Eugene	154.2050	
Ch5	State Fire Mutual Aid	Eugene	154.2800	Simplex
Ch6	Training / Drills	Eugene	154.2950	Simplex
Ch7	Fireground	Eugene	155.1600	
Ch8	MEDCOM / HEARS (Primary)	Eugene	155.3400	
Ch9	MEDCOM / HEARS (Secondary)	Eugene	155.4000	
Ch10	Fireground	Eugene	153.8300	Simplex

154.1900 154.2500 154.4450

Eugene Fire Department

Station Locations & Apparatus

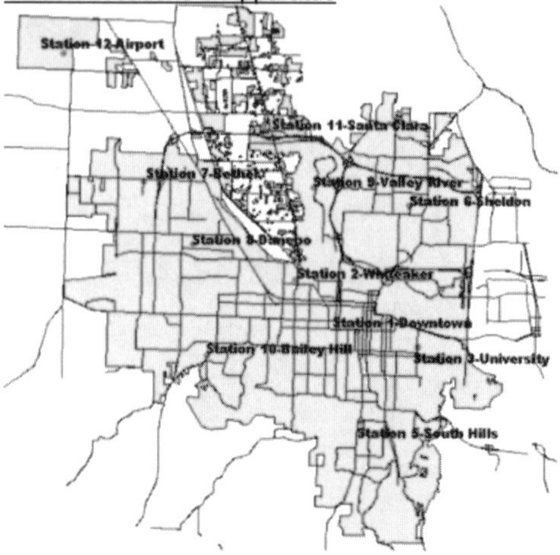

1 777 Pearl St., Downtown
E1, E11, Trk 1, BC1, Water R
2 1725 W. 2nd Ave, Whiteaker
E2, Trk 2, M2, M12, BC2
3 1695 Agate, University: E3
5 80 East 33rd Ave, South Hills
E5, WT5, M5
6 2435 Willakenzie Rd, Sheldon:
E6, Trk 6, HM6, EOC
7 4664 Barger Dr, Bethel: E7, M7
8 500 Berntzen, Danebo
E8, Trk 8, M8
9 697 Good Pasture Isl, Vly Rvr
E9, WT9, M9, M19
10 2002 Bailey Hills Rd: E10, M10
11 119 Santa Clara Ave: E11
12 28827 Douglas Dr., Airport
Crash 1, Crash 2

Eugene Police Department

Ch1	Dispatch / Operations	[Dig-P25]	Eugene	460.3000	465.3000
Ch2	Data / Information Access	[Dig-P25]	Eugene	460.0500	465.0500
Ch3	County NET	[Dig-P25]	Eugene	460.1750	465.1750
Ch4	Detectives	[Dig-P25]	Eugene	460.3500	465.3500
Ch5	Tactical	[Dig-P25]	Eugene	460.2500	465.2500
Ch6	Multi-Use (Tactical)	[Dig-P25]	Eugene	460.4000	465.4000
Ch7	Multi-Use (Tactical)	[Dig-P25]	Eugene	460.5250	465.5250
Ch8	University of Oregon PD Ch1		Eugene	460.4500	465.4500
Ch9	University of Oregon PD Ch2		Eugene	460.3750	
	SWAT Team		Eugene	453.3250	Simplex
	SWAT Team		Eugene	458.3250	Simplex
Eugene City Government			Eugene	453.7500	458.7500
Florence Police Department			Florence	155.6550	158.7900
Hazeldell Rural Fire District – Dispatched by Oakridge Fire					
Junction City Police Department			Junction City	153.7550	Simplex
	Dispatch	[Dig-P25]	Junction City	453.5125	458.5125
Junction City Public Works			Junction City	154.0550	Simplex
Lane County Fire Districts			Countywide	154.0100	158.7750
	Dispatch		Countywide	154.4300	Simplex
	Fireground I		Countywide	154.2050	Simplex
Lane County Government			Countywide	154.9950	Simplex
	Transit District		Countywide	453.4250	458.4250
	Transit District		Countywide	453.5750	458.5750
	Transit District		Countywide	453.9750	458.9750
	Facility Maintenance		Countywide	453.4750	458.4750
	Unknown Use		Countywide	154.0550	Simplex
	Juvenile Justice Center		Eugene	451.9000	Simplex
Lane County Sheriff's Department			Countywude	460.6250	465.6250
Ch1	Dispatch / Operations		Countywide	460.2250	465.2250
Ch2	Data / Information		Countywide	460.4250	465.4250
Ch3	Blue River Channel		Countywide	460.5500	460.5500
Ch4	County NET		Countywide	460.1750	465.1750
Ch5	Tactical		Countywide	460.2000	465.2000

Ch6	Corrections Dept. / Jail		Countywide	460.1500	
Ch7	Search & Rescue (Statewide)		Countywide	155.8050	
Ch8	Cottage Grove Police / Fire		Countywide	460.3250	
Ch9	Junction City PD / Coburg PD		Countywide	153.7550	
	Repeater		Countywide	460.2750	465.2750
	Repeater		Countywide	460.5000	465.5500
	Unknown Use		Countywide	453.6625	Simplex

453.6875 453.8125 458.6625 458.6875 458.8125 158.9250

Lifeflight Network			Eugene	462.9750	467.9750
Lowell Rural Fire Protection District			Lowell	154.8900	159.1575
Metro West Ambulance			Eugene	156.1650	Simplex

Mohawk Valley Fire District: See Lane County Fire Districts

	Operations Repeater		Marcola	154.3850	150.7750

Station Locations & Apparatus

141	Marcola Road, Marcola	E11, E42, R18
142	Old Marcola Rd	E21, WT26, R28
143	Parsons Creek Rd	M31
144	Nadeau Rd	E41, WT46, R48
145	Black Canyon Rd	E51, WT56

McKenzie Fire Department			McKenzie	154.4150	
Oakridge City Government			Oakridge	154.1150	154.9500
			Oakridge	155.6100	Simplex
Oakridge Fire Department					
Ch1	Dispatch	[PL 91.5]	Oakridge	154.2950	
Ch2	Fireground / Operations	[PL 91.5]	Oakridge	155.1600	
Oakridge Police Department					
	Dispatch & Operations		Oakridge	155.9550	158.9400
	Tactical		Oakridge	453.8000	Simplex
	UHF Repeater in Lowell		Oakridge	453.8000	458.8000
Oregon Dept. of Forestry		[PL 162.2]	Nebo/Farview	151.2200	159.3975
		[PL 131.8]	Noti	151.1825	159.3075
Oregon State Police					
St 25	Central Lane County	[PL 156.7]	Springfield	154.8450	155.5200
St 25A	East Lane County	[PL 156.7]	Cottage Grove	154.9050	155.4600
St 29	West Lane County	[PL 156.7]	Florence	154.7850	159.2100
Pleasant Hill Fire Department			Pleasant Hill	154.0700	
South Lane Fire District			Cottage Grv	154.4000	Simplex
	County Dispatch		Cottage Grv	154.4300	Simplex
Springfield Fire Department			Springfield	155.7600	Repeater
			Springfield	159.2025	155.9775

Station Locations & Apparatus

81	N 5th St & Olympic	E811, M819
82	Main St & N 48th St	E821, L824
83	N 28th St & Centennial	E831, L834, BC803, M839
84	Main St & N 69th St	E841, M849
85	Harlow & Pheasant	E851, M859

Springfield Police Department					
Ch1	Dispatch	[Dig-P25]	Springfield	460.1250	465.1250
Ch2	Information / Records	[Dig-P25]	Springfield	460.4750	465.4750
Suislaw Fire Department			Suislaw	451.0250	
Western Lane Ambulance District			Florence	155.8950	153.8600
Westfir Fire Department – Dispatched by Oakridge Fire				154.4900	Simplex
			Westfir	154.3250	Simplex
			Westfir	154.2350	Simplex

University of Oregon Police Department

Ch1	Dispatch	Eugene	460.4500	465.4500
Ch2	Tactical	Eugene	460.3750	

Other Fire District Numbering

21	Cottage Grove	
31	Junction City FD	
31	Harrisburg FD	
51	Lane Rural FD	
61	Santa Clara FD	
71	Creswell FD	
91	Coburg FD	N Coburg Rd
111	Veneta FD	
121	Westfir FD	
131	Goshen FD	First Street
151	Oakridge FD	Hwy 58
161	McKenzie FD	
171	Monroe FD	
181	Crow Valley	
191	Pleasant Hill	Hwy 58
192	Pleasant Hill	Jasper/Lowell Rd
201	Elmira Noti	
211	Dexter FD	
221	Walton FD	
231	Lowell FD	N Pioneer St
232	Lowell FD	Jasper Lowell Rd
241	ODF	
251	Lorane FD	Old Lorane Rd
261	Lake Creek FD	
271	Blue River VFD	
281	McKenzie Bridge	
291	Hazeldell FD	
301	Upper McKenzie	

Miscellaneous

Bohemia Mine Association	Cottage Grv.	151.5950	160.0500
Cottage Grove Hospital	Cottage Grv	155.3400	Simplex
Cottage Grove State Airport	Cottage Grv.	122.8000	AM
Emerald People's Utility District	Springfield	451.0750	456.0750
	Springfield	451.1000	456.1000
Emerald Valley Golf Club	Creswell	452.1875	Simplex
457.1875 457.5250 457.5500 457.5750			
Eugene Airport	Eugene	462.8000	Simplex
Eugene Good Samaritan Health Center		461.6000	Simplex
	Eugene	463.4250	Simplex
Eugene School District	Eugene	463.5125	468.5125
	Eugene	463.4875	468.4875
	Eugene	461.5375	466.5375
Eugene Water & Electric	Eugene	451.1375	456.1375
Florence Municipal Airport	Florence	122.8000	AM
Hobby Field Airport	Creswell	122.7000	AM
Lake Woahink Seaplane Base	Florence	122.9000	AM
Lane County Community College	Eugene	453.3000	458.3000
Public Safety Department	Eugene	461.1500	466.1500
Lane Electric Cooperative	Countywide	463.3125	468.3125
Operations Repeater	Countywide	464.0875	469.0875
Operations Repeater	Countywide	464.6625	469.6625
Operations Repeater	Countywide	464.8625	469.8625
Operations Repeater	Countywide	464.0875	469.0875
Mahlon Sweet Field Airport (EUG)	Eugene	122.9500	AM
Control Tower	Eugene	118.9000	371.9000
ATIS – Airport Terminal Info Service	Eugene	125.2250	AM
Ground Control	Eugene	121.7000	269.5000
Cascade App / Dep – N & E	Eugene	119.6000	348.7000
Cascade App / Dep – S & W	Eugene	120.2500	298.9000
Clearance Delivery	Eugene	121.7000	269.5000

Seattle Center Approach / Departure	Eugene	125.8000	291.7000
United Airlines	Eugene	460.8000	465.8000
McKenzie Bridge State Airport	McKenzie Br.	122.9000	AM
McKenzie Willamette Reg Med Cen	Springfield	155.2350	Simplex
Paging System	Springfield	152.0075	Simplex
Oakridge State Airport	Oakridge	122.9000	AM
Oregon Country Fair	Eugene	461.3000	466.3000
Operations Repeater	Eugene	464.5250	469.5250
460.8875 463.2625 463.8125 464.7625 464.8750			
464.9125 468.2625 468.8125 469.7625 469.9125			
Sacred Heart Medical Center	Eugene	155.3400	Simplex
	Eugene	155.4000	Simplex
Operations Repeater	Eugene	460.7750	465.7750
Siltcoos Lake Seaplane Base	Florence	122.9000	AM
Skywest Airlines	Eugene	460.6500	465.6500
South Lane School District	Eugene	461.4500	Simplex
461.1125 466.1125			
University of Oregon	Eugene	453.6500	458.6500
Willamette Pass Ski Park	Crescent Lake	464.3250	469.3250

Weather Information

National Weather Service – NOAA Weather Radio

Conditions & Forecasts – Eugene	162.4000	Simplex
Conditions & Forecasts – Newport (NW Lane)	162.5500	Simplex
Conditions & Forecasts – Florence (West Lane)	162.5000	Simplex

Lane County SAME Alert Code: 041039

LINCOLN COUNTY

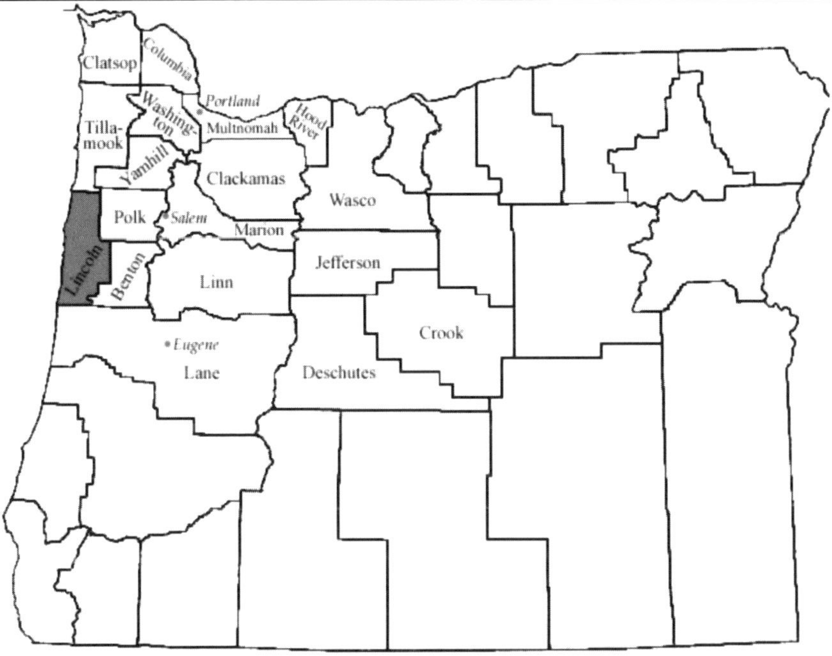

Government & Public Safety

Depoe Bay City Government		Depoe Bay	154.1000	Simplex
Kernville Gleneden Beach Lincoln Beach Water District				
Operations Simplex		Glenden Bch.	158.9550	Simplex
North Lincoln Fire Protection District				
	[PL 141.3]	Lincoln City	154.3700	154.0100
		Lincoln City	154.4000	154.0700
		Lincoln City	158.7600	Simplex

Protects: Lincoln City, Rose Lodge, Otis, Neotsu, Roads End, Kernville, Cutler City, Schooner, Drift Creek

Stations:

1200	Rose Lodge – Highway 18	E1203, E/T1421, R1231
1300	Otis – Highway 18	E1322, Quick Attack 1313, R1335
1400	Lincoln City – 2525 NW 101	E1405, Trk1455, R1438
1500	DeLake	E1507, RA1537
1600	Lincoln City – 4520 SE 101	E1608, E1604, R1632, Beach Rescue 1533
1700	Kernville – Hwy 229	E1702, WT1732

Lincoln City Police Dept.	[PL 156.7]	Lincoln City	155.5350	
		Lincoln City	155.7750	
		Lincoln City	154.8900	155.6250
		Lincoln City	155.0700	159.0150
Lincoln County EMS		Countywide	155.4000	Simplex
MedNet	[PL 162.2]	Countywide	155.3400	Simplex

Agency / Channel	PL	Area	Frequency	Input / Mode
Lincoln County Fire Departments				
Dispatch	[PL 156.7]	Countywide	154.4300	153.9500
			154.1000	
			154.3250	
Lincoln County Sheriff's Office		Countywide	155.1450	Simplex
Dispatch (+ Newport PD)	[PL 156.7]	Countywide	155.0700	
			154.9650	Repeater
Jail Operations			154.8750	Simplex
Search & Rescue		Countywide	155.8050	Simplex
Newport Public Works Department		Newport	154.1000	Simplex
		Newport	155.0250	Simplex
Newport Fire Department		Newport	154.1750	Simplex
Newport Police Department				
Dispatch (+ Newport PD)	[PL 156.7]	Countywide	155.0700	
Talkaround		Newport	155.6250	Simplex
Oregon State Police				
Sta 26 Central & North Lincoln	[PL 156.7]	Newport	154.9350	155.5800
Sta 29 Far South Lincoln County	[PL 156.7]	Florence	154.7850	159.2100
Sta 23 East Lincoln County	[PL 156.7]	Albany	154.6650	159.0300
Oregon Dept. of Forestry	[PL 167.9]	Euchre	151.1675	159.3225
Pacific West Ambulance		Countywide	155.1750	Simplex
		Countywide	155.2950	Simplex
Seal Rock Fire District		Seal Rock	154.1900	150.7900
		Seal Rock	154.2800	Simplex
Seal Rock Water District		Seal Rock	154.1000	Simplex
Siletz City Government		Siletz	154.1000	Simplex
Siletz Confederated Indian Tribes				
Forestry Management		Siletz	171.7250	Simplex
Forestry Management		Siletz	172.7750	Simplex
Toledo City Government		Toledo	154.8900	159.1575
		Toledo	158.8200	Simplex
Toledo Fire Department		Toledo	154.1450	Simplex
Toledo Police Department	[PL 156.7]	Toledo	154.8300	

Miscellaneous

Agency	Location	Frequency	Frequency/Mode
Central Lincoln People's Utility District			
	South Beach	153.5300	158.2500
	South Beach	153.5750	158.9750
	Countywide	153.4550	159.9750
	Countywide	153.4850	159.9900
	Countywide	153.6350	Repeater
Chinook Winds Gaming Center (Confederated Tribes of Siletz Indians)			
Operations Repeater	Lincoln City	452.6250	457.6250
Operations Simplex	Lincoln City	462.1375	Simplex
461.3125 461.6125 461.9875	Lincoln City	462.0125	Simplex
462.1375 463.2375 463.3125	Lincoln City	463.5625	Simplex
463.8125 463.9375 464.0125	Lincoln City	464.1125	Simplex
464.4875 464.6875 464.9375	Lincoln City	466.3125	Simplex
466.6125 466.9875 467.0125	Lincoln City	467.1375	Simplex
468.2375 468.3125 468.5625	Lincoln City	468.8125	Simplex
468.9375 469.0125 469.1125	Lincoln City	469.4875	Simplex
469.6875	Lincoln City	469.9375	Simplex
Embarcadero Resort	Embarcadero	159.7800	Simplex
Good Samaritan Hospital	Newport	152.2700	157.5300
	Toledo	152.3450	157.6050
	Countywide	152.4050	157.6650
Inn at Spanish Head	Lincoln City	463.3750	468.3750
	Lincoln City	464.7500	469.7500
Lincoln County School District	Countywide	463.2250	468.2250
Lincoln County Transit	Lincoln	464.8000	469.8000
Newport Municipal Airport	Newport	122.8000	AM
Seattle Center Approach / Departure	Newport	125.8000	291.7000
Northwest Natural Gas Company	Newport	153.4100	Simplex
	Lincoln City	153.4400	Simplex
Oregon Coast Aquarium	Newport	461.2250	466.2250
Oregon State University			
Hatfield Marine Science Center	Newport	467.7875	Simplex
Pacific Communities Hospital	Newport	461.0750	466.0750
Salishan Spa & Golf Resort	Gleneden Bch	150.8150	Simplex
	Gleneden Bch	151.7150	Simplex
	Gleneden Bch	151.8050	Simplex
Siletz Bay State Airport	Gleneden Bch	122.7000	AM
Southwest Lincoln County Water Dist	Waldport	153.5000	Simplex
Toledo State Airport	Albany	122.9000	AM
Willamette & Pacific Railroad	Toledo	160.2150	Repeater
160.3650 160.8450 160.8750	Toledo	160.7700	Repeater
161.0850 161.1150 161.1750	Toledo	161.2150	Simplex

Weather Information

National Weather Service – NOAA Weather Radio

Conditions & Forecasts – Newport	162.5500	Simplex
Conditions & Forecasts – Florence (South Lincoln)	162.5000	Simplex

Lincoln County SAME Alert Code: 041041

LINN COUNTY

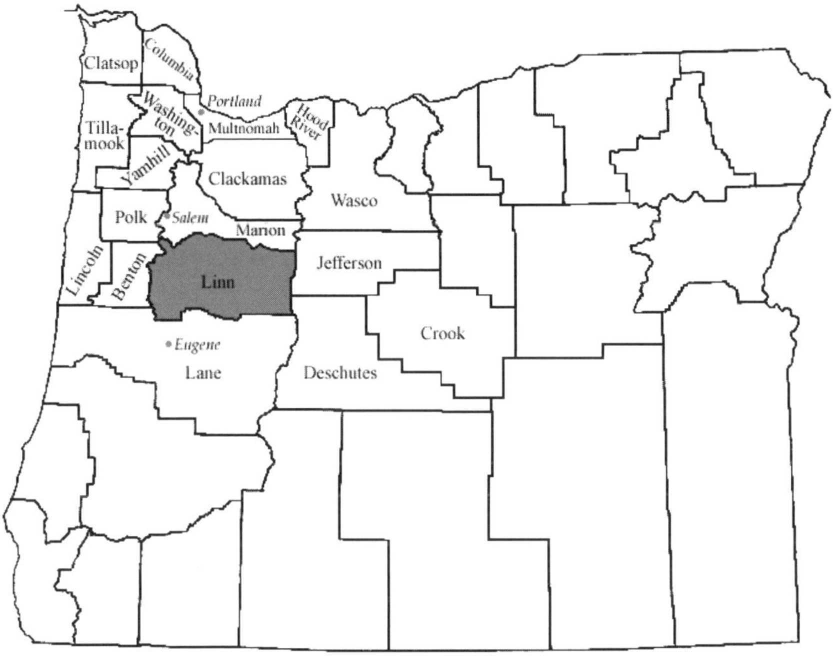

Government & Public Safety

Albany City Government		Albany	453.0375	Simplex
		Albany	458.0375	Simplex
		Albany	153.8750	Simplex
Albany Police Department		Albany	155.0100	Simplex
Dispatch	[PL 156.7]	Albany	153.8150	158.9100
Lebanon Community Hospital		Lebanon	155.2200	Simplex
Paging System		Lebanon	152.0075	Simplex
		Lebanon	155.3400	Simplex
Lebanon Fire District		Lebanon	154.3100	Simplex
Lebanon Police Department		Lebanon	155.0100	Simplex
Operations Simplex		Lebanon	155.5350	Simplex
Operations Repeater		Lebanon	460.0250	465.0250
Linn County Fire / EMS				
Ch1 Dispatch		Countywide	154.3100	
Ch2 Fireground West		Countywide	154.1300	
Ch3 Fireground East		Countywide	153.9950	
Ch4 State Fire Interagency		Countywide	154.2800	Simplex
Corvallis Mountain Rescue		Countywide	155.1600	Simplex
Statewide Search & Rescue		Countywide	155.8050	

Agency / Channel	Tone	Location	Frequency	Input / Mode
Linn County Sheriff's Department		Countywide	155.8500	
Ch1 Dispatch / Operations	[PL 156.7]	Countywide	154.7100	
Ch2 Tri County Net (w/ Benton & Lincoln)		Countywide	155.0100	Simplex
Jail Operations		Countywide	155.7825	158.9775
Lyons Rural Fire District		Lyons	154.1600	Simplex
		Lyons	154.3100	Simplex
		Lyons	154.4150	Simplex
Mill City Rural Fire District		Mill City	154.1600	Simplex
Fire Mutual Aid		Mill City	154.2800	Simplex
Operations Simplex		Mill City	154.4150	Simplex
Oregon State Police				
St 21A East Linn County	[PL 156.7]	Mill City Sta.	154.8600	155.4450
St 23 Far West Linn County	[PL 156.7]	Albany Station	154.6650	159.0300
St 25B Far Southeast Linn	[PL 156.7]	Blue River Sta.	154.7700	155.3100
Oregon Dept. of Forestry	[PL 167.9]	Scott / Coffin	151.2425	159.4050
Scio City Government		Scio	153.7400	Simplex
Scio Fire District		Scio	154.1300	Simplex
		Scio	154.3100	Simplex
Sweet Home Police Department		Sweet Home	158.7825	Simplex
Dispatch		Sweet Home	154.8150	158.7825
Sweet Home Public Works		Sweet Home	156.0000	Simplex
		Sweet Home	453.0875	458.0875

Miscellaneous

Agency / Channel	Location	Frequency	Input / Mode
Albany Municipal Airport			
CTAF / UNICOM	Albany	122.8000	AM
Cascade Approach / Departure	Albany	127.5000	AM
Seattle Center Approach / Departure	Albany	125.8000	291.7000
Greater Albany Public School District	Albany	151.7150	Simplex
Heritage Mall	Albany	457.3000	Simplex
Operations Repeater	Albany	464.3250	469.3250
Hoodoo Ski Bowl	Sisters	157.6200	152.3600
	Sisters	157.5600	Simplex
Lebanon Community School District	Sweet Home	856.9625	811.9625
Lebanon High School	Lebanon	463.4250	468.4250
461.3250 462.2250	Lebanon	464.9250	Simplex
Lebanon State Airport	Lebanon	122.8000	AM
Linn-Benton Community College	Albany	464.8000	Simplex
461.9000 462.0000 464.0250	Albany	464.3250	Simplex
Santiam Junction State Airport	Santiam Jct.	122.9000	AM
Sweet Home School District	Sweet Home	152.4200	Simplex

Weather Information

National Weather Service – NOAA Weather Radio

Channel	Frequency	Mode
Conditions & Forecasts – Eugene	162.4000	Simplex
Conditions & Forecasts – Salem	162.4750	Simplex

Linn County SAME Alert Code: 041043

MARION COUNTY

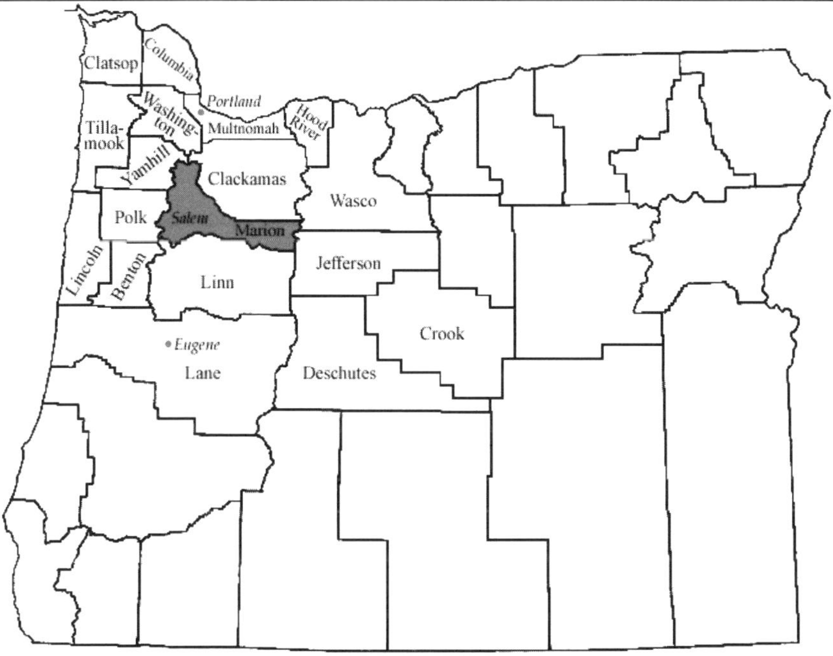

Government & Public Safety

Aumsville Fire District: See Marion County Fire Departments

Jefferson City Government			Jefferson	154.0400	Simplex
Keizer Police Department			Keizer	154.0400	154.9950
	Dispatch		Salem	TG: 57840	Salem
	Tac 1		Salem	TG: 57808	Salem
	Tac 2		Salem	TG: 57776	Salem
	Tac 3		Salem	TG: 57744	Salem
	Tac 4 – Talk-around		Salem	TG: 57712	Salem
	Conventional Repeater	[PL 114.8]	Keizer	857.4875	812.4875
Marion County Fire Departments					
North-1	Dispatch	[PL 123.0]	Countywide	154.1600	
North-2	Working Incidents	[PL 123.0]	Countywide	159.3450	
North-3	Working Incidents		Woodburn	LTR Trunking (below)	
North-4	Working Incidents		Woodburn	LTR Trunking (below)	
Central-1	Alerts / Working		Countywide	154.3850	
Central-2	Working Incidents		Countywide	153.7700	
Central-3	Polk County Alerts	[PL 110.9]	Countywide	154.2350	
Central-4	Polk County Working Incidents		Countywide	154.4450	
South-1	Alerts	[PL 123.0]	Countywide	154.4150	
South-2	Working Incidents	[PL 123.0]	Countywide	154.0100	
South-3	Working Incidents	[PL 123.0]	Countywide	154.2500	
South-4	Tac-1 / Working	[PL 123.0]	Countywide	154.2950	
	Statewide Mutual Aid		Countywide	154.2800	Simplex
HEAR	Hospital / Ambulance Notification		Countywide	155.3400	

	Statewide Fire Training	Countywide	153.8300	

Fire Station Locations & Apparatus

30	Chemeketa	E308
31	Cordon Rd	E315, M31, BR318
32	Cordon Rd	E325, E327, RS323, M32, E323
33	Pratum Av NE	E335, WT339, S334
34	84th Ave NE	E345, WT349
35	BLM Salem	WT355, WT357
36	Keizer	E365, L368, R363, M36
71	Brooklake Rd	E715, M71, BR718
72	Wheatland Rd	E725, WT729, BR728
73	Labish Cnt Rd	E735, WT739, BR738

Marion County Government Mehama 453.3750 458.3750

Marion County Sheriff's Department

Ch1	Dispatch / Operations	[PL 100.0]	Countywide	154.7400	
Ch3	Alternate		Countywide	154.8900	155.7300
	County Intersystem		Countywide	460.5000	
	Search & Rescue Explorers		Countywide		

155.1600 155.2200 155.2800

Mount Angel Fire District: See Marion County Fire Departments

Mount Angel Police	[PL 100.0]	Mount Angel	154.8150	158.8950
Oregon Department of Corrections		Salem	460.1250	465.1250
460.4000 460.3500 460.5500		Salem	453.7250	458.7250
Oregon Dept. of Forestry	[PL 179.9]	Mount Horeb	151.2050	159.4375

Oregon State Police

St 21	West Marion	[PL 156.7]	Salem Station	155.9100	159.1500
St 21A	East Marion	[PL 156.7]	Mill City Sta.	154.8600	155.4450

Salem City Government

IF THESE FREQUENCIES CHANGE, FREE UPDATES ARE AT SCANOREGON.COM

TRUNKED RADIO SYSTEM (Motorola SmartNet): TrunkTracker Setting: Type II 800

A Tower Site Frequencies: Salem

856.7125 857.7125 859.4875 860.4875 866.1250 868.1000

B Tower Site Frequencies: Salem

856.4875 857.4875 858.4875 867.8125 867.5500

C Tower (Conventional / Non-Trunked) Salem 866.4250 821.4250

All talkgroups accessible on either tower, but are not simulcast

Building Inspectors	Salem	TG: 59728	Salem
Building Services Division	Salem	TG: 59792	Salem
Building Services Division	Salem	TG: 59344	Salem
Cherriots / Bus-1	Salem	TG: 57836	Salem
Cherriots / Bus-2	Salem	TG: 57968	Salem
Cherriots / Bus-3 Supervisors	Salem	TG: 58032	Salem
Community Enforcement	Salem	TG: 59696	Salem
Housing Authority	Salem	TG: 60080	Salem
Parks Department	Salem	TG: 59760	Salem
457.5250 457.5500 457.5750 457.6000			
Parking Enforcement	Salem	TG: 59696	Salem
Pollution Control Department	Salem	TG: 59984	Salem
Public Library	Salem	453.7375	Simplex
Public Library	Salem	458.7375	Simplex
Public Works	Salem	TG: 59856	Salem
Sewer Department	Salem	TG: 59920	Salem
Street Maintenance	Salem	TG: 59952	Salem
Street Maintenance	Salem	TG: 60496	Salem
Water Department	Salem	TG: 59888	Salem
Water Department	Salem	TG: 60176	Salem

Salem Fire Department

01	A Dispatch	Salem	TG: 58416	Salem
02	A Ops 2	Salem	TG: 58448	Salem
03	A Ops 3	Salem	TG: 58480	Salem
04	A Ops 4	Salem	TG: 58512	Salem
05	A Ops 5	Salem	TG: 58544	Salem
06	A Ops 6 – McNary Field Incidents	Salem	TG: 58576	Salem
07	A Ops 7 – Training	Salem	TG: 58608	Salem
08	A Ops 8 – F BLS	Salem	TG: 58640	Salem
09	A Ops 9 – Drills	Salem	TG: 58704	Salem
10	A Command 1	Salem	TG: 58800	Salem
11	A Command 2	Salem	TG: 58832	Salem
12	C-1 – Conventional Repeater	Salem	866.4250	821.4250

Station Locations

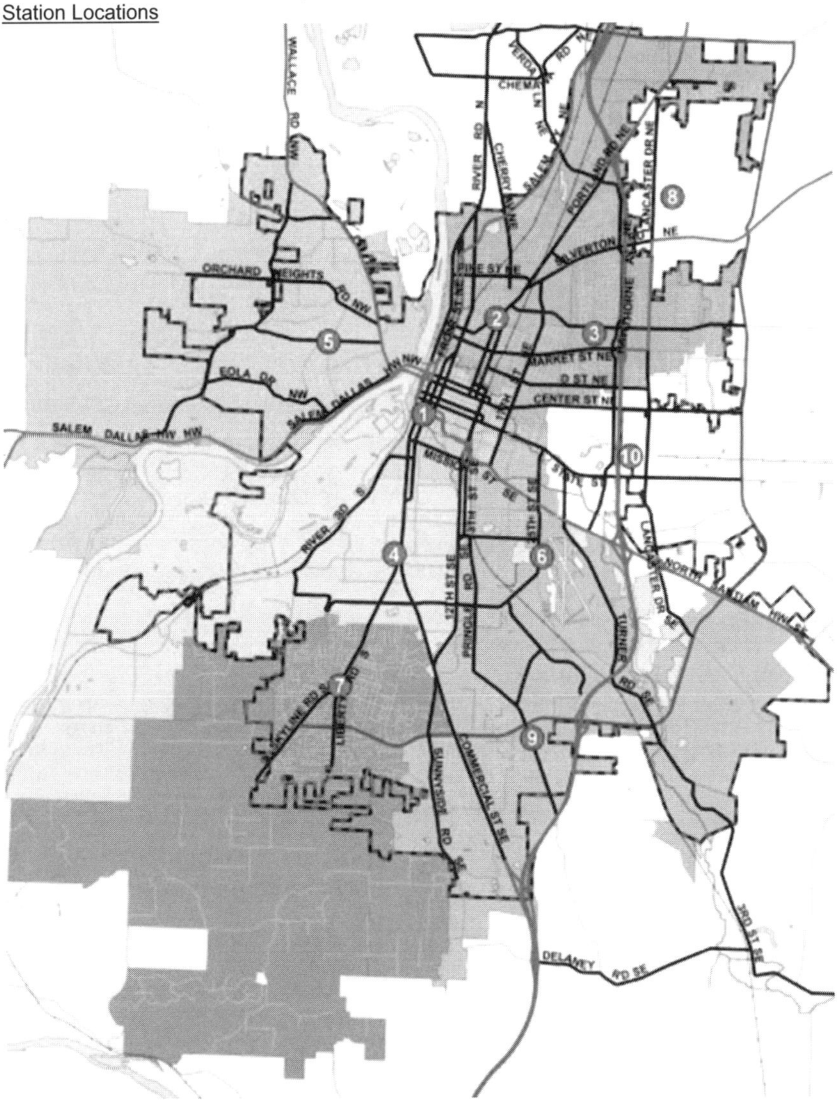

Station-1	370 Trade Street SE	E1 L1 M1 WT1 R1
Station-2	875 Madison Street NE	E2 M2 L2 BR2
Station-3	1884 Lansing NE	E3 M3
Station-4	200 Alice S	E4 M4 L4 BC
Station-5	1520 Glen Creek Rd. NW	E5 BR5
Station-6	2740 25th St SE, Airport	E6 M6 WT6 ARFF
Station-7	4730 Liberty Road S	E7 WT7 BR7
Station-8	4000 Lancaster Dr. NE	E8
Station-9	5080 Battlecreek Rd. SE	E9

Salem Police Department

Ch1	Operations Repeater	[PL 192.8]	Salem	460.4750	465.4750
Ch2	Operations Direct	[PL 192.8]	Salem	460.4750	Simplex
Ch3	Operations Repeater	[PL 192.8]	Salem	460.0750	465.0750
Ch4	Operations Direct	[PL 192.8]	Salem	460.0750	Simplex
Ch5	Programmed, Not in Use	[PL 192.8]	Salem	460.1750	465.1750
Ch6	Alternate Simplex	[PL 192.8]	Salem	460.1750	Simplex

Santiam Ambulance

			Stayton	155.2950	Simplex

Silverton Fire District

			Silverton	154.1600	Simplex
	OPS 1 – Working	[PL 100.0]	Silverton	155.1600	Simplex
	OPS 2 – Working		Silverton	155.1150	Simplex

154.1525 154.1675

Fire Station Locations & Apparatus

40	North 1st	E406, R404, Amb 413
41	McClaine St	E415, E408, WT418, WT419
42	N Abiqua Rd	E425
43	Riches Rd	E436
45	3rd Street	E445, WT449
46	Crooked Fngr	E456, E457, WT469
48	Scotts Mill	E485, WT489, R484

Silverton Police Department

Ch1	Dispatch by MAPD	[PL 100.0]	Mount Angel	154.8150	158.8950
Ch2	Tactical		Silverton	155.5350	Simplex

Stayton Fire District: See Marion County Fire Departments

Fire Station Locations & Apparatus

81	Main	E815, WT809, Trk 806, R803
82	Mehama	E825, WT829, BR828
83	Marion	E835, WT839
84	Mt Pleasant	E845, WT849, BR848
85	Elkhorn	E855, WT859, BR858

Stayton Police Department

Ch1	Dispatch	Stayton	154.8300	Simplex
Ch2	Tactical	Stayton	155.5350	Simplex

Sublimity Rural Fire District

Fire Station Locations & Apparatus

51	Parker St	E515, E516, WT518, R513
52	Drift Creek Rd	E525

Woodburn Ambulance

Woodburn	155.2050	Simplex

Woodburn Fire District: See Marion County Fire Departments

TRUNKED RADIO SYSTEM (LTR)

	LTR Trunked – LCN 01	Woodburn	453.4625	LTR
	LTR Trunked – LCN 05	Woodburn	460.6125	LTR
	LTR Trunked – LCN 09	Woodburn	453.6125	LTR
	LTR Trunked – LCN 13	Woodburn	453.8625	LTR
[#]	Dispatch (Linked to 154.160)	Woodburn	TG: 0-09-012	
[#]	Tactical (Linked to 159.345)	Woodburn	TG: 0-09-011	

Fire Station Locations & Apparatus

20	Newberg Hwy	E205, WT208, R203
21	James St	E216
22	Fourth St	E225
23	River Rd NE	E235, WT239
24	Butteville Rd	E245

Woodburn Police Department	Woodburn	155.1300	158.7750
Woodburn Public Works	Woodburn	154.1150	Simplex

Other Marion County Fire Station Locations

59	Detroit/Idanha	E595, E596, WT598
60	Jefferson	E605, WT609, R603
64	Talbot	E645
65	Aumsville – Church St	E656, E657, WT659, R653
66	US BLM	E660, E661, E662, E663
67	Aumsville – Shaw	E676, WT679
68	Gates	E685, WT689
75	St Paul – 2nd	E756 E766 WT759 769
79	Mill City	E795, WT798
87	Hubbard FPD	E876, WT876, R873
90	Aurora FPD	E914, E915, BR918, R903, WT909, WT919, 901
92	Donald FPD	E926, E927, WT929, Pumper 924
94	Monitor FPD	E945, WT949, R943
95	Turner FPD	E955, WT959, R953
96	Turner RFPD	E965

Miscellaneous

Albany & Eastern Railroad	Countywide	161.4750	
American Red Cross Disaster Relief	Countywide	453.5750	458.5750
Aurora State Airport	Aurora	122.7000	AM
Aurora Jet Center	Aurora	122.8500	AM
Portland Approach / Departure	Aurora	126.0000	AM
Clearance Delivery	Aurora	119.9500	AM
Chemeketa Community College	Salem	153.4475	158.1525
157.5075 157.6200 154.5400	Salem	152.2850	157.5450
Columbia Helicopters	Aurora	122.7750	AM
Davis Airport	Gates	122.9000	AM
Lancaster Mall			
Ch1		464.7750	
Ch2		452.2750	
Ch3		461.0250	
Lenhardt Airpark	Hubbard	122.9000	AM
Marion County School District	Salem	463.2000	468.2000
461.4875 461.7125 461.9375 461.9625 463.4125			
463.5375 463.8375 463.9125 464.0875 465.8875			
Oregon State Hospital	Salem	153.9050	Simplex
Operations Simplex	Salem	151.8050	Simplex
Operations Simplex	Salem	467.7625	Simplex
Oregon State Lottery	Salem	453.4750	458.4750
Portland General Electric [PL 203.5]	Countywide	153.5750	173.2250
	Countywide	153.3800	160.0500

Portland & Western Railroad			
Road Channel	Countywide	160.7700	
Road Channel / Switching	Countywide	160.8450	
PBX – Bald Peak	Countywide	160.2150	
Salem Hospital	Salem	463.6500	468.6500
Memorial Unit	Salem	469.6875	Simplex
Regional Health Services	Salem	155.3850	Simplex
Salem McNary Field Airport			
Tower	Salem	119.1000	257.2000
Ground Control	Salem	121.9000	AM
ATIS (Airport Info)	Salem	124.5500	AM
Seattle Center Approach / Departure	Salem	125.8000	291.7000
Oregon National Guard Aviation Unit	Salem	040.9000	
Oregon National Guard Aviation Unit	Salem	241.6000	AM
135.0000 139.1375 139.1875 139.2125 139.2375			
139.2625 139.3875 141.0500 141.1000 141.1500			
141.2000 142.8750 142.9250 148.5250 150.6250			
UNICOM	Salem	122.9500	AM
Silverton Public School	Silverton	152.3450	157.6050
Stayton Middle School	Stayton	151.8350	Simplex
Union Pacific Railroad – Brooklyn Subdivision			
Dispatch 68/70	Countywide	161.5500	
PBX – Prospect Hill	Countywide	160.8900	
Willamette Valley Railroad	Countywide	161.4750	

Weather Information

National Weather Service – NOAA Weather Radio

Conditions & Forecasts – Salem	162.4750	Simplex
Conditions & Forecasts – Portland	162.5500	Simplex

Marion County SAME Alert Code: 041047

MULTNOMAH COUNTY

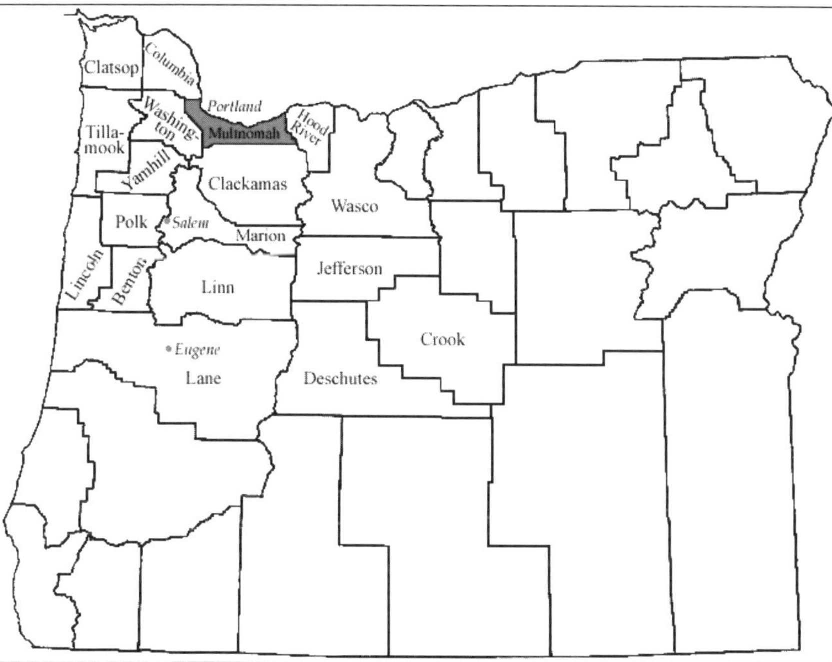

Government & Public Safety

American Medical Response NW		Portland	855.3025	810.3625
Ch1	Operations Repeater	Portland	155.2350	150.8050
Ch2	Operations Simplex	Portland	155.2200	Simplex
	County Contract Dispatch	Portland	TG: 19088	Portland
		Portland	855.4125	810.4125

AMR Northwest is a private ambulance company with a contract to provide Advanced Life Support and Basic Life Support medical transportation in Multnomah county.

Bonneville Power Administration				
	RAMS 1 – Security	Areawide	TG: 36816	Portland
	RAMS 2 – Security	Areawide	TG: 36848	Portland
	RAMS 3 – Security	Areawide	TG: 36880	Portland
Coast Guard, United States				
	Marine 13	Portland	TG: 29520	Portland
	Portland / Astoria 26	Portland	TG: 29552	Portland
	Rainier 28	Portland	TG: 29584	Portland
	Main	Portland	TG: 16208	Portland

Corbett Fire District

See Portland Fire Bureau Radio System

ICS-1	Alternate	Corbett	TG: 15344	Portland
ICS-2	Alternate	Corbett	TG: 15376	Portland
ICS-3	Alternate	Corbett	TG: 15408	Portland

Fire Station Locations & Apparatus

61	E Crown Point Hwy, Springdale	E62, WT67, R65, S69, BR66
62	E Crown Point Hwy	E61, BR64
63	SE Gordon Creek Rd	E63, WT68

Fairview City Public Works

Public Works 1	Fairview	TG: 400	Portland
Public Works 2	Fairview	TG: 432	Portland

Fairview Police Department

Dispatch (Shared) on MCSO Net-4	Fairview	TG: 20816	Portland
Tactical	Fairview	TG: 560	Portland
Training	Fairview	TG: 592	Portland
ICS	Fairview	TG: 624	Portland
Inter-Department Talk	Fairview	TG: 464	Portland

Gresham Fire Rescue and Emergency Services

See Portland Fire Bureau Radio System

Gresham FD Talk	Gresham	TG: 1456	Portland

Fire Station Locations

71	1331 NW Eastman Pkwy.	E71, T71, E77
72	500 NE Kane Road	E72, SQ72, E78
73	2301 SW Pleasant View	E73, HM73
74	1520 NE 192nd Avenue	E74, ALS Units
75	600 SW Cherry Park Road	E75, BR75
76	7144 SE 302nd Avenue	E76

Gresham City Government

Gresham City Government	Gresham	037.1000	Simplex
Unknown Use Simplex	Gresham	037.2600	Simplex
Gresham ICS - 1	Gresham	TG: 2288	Portland
Gresham ICS - 2	Gresham	TG: 2320	Portland
Gresham ICS - 3	Gresham	TG: 2352	Portland
Gresham ICS - 4	Gresham	TG: 2384	Portland
GEM 1 (Emergency Management)	Gresham	TG: 1744	Portland
GEM 2	Gresham	TG: 1776	Portland
GEM 3	Gresham	TG: 1808	Portland
GEM 4	Gresham	TG: 1840	Portland
GEM 5	Gresham	TG: 1872	Portland
GEM 6	Gresham	TG: 1904	Portland
Environmental Services Dispatch	Gresham	TG: 496	Portland
Storm Sewer Maintenance	Gresham	TG: 848	Portland
Sanitary Sewer Maintenance	Gresham	TG: 688	Portland
Water Department	Gresham	TG: 1264	Portland

Gresham Police Department

Dispatch (Shared)	Gresham	TG: 20816	Portland
Detectives / Investigations 1	Gresham	TG: 2000	Portland
Detectives / Investigations 2	Gresham	TG: 2032	Portland
Records Access / Service Net	Gresham	TG: 2576	Portland
Tac 1 (Tactical)	Gresham	TG: 2128	Portland
Tac 2 (Tactical)	Gresham	TG: 2160	Portland
Tac 3 (Tactical) / Citywide	Gresham	TG: 2192	Portland
SERT – EOD	Gresham	TG: 2608	Portland
SERT – Negotiators	Gresham	TG: 2640	Portland
MSS Training	Gresham	TG: 2512	Portland

Gresham Police Department – Beat Map

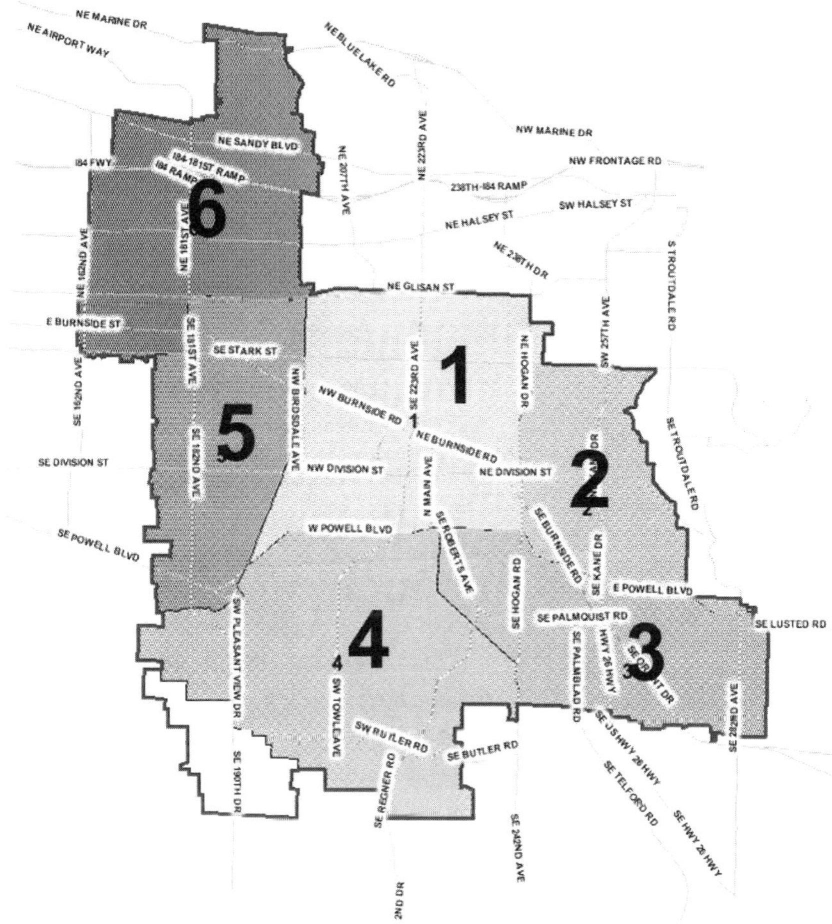

Life Flight				
Dispatched via NEXTEL Phones				
Operations	[PL 146.2]	Hillsboro	463.1750	468.1750
Unknown Use		Portland	155.3400	
Unknown Use		Portland	155.4000	
EMS Operations		Portland	462.9500	467.9500
EMS Operations		Portland	462.9750	467.9750
Air Operations		Portland	123.0500	AM
TRS Link (Unused)		Portland	TG: 16880	Portland
Maritime Fire & Safety Association		Countywide	150.9800	154.5850
Operations Repeater		Countywide	159.4800	158.4450
Metro West Ambulance		Portland	452.2375	457.2375
463.2375 462.2375 462.3125				
Non-Emergency Ops	[PL 192.8]	Portland	151.8650	Simplex
Emergency Operations	[PL 192.8]	Portland	155.3550	
Multnomah County Government				
Animal Control		Countywide	153.8600	
Animal Control		Countywide	TG: 19920	Portland

Human Services Department	Countywide	153.7400	

Multnomah County Sheriff's Office

Administration	Countywide	TG: 20528	Portland
Civil Division	Countywide	TG: 20048	Portland
Correctional Facility	Portland	TG: 20784	Portland
Courthouse Jail Security	Portland	TG: 20112	Portland
Detectives	Portland	TG: 21104	Portland
Dispatch / Operations (Shared – Net 4)	Countywide	TG: 20816	Portland
Justice Center	Countywide	460.5500	465.5500
Multnomah County ICS - 1	Countywide	TG: 20208	Portland
Multnomah County ICS - 2	Countywide	TG: 20240	Portland
Records Access	Countywide	TG: 20848	Portland
River Patrol	Countywide	TG: 20880	Portland
Search & Rescue	Countywide	TG: 20944	Portland
SIU - Tac 1	Portland	TG: 20560	Portland
SIU - Tac 2	Portland	TG: 20592	Portland
Tac 1 (Tactical)	Countywide	TG: 21008	Portland
Tac 2 (Tactical)	Countywide	TG: 20624	Portland
Traffic Detail – DUI Enforcement	Portland	TG: 20016	Portland
Training	Portland	TG: 21168	Portland
Warrant Detail	Countywide	TG: 21232	Portland

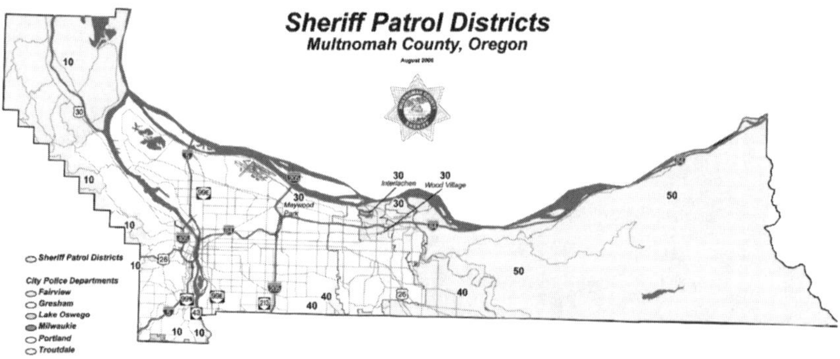

Oregon Department of Transportation

2A	Dispatch 1 (Comet)	Portland	TG: 17168	Portland
2B	Dispatch (VHF Link)	Portland	TG: 17008	Portland
	TAC 1	Portland	TG: 17072	Portland
	TAC 2	Portland	TG: 17680	Portland
	TAC 3	Portland	TG: 17712	Portland
	TAC 4	Portland	TG: 17744	Portland

Oregon Liquor Control Commission

Oregon Liquor Control Commission	Portland	TG: 27760	Portland
Net 2	Portland	TG: 27792	Portland

Oregon State Police

Sta 10 Channels 9, 10, 11, 12	[PL 179.9]	Portland	154.9350	159.0900
Link to Ch9		Portland	TG: 17392	Portland
Link to Ch12		Portland	TG: 17424	Portland
Link to Ch10		Portland	TG: 17936	Portland
Link to Ch11		Portland	TG: 18064	Portland
TAC 1		Portland	TG: 17648	Portland
TAC 2		Portland	TG: 17968	Portland

Port of Portland

See Port of Portland in the Industries Section

Portland International Airport

See Port of Portland in the Industries Section & Airports Section

Portland Bureau of Communications & Networking (ComNet)

TRUNKED RADIO SYSTEM (Motorola SmartZone) ("Portland" Below)

TrunkTracker Setting: Type II 800

IF THESE FREQUENCIES CHANGE, FREE UPDATES ARE AT SCANOREGON.COM

Note: Virtually all talkgroups EXCEPT Port of Portland related talkgroups can be heard on the Portland Simulcast Zone at all times. Port of Portland related talkgroups can generally only be heard on the Portland Airport Intellirepeater, and non-Port related traffic is generally not heard on the Portland Airport Intellirepeater. All other Intellirepeaters can generally be used by by any talkgroup with radios affiliated in that area at the time, but a specific talkgroup is generally not active on Intellirepeaters unless a radio using that talkgroup is presently affiliated with the particular Intellirepeater.

Simulcast Zone: Portland

Council Crest, Willalatin Tank, Mount Scott, Lookout Point, Prune Hill

858.7125* 858.7375* 859.2625* 859.4625* 856.2625
856.4625 856.7375 857.2625 857.4375 857.4625
857.7375 858.2625 858.4375 858.4625 859.4375
859.7125 860.2625 860.4375 860.4625 860.7125
867.3125 867.6625 868.3125 868.6375

Note: Most comms on Simulcast Zone. Intellirepeaters below are fill-in only.

Goat Mountain Intellirepeater: Estacada
854.2875* 854.7375* 856.5375 858.5375 859.5375

Biddle Butte Intellirepeater: Washougal, WA
859.0875* 854.3125* 858.6625 863.7625 854.5875

Timberline Lodge Intellirepeater: Government Camp
860.8125* 860.8625* 855.7875 857.9125 859.8375

Walter's Hill Intellirepeater: Gresham
868.5750* 868.9000* 866.3500 866.6875 866.9000
867.1000 867.3875 867.8000 868.1625 868.6750

Forest Heights Intellirepeater: Portland
868.4625* 868.7125* 866.4625 866.9875 867.9125

Cornelius Pass Intellirepeater: Portland
868.3500* 868.8500* 866.2750 866.7125 867.7500

Portland Airport Intellirepeater: Portland
868.3500* 868.8500* 866.2750 866.7125 867.7500

Arrowood Intellirepeater: Portland
868.2750* 868.6000* 867.7750 867.8750 868.1875

Bull Run Reservoir Intellirepeater: Sandy
867.2875* 868.8250* 866.3375 866.3750 866.7250

Multnomah A – Systemwide Common	Portland	TG: 176	Portland
Multnomah B – Systemwide Common	Portland	TG: 208	Portland
Multnomah C – Systemwide Common	Portland	TG: 240	Portland
Communications Maintenance Back-Up	Portland	155.6850	Repeater

Portland Bureau of Emergency Communications

Law Enforcement Common Channels

Law 1	Portland	TG: 8656	Portland
Law 2	Portland	TG: 368	Portland
Comm 1	Portland	TG: 20752	Portland
Comm 2	Portland	TG: 336	Portland
Citywide Calling	Portland	TG: 3248	Portland
Citywide Tactical	Portland	TG: 3280	Portland

Portland Bureau of Environmental Services

Dispatch	Portland	TG: 3536	Portland
Operations 1	Portland	TG: 3760	Portland
Operations 2	Portland	TG: 3792	Portland

Operations 3	Portland	TG: 4112	Portland
Operations 4	Portland	TG: 3920	Portland
Operations 5	Portland	TG: 3824	Portland
Operations 6	Portland	TG: 3856	Portland
Operations 7	Portland	TG: 3888	Portland
TAC 1	Portland	TG: 3568	Portland
TAC 2	Portland	TG: 3600	Portland

Portland Bureau of Maintenance

Dispatch	Portland	TG: 5232	Portland
Tac 1	Portland	TG: 4848	Portland
Tac 2	Portland	TG: 4880	Portland
Tac 3	Portland	TG: 4912	Portland
Tac 4	Portland	TG: 4944	Portland
Tac 5	Portland	TG: 4976	Portland
Tac 6	Portland	TG: 5008	Portland
Tac 7	Portland	TG: 5040	Portland
Tac 8	Portland	TG: 5072	Portland
Tac 9	Portland	TG: 5104	Portland
Tac 10	Portland	TG: 5136	Portland
Drainage Systems	Portland	TG: 5168	Portland
Streets – Cleaning	Portland	TG: 5296	Portland
Sewers – Cleaning	Portland	TG: 5360	Portland
Sewers – Repair	Portland	TG: 5392	Portland
Sidewalks	Portland	TG: 5424	Portland
Streets – Paving Operations	Portland	TG: 5456	Portland
Streets – Repair	Portland	TG: 5488	Portland
Traffic Structures (i.e. Bridges)	Portland	TG: 5520	Portland
Electrical Repairs (Signals)	Portland	TG: 5552	Portland
TCD – Traffic Control Devices Group	Portland	TG: 5584	Portland

Portland Bureau of Park & Recreation

Athletics – 2	Portland	TG: 5684	Portland
West & Northwest Districts	Portland	TG: 5712	Portland
North & Northeast Districts	Portland	TG: 5744	Portland
Southeast District	Portland	TG: 5776	Portland
District Dispatch	Portland	TG: 5840	Portland
Facilities & Equipment	Portland	TG: 5904	Portland
Operations	Portland	TG: 5936	Portland
Facilities / Structures – 1	Portland	TG: 5968	Portland
Facilities / Structures – 2	Portland	TG: 6000	Portland
Forestry	Portland	TG: 6032	Portland
Golf – Facilities Foreman	Portland	TG: 6096	Portland
Golf – Facilities Group	Portland	TG: 6128	Portland

Portland Bureau of Transportation Engineering

Portland Bureau of Transportation Engineering		151.0700	Simplex
Arterials	Portland	TG: 30032	Portland
Collectors	Portland	TG: 30064	Portland
Development	Portland	TG: 30096	Portland
Traffic Signals	Portland	TG: 30192	Portland
Survey Team 1	Portland	TG: 30800	Portland
Survey Team 2	Portland	TG: 30832	Portland
Survey Team 3	Portland	TG: 30864	Portland
Survey Team 4	Portland	TG: 30896	Portland
Survey Team 5	Portland	TG: 30928	Portland
Survey Team 6	Portland	TG: 30960	Portland
Survey Team 7	Portland	TG: 30992	Portland
Survey Team 8	Portland	TG: 31024	Portland
Survey Team 9	Portland	TG: 31056	Portland
Survey Team 10	Portland	TG: 31088	Portland
Survey Team 11	Portland	TG: 31120	Portland

	Street Lighting	Portland	TG: 15056	Portland

Portland Bureau of Water

	Dispatch	Portland	TG: 14544	Portland
	Tac 1	Portland	TG: 14000	Portland
	Tac 2	Portland	TG: 14032	Portland
	Tac 3	Portland	TG: 14128	Portland
	Tac 4	Portland	TG: 14160	Portland
	EOC 1	Portland	TG: 14320	Portland
	EOC 2	Portland	TG: 14352	Portland
	Operations & Engineering	Portland	TG: 13872	Portland
	Sandy River	Portland	TG: 13904	Portland
	Electricians	Portland	TG: 13744	Portland
	Engineering	Portland	TG: 13776	Portland
	Laboratory	Portland	TG: 13808	Portland
	Northwest District	Portland	TG: 13936	Portland
	Southwest District	Portland	TG: 13968	Portland
	Northeast District	Portland	TG: 14064	Portland
	Southeast District	Portland	TG: 14096	Portland
	Public Works – Tac 1	Portland	TG: 14192	Portland
	Public Works – Tac 2	Portland	TG: 14224	Portland
	Survey 1	Portland	TG: 14416	Portland
	Survey 2	Portland	TG: 14448	Portland
	Survey 3	Portland	TG: 14480	Portland
	Control Board	Portland	TG: 14512	Portland
	Meter Shop	Portland	TG: 14288	Portland
	Construction	Portland	TG: 14256	Portland
	Hiyu Mountain Repeater	Hiyu Mtn.	867.3500	822.3500

Portland City Government

	Air Operations	Portland	123.0500	AM
	Air Search & Rescue	Portland	122.9000	AM
	Air Search & Rescue	Portland	123.1000	AM
	Parking Enforcement – Administration	Portland	TG: 14928	Portland
	Parking Enforcement – Dispatch	Portland	TG: 14992	Portland
	Parking Enforcement – Records	Portland	TG: 15024	Portland

Portland City Streetcar

A1	Portland Streetcar – Dispatch	Portland	TG: 38416	Portland
A2	Portland Streetcar – Tac 1	Portland	TG: 38448	Portland
A3	Portland Streetcar – Tac 2	Portland	TG: 38480	Portland

Portland Emergency Medical Services

	Dispatch - Multnomah County	Portland	462.9500	467.9500
	Dispatch - Washington County	Portland	462.9750	467.9750
	Operations - Washington County	Portland	463.0250	468.0250
	Operations - Washington County	Portland	463.0500	468.0500
	Operations - Multnomah County	Portland	463.1250	468.1250
	Operations - Life Flight [PL 146.2]	Portland	463.1750	468.1750
	EMS Dispatch	Portland	TG: 19088	Portland
	Medical Resource Hospital*	Portland	TG: 19152	Portland
	*Regional Online Medical Control at OR Health & Science University			
	Ambulance/Hospital (HEAR) VHF	Portland	155.3400	Simplex
	Trauma System Entry	Portland	TG: 19312	Portland
	Multi Casualty Incident Working Chan	Portland	TG: 19120	Portland
	Amb/Hosp: Emanuel Hospital	Portland	TG: 19440	Portland
	Amb/Hosp: Good Samaritan Hospital	Portland	TG: 19472	Portland
	Amb/Hosp: Mt. Hood Medical Center	Portland	TG: 19536	Portland
	Amb/Hosp: Kaiser Sunnyside Hospital	Portland	TG: 19728	Portland
	Amb/Hosp: OR Health Science	Portland	TG: 19856	Portland
	Amb/Hosp: Portland Adventists Hosp.	Portland	TG: 19568	Portland
	Amb/Hosp: Providence Hosp. Portland	Portland	TG: 19632	Portland

	Amb/Hosp: Providence Hosp. Milwaukie	Portland	TG: 19664	Portland
	Amb/Hosp: St. Vincent's Hospital	Portland	TG: 29936	Portland
	Amb/Hosp: Veteran's Admin. Hospital	Portland	TG: 19600	Portland
	Amb/Hosp: Willamette Falls Hospital	Portland	TG: 19888	Portland

Portland Emergency Operations Center

Ch1	EOC 1	Portland	TG: 28976	Portland
Ch2	EOC 2	Portland	TG: 29008	Portland
Ch3	EOC 3	Portland	TG: 29040	Portland
Ch4	EOC 4	Portland	TG: 29072	Portland
Ch5	EOC 5	Portland	TG: 29104	Portland
Ch6	EOC 6	Portland	TG: 29136	Portland
Ch7	EOC 7	Portland	TG: 29168	Portland
Ch8	EOC 8	Portland	TG: 29200	Portland
Ch9	EOC 9	Portland	TG: 29232	Portland
Ch10	EOC 10	Portland	TG: 29264	Portland
Ch11	EOC 11	Portland	TG: 29296	Portland
Ch12	EOC 12	Portland	TG: 29328	Portland
Ch13	EOC 13	Portland	TG: 29360	Portland
Ch14	EOC 14	Portland	TG: 29392	Portland
Ch15	EOC 15	Portland	TG: 29424	Portland

Portland Fire & Rescue Bureau

B1	Tap Out – Dispatch / Alarms		Portland	TG: 28944	Portland
A1	Operations 1 – Minor Incidents		Portland	TG: 4528	Portland
A2	Operations 2 – Portland Working Fires		Portland	TG: 4560	Portland
A3	Operations 3 – Portland Working Fires		Portland	TG: 4592	Portland
A4	Operations 4 – Gresham Working Fires		Portland	TG: 4624	Portland
A5	Operations 5 – Gresham Working Fires		Portland	TG: 4656	Portland
A6	Operations 6 – Major Medical Calls		Portland	TG: 4688	Portland
A7	Operations 7 – Major Medical Calls		Portland	TG: 4720	Portland
A8	Operations 8 – Training / Muli-Use		Portland	TG: 1520	Portland
A9	Operations 9 – Training / Muli-Use		Portland	TG: 1552	Portland
A10	Operations 10 – Training / Muli-Use		Portland	TG: 19184	Portland
A11	Operations 11 – Training / Muli-Use		Portland	TG: 19216	Portland
A12	Operations 12 – Training / Muli-Use		Portland	TG: 19248	Portland
A13	Simplex 1	[DPL 074]	Portland	866.4375	Simplex
A14	Simplex 2	[DPL 114]	Portland	866.0375	Simplex
A15	Simplex 3	[DPL 131]	Portland	866.9500	Simplex
A16	Simplex 4	[DPL 143]	Portland	866.1750	Simplex
	MAYDAY Channel	[DPL 025]	Portland	866.3875	Simplex
	COMM 1 – Interagency		Portland	866.5125	Simplex
	COMM 2 – Interagency		Portland	867.0125	Simplex
	COMM 3 – Interagency		Portland	866.7750	Simplex
	COMM 4 – Interagency		Portland	866.3250	Simplex
	Staging		Portland	TG: 4752	Portland
	Administration		Portland	TG: 4784	Portland
	Training		Portland	TG: 4816	Portland
	PFR Talk – Alternate		Portland	TG: 4304	Portland
	Arson Task Force		Portland	TG: 4336	Portland
	Investigations		Portland	TG: 4432	Portland
	Logistics		Portland	TG: 4464	Portland
	Public Information / FMO		Portland	TG: 4368	Portland
	State Fire Net Link (154.280)		Portland	TG: 18480	Portland
	Back-Up Repeater 1		Portland	866.5500	821.5500
	Back-Up Repeater 2		Portland	866.5750	821.5750

Portland Fire Bureau – Fire Station Map

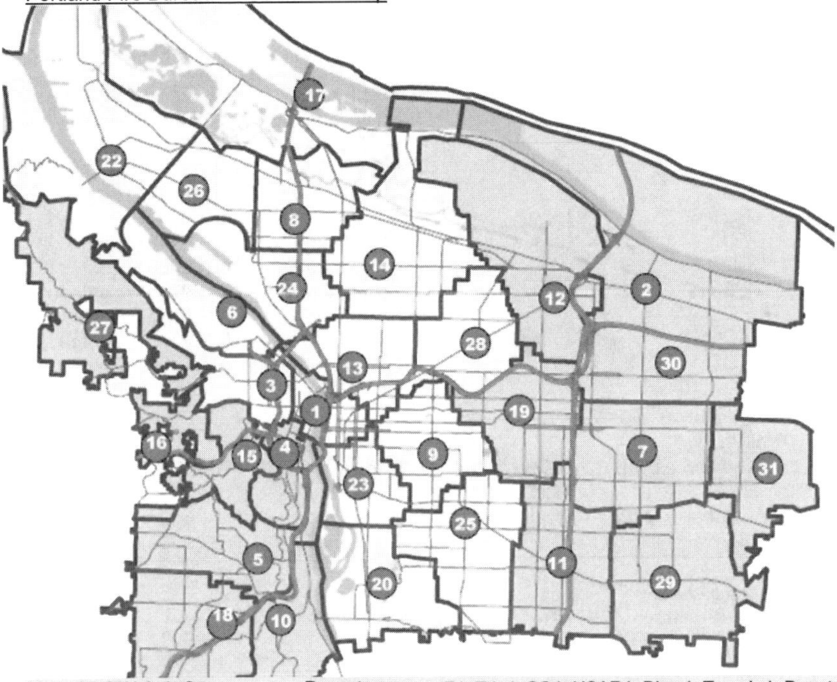

1	55 SW Ash St	Downtown	E1, Trk 1, SQ1, USAR1, Dive 1, Trench 1, Deputy
2	4800 NE 122nd Ave.	Parkrose	E2, Trk 2, E2R, Air 2
3	1715 NW Johnson St	NW Pearl	E3, Trk 3, Jeep 3
4	511 SW College Street	PSU	E4, Trk 4, E4R
5	1505 SW DeWitt Street	Hillsdale	E5, R99
6	3660 SW Front Avenue	NW Industrial	E6, Fireboat 6, Fireboat 6R, Rescue Boat 6
7	1500 SE 122nd Avenue	Mill Park	E7, Trk 7, HM7, Recon 7, BC3
8	7134 N. Maryland Ave.	Kenton	E8, Trk 8, Air 8
9	1706 SE 39th Ave.	Hawthorne	E9, E14R
10	451 SW Taylors Ferry	Burlingame	E10, Air 10, Water Tender 10
11	5707 SE 92nd Avenue	Lents	E11
12	8645 NE Sandy Blvd.	Sumner	E12, Trk 12R
13	926 NE Weidler Street	Lloyd Center	E13, Trk 13, E13R
14	1905 NE Killingsworth	Vernon	E14, E14R
15	1920 SW Spring St	SW Hills	E15
16	1715 SW Skyline Blvd	Sylvan Village	E16, Urban Interface 16
17	848 N Tomahawk Drive	Hayden Island	E17, Fireboat 17, Rescue Boat 17, E32R
18	8720 SW 30th Ave	Multnomah Vlg	E18, Brush 18, E18R
19	7301 East Burnside St	Mt. Tabor	E19
20	2235 SE Bybee Road	Sellwood	E20, Trk 20R, Urban Interface 20
22	7205 North Alta St.	St. Johns	E22, Trk 22, BR22, R-Boat 22, U22, ATV
23	2915 SE 13th Place	Lwr Eastside	E23, Trk 19R, BC4
24	4515 Maryland Ave	Swan Island	E24, E24R, SQ24, BC2
25	5211 SE Mall Street	Woodstock	E25, Trk 25
26	5247 North Lombard St	U of Portland	E26, Air 8
27	3130 NW Skyline Blvd	Forest Park	E27, BR27
28	5540 NE Sandy Blvd	Rose City	E28, Air 28
29	13310 SE Foster Road	Gilbert	E29, BR29
30	13313 NE San Rafael	Gateway	E30, Brush 30, E30R
31	1927 SE 174th St	Rockwood	E31

Portland Office of Emergency Management

	EOC – Metro	Portland	TG: 3312	Portland
	EOC 1 – Police	Portland	TG: 28976	Portland
	EOC 2 – Operations	Portland	TG: 29008	Portland
	EOC 3 – Water	Portland	TG: 29040	Portland
	EOC 4 – Fire	Portland	TG: 29072	Portland
	EOC 5 – Sewer	Portland	TG: 29104	Portland
	EOC 6	Portland	TG: 29136	Portland
	EOC 7	Portland	TG: 29168	Portland
	EOC 8	Portland	TG: 29200	Portland
	EOC 9	Portland	TG: 29232	Portland
	EOC 10	Portland	TG: 29264	Portland
	EOC 11	Portland	TG: 29296	Portland
	EOC 12	Portland	TG: 29328	Portland
	EOC 13	Portland	TG: 29360	Portland
	EOC 14	Portland	TG: 29392	Portland
	EOC 15	Portland	TG: 29424	Portland

Portland Police Bureau

Central Precinct

A1	**Central – Dispatch / Operations**	Portland	TG: 7472	Portland
B1	Central – Tac 1 (Tactical)	Portland	TG: 7536	Portland
C1	Central – Tac 2 (Tactical)	Portland	TG: 7600	Portland
D1	Central – Tac 3 (Tactical)	Portland	TG: 7568	Portland
	Central – Tac 4 (Tactical)	Portland	TG: 7632	Portland

Southeast Precinct

A2	**Southeast – Dispatch / Operations**	Portland	TG: 9456	Portland
B2	Southeast – Tac 1 (Tactical)	Portland	TG: 9520	Portland
C2	Southeast – Tac 2 (Tactical)	Portland	TG: 9584	Portland
D2	Southeast – Tac 3 (Car to Carl)	Portland	TG: 9552	Portland
	Southeast – Tac 4 (Tactical)	Portland	TG: 9616	Portland

East Precinct

A3	**East – Dispatch / Operations**	Portland	TG: 12752	Portland
B3	East – Tac 1 (Tactical)	Portland	TG: 12848	Portland
C3	East – Tac 2 (Tactical)	Portland	TG: 12912	Portland
D3	East – Tac 3 (Tactical)	Portland	TG: 12880	Portland
	East – Tac 4 (Tactical)	Portland	TG: 12944	Portland

North & Northeast Precincts

A4	**North & NE – Dispatch / Operations**	Portland	TG: 10704	Portland
B4	North – Tac 1 (Tactical)	Portland	TG: 10800	Portland
C4	North – Tac 2 (Tactical)	Portland	TG: 10864	Portland
D4	North – Tac 3 (Tactical)	Portland	TG: 10832	Portland
	North – Tac 4 (Tactical)	Portland	TG: 10896	Portland
B5	Northeast – Tac 1 (Tactical)	Portland	TG: 30640	Portland
C5	Northeast – Tac 2 (Tactical)	Portland	TG: 30672	Portland
D5	Northeast – Tac 3 (Tactical)	Portland	TG: 30704	Portland
	Northeast – Tac 4 (Tactical)	Portland	TG: 30736	Portland

Regional Organized Crime & Narcotics Task Force (ROCN)

	ROCN – Tac 1	Portland	TG: 11120	Portland
	ROCN – Tac 2	Portland	TG: 11152	Portland
	ROCN – Tac 3	Portland	TG: 11184	Portland
	ROCN – Tac 4	Portland	TG: 11216	Portland

School Police Detail

	School Police – Dispatch	Portland	TG: 27536	Portland
	School Police – Tac-1	Portland	TG: 27600	Portland
	School Police – Tac-2	Portland	TG: 27632	Portland
	School Police – Tac-3	Portland	TG: 27664	Portland

Special Emergency Response Team (SERT) / Special Weapons & Tactics (SWAT)

	SERT – Tac 1	Portland	TG: 11472	Portland

SERT – Tac 2	Portland	TG: 11504	Portland
SERT – Tac 3	Portland	TG: 11536	Portland
SERT – Tac 4	Portland	TG: 11568	Portland
Tactical Operations Division (TOD)			
TOD – Tac 1	Portland	TG: 13072	Portland
TOD – Tac 2	Portland	TG: 13104	Portland
Traffic Division			
Traffic – Operations	Portland	TG: 13456	Portland
Traffic – Tac 1 (Tactical)	Portland	TG: 13328	Portland
Traffic – Tac 2 (Tactical)	Portland	TG: 13360	Portland
Traffic – Administration	Portland	TG: 13168	Portland
Miscellaneous			
Service Desk / Records Checks	Portland	TG: 11248	Portland
Special Event 1	Portland	TG. 11600	Portland
Special Event 2	Portland	TG: 11632	Portland
Special Event 3	Portland	TG: 11664	Portland
Training – 1	Portland	TG: 12080	Portland
Training – 2	Portland	TG: 12112	Portland
Training – 3	Portland	TG: 12144	Portland

Portland Police Bureau – Precincts Map

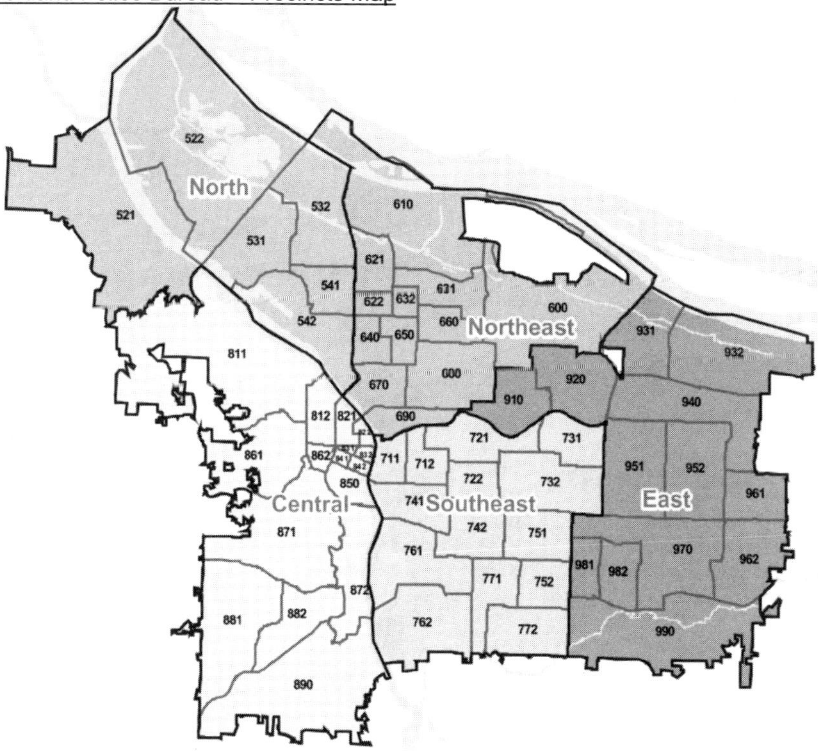

Portland Police Bureau – Districts Map

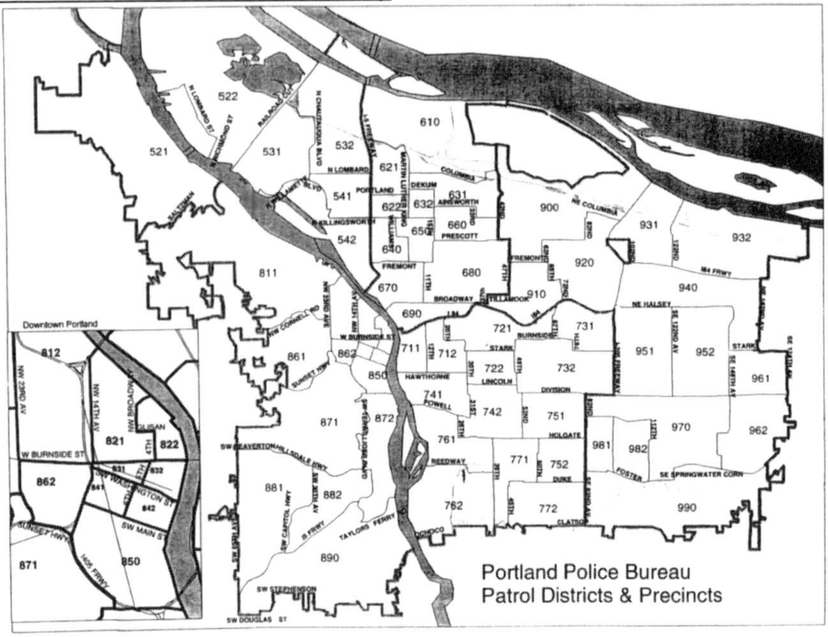

Portland Police Bureau Coverage Areas & Beats by Precinct

Precinct	Beats	Coverage
North Precinct	500	West of I-5; North of Willamette River
Northeast Precinct	600	North of I-84; East of I-5; West of 42nd Ave (except Port)
Southeast Precinct	700	South I-84; E of Willamette River; West of I-205; W of 82nd Ave
Central Precinct	800	West of Willamette River; South of NW Saltzman Road
East Precinct	900	East of 82nd Ave & I-205; N of I-84 from 42nd Ave (except Port)

Portland Police Bureau Radio Codes

Code	Meaning	Code	Meaning	Code	Meaning
10-0	Officer in Peril	10-42	Switch to…	10-72	Special Assign.
10-1	Receiving Well	10-43	Call Precinct	10-73	Call Follow-up
10-2	Not Receiving Well	10-45	Open Door/Gate	10-74	Fixing Equipment
10-4	OK / Acknowledge	10-46	Confidential Info	10-75	Surveillance OPS
10-6	Busy	10-47	Minor Injury	10-79	End of Shift
10-7	Out of Service	10-48	Severe Injury	10-82	Meal Break
10-8	In Service	10-49	Fatal Injury	10-83	Writing a Report
10-11	Beginning Shift	10-51	Sensitive Info	10-84	Vehicle Check
10-19	Return to …	10-52	Felony Warrant	10-86	Interview
10-20	Location	10-53	Warrant -Misd.	Code 0	Ofcr Needs Help!
10-21	Telephone	10-54	Confirmed Warrant	Code 1	No Hurry
10-23	On the scene	10-55	Dangerous Subject	Code 2	Hurry, No Lights
10-27	License Info	10-61	In Custody	Code 3	Lights & Siren
10-28	Registration Info	10-62	Prisoner Transport	Code 4	Situation OK
10-29	Warrant Check	10-63	Civilian Transport	Code 5	Surveillance Op.
10-30	Use Caution	10-64	Finished Transport		
10-34	Time Check	10-71	Warrant Service		

Tri County Metropolitan Transportation District (Tri-Met)

A1	Rail – Ruby Junction Yard	Portland	TG: 28304	Portland
A2	Rail – East Side Trains	Portland	TG: 28208	Portland
	Note: A2 is 82nd Ave east, including Red & Green lines			
A3	Rail – Central Trains	Portland	TG: 28272	Portland
	Note: A3 is 82nd Ave west to Rose Quarter, including Yellow Line			

A4	Rail – Central Business District Trains	Portland	TG: 28336	Portland
	Note: A4 is from Steel Bridge to PGE Park, including the Mall Alignment			
A5	Rail – West Side Trains	Portland	TG: 28240	Portland
	Note: A5 is PGE Park to the end of the line			
A6	Rail – Elmonica Rail Yard	Portland	TG: 28176	Portland
A7	Rail – Station Agent (Manpower)	Portland	TG: 28368	Portland
A8	Rail – Fare Inspectors – Tactical	Portland	TG: 28112	Portland
A9	R-Check – Repeat Offenders	Portland	TG: 28464	Portland
A10	Customer Service	Portland	TG: 27952	Portland
A11	ICS 1 – Incident Command	Portland	TG: 27888	Portland
A12	ICS 2 – Incident Command	Portland	TG: 28624	Portland
A13	ICS 3 – Incident Command	Portland	TG: 28656	Portland
A14	Bus – Tac 2 – Tactical	Portland	TG: 27856	Portland
A15	Bus – Tac 1 – Tactical	Portland	TG: 28496	Portland
A16	Bus – Dispatch	Portland	TG: 28432	Portland
B2	Rail – Tac East – Security	Portland	TG: 28400	Portland
B3	Rail – Tac Central – West Portal	Portland	TG: 28048	Portland
B4	Rail – Tac CBD – Fare Inspector Supv.	Portland	TG: 28080	Portland
B5	Rail – Tac West – Tunnel Group	Portland	TG: 28016	Portland
B6	Portland Streetcar – Dispatch	Portland	TG: 38416	Portland
B7	Portland Streetcar – Tac 1	Portland	TG: 38448	Portland
B8	Portland Streetcar – Tac 2	Portland	TG: 38480	Portland
B9	Multnomah A – Systemwide Common	Portland	TG: 176	Portland
B10	Multnomah B – Systemwide Common	Portland	TG: 208	Portland
B11	Multnomah C – Systemwide Common	Portland	TG: 240	Portland
B12	Oregon Ops 1 (Interagency) Fire 1	Statewide	866.3250	Simplex
B13	Oregon Ops 2 (Interagency) Law 1	Statewide	866.3875	Simplex
B14	Oregon Ops 3 (Interagency) Local Gov't	Statewide	866.7500	Simplex
B15	Oregon Ops 4 (Interagency) Fire 2	Statewide	866.7750	Simplex
B16	Oregon Ops 5 (Interagency) Law 2	Statewide	866.8000	Simplex
C2	Rail Maintenance – East	Portland	TG: 28144	Portland
C3	Rail Maintenance – Central	Portland	TG: 27984	Portland
C4	Rail Maintenance – Central Business D.	Portland	TG: 27920	Portland
C5	Rail Maintenance – West	Portland	TG: 27824	Portland
C6	Alternate	Portland	TG: 28688	Portland
Ch1	Bus Operations 1	Portland	452.7500	457.7500
Ch2	Bus Operations 2	Portland	452.7000	457.7000
Ch3	Bus Operations 3	Portland	452.3750	457.3750
Ch4	Bus Operations 4	Portland	453.7750	457.7750
Ch5	Paratransit Service	Portland	452.4250	457.4250
Ch6	Paratransit Service	Portland	452.7750	457.7750
	Spare	Portland	453.1500	458.1500
	Willamette Shore Trolley	Portland	460.5250	465.5250
	Low Power 1	Portland	452.8125	Simplex
	Low Power 2	Portland	452.7625	Simplex
	Unused Frequencies	Portland	452.8750	457.8750

Tri-Met Radio Codes: (Note: Tri-Met gradually eliminating use of Codes)

10-4	Affirmative	10-20	Location	10-74	Fixing Equipment
10-8	In Service/Clear	10-21	Call by Telephone	10-79	End of Shift
10-11	On Property	10-22	Call by Radio	10-81	Coffee Break
10-12	Bus Trade	10-23	On the scene	10-82	Meal Break
10-13	Road Call	10-24	Vehicle Blocking	10-83	Writing a Report
10-14	Fill	10-25	Bus vs Vehicle	33-B	Bomb Threat
10-15	Driver Intoxication	10-26	Bus vs Pedestrian	Code 4	Situation OK
10-16	Disturbance	10-27	Falling Accident	R-Check:	Repeat Offender Check
10-17	Passenger/Oper.	10-34	Time Check		
10-18	Sick Passenger	10-63	Transport		
10-19	Return to office	10-72	Special Assign.		

Troutdale City Government		Troutdale	154.9950	
Parks Department		Troutdale	TG: 15184	Portland
Public Works Department		Troutdale	TG: 15216	Portland
Sewage Treatment		Troutdale	TG: 15248	Portland
Troutdale Police Department				
Disp Dispatched on MCSO Net 4		Troutdale	TG: 20816	Portland
Tac 1 Troutdale ICS 1		Troutdale	TG: 15152	Portland
Tac 2 Troutdale ICS 2		Troutdale	TG: 31920	Portland
Administration		Troutdale	TG: 15280	Portland
Police Command		Troutdale	TG: 15312	Portland
Police Chat Channel	[PL 127.3]	Troutdale	154.9950	Simplex
Tualatin Valley Ambulance		Tualatin Vly.	155.3250	
Veterans Affairs Hospital Police		Portland	TG: 6480	Portland
Wood Village City Government		Wood Village	158.9550	

Portland Area EMS Codes

All Codes are a Nature Code (Text: XX) + Response Code (numeric: #)

EMS Nature Codes

AB#	Abdominal Complaint	DR#	Drowning	SK#	Sick Person
AL#	Allergies / Reaction	EL#	Electrocution	ST#	Stroke
AN#	Animal Bites	EY#	Eye Problems	SZ#	Seizure / Convulsions
AS#	Assault/GSW/Stab	FA#	Fall	TA#	Traffic Accident
BE#	Behavior Problem	HE#	Headache	TR#	Traumatic Injury
BK#	Back Pain	HC#	Hot / Cold	TZ#	Tazer Probe Remove
BL#	Bleeding	IN#	Inhalation Poisoning	UK#	Unknown Problem
BR#	Breathing Difficulties	MN#	Man Down	UN#	Unconscious / Faint
BU#	Burns	OB#	Pregnancy / Childbirth / Miscarriage		
CH#	Chest Pains / Heart				
DI#	Diabetic Problem	OD#	Overdose / Poisoning		

EMS Response Code (except 0, Lower number indicates more serious)

0	Non-Emergency (No Response)	8	ALS Ambulance only, Code-1
1	First Responder + ALS Fire & Amb	9	Fire or Detox unit only to check
3	First Responder + ALS Ambulance	TA1P	Pin in Accident with Extrication
4	ALS Ambulance only, Code-3	TA1R	Rollover Accident
5	First Responder Code-3, ALS Amb 1	DR1R	Drowning in the River
7	Closest Unit Code-3, Backup Code-1		

DISPOSITION CODES

Q - DISPATCHER TERMINATED
1 - Cancel
2 - Re-Assigned

R - REPORT WRITTEN
1 - Original
2 - Supplemental

S - UNABLE TO LOCATE
1 - Person
2 - Location
3 - Subject Left Loc.

T - PREMISE CHECKED
1 - False Alarm
2 - Found Secure
3 - Made Secure

W - CALL SERVICED
1 – Assign. Complete
2 - Person Assisted
3 - Person Advised
4 - Hazard Removed
5 - Delivered
6 - Peace Restored
7 - Nuisance Abated
8 - Other Unit / Agency

X - SUSPICIOUS ACTIVITY
1 - Person Check OK
2 - Vehicle Check OK

Y - ACTION TAKEN
1 - FIR Complete
2 - Warning Issued
3 - Citation Issued

Z - WARRANT SERVICE
1 - Served
2 - Not Served

Columbia Regional Information Sharing System CRISS Codes

25 - ASSIST OFCR
- A - In Peril
- B - Code 3
- C - Hazardous
- D - Code 1
- E - Meet Officer
- F - Other Agency
- G - Security Officer

26 - CHECK FOR
- A - Police
- B - Ambulance
- C - Fire
- D - Tow
- E – Utility Co.

27 - ALARM
- A - Holdup
- B - Silent
- C - Burglary
- D - ADT
- E - Fire

28 - ROBBERY
- A - Armed
- B - Strong-arm
- C - Other Weapon
- D - Purse-Snatch

29 - ASSAULT
- A - Shooting
- B - Cutting
- C - Other Weapon
- D - No Weapon

30 - MENTAL
- A - Violent
- B - Non-Violent

31 - DISTURBANCE
- A - Domestic
- B - Screaming
- C - Fight
- D - Drunk
- E - Commotion
- F - Juvenile
- G - Loud Party
- H - Sound System

32 - UNWANTED
- A - Ex-Spouse
- B - Stranger
- C - Customer

33 - THREAT
- A - Person
- B - Phone - Bomb
- C - Phone - Other
- D - Note/Letter

34 - HAZARD
- A - Explosive
- B - Smoke
- C - Odor of Gas
- D - Leaking Fuel
- E - Wire Down
- F – Tree Down
- G - Hole
- H - Debris

35 - SUSPICIOUS
- A - Prowler - Res.
- B - Prowler - Com
- C - Prowler - Veh.
- D - Accosting Kid
- E - Stranger

36 - BURGLARY
- A - Residential
- B - Commercial

37 - PREMISES
- A - Noise Inside
- B - Open Door
- C - Open Window
- D - Lights
- E - Lights Off
- F - Vandalism

38 - SUSPICIOUS
- A - Parked Car
- B - Moving Car
- C - Other Car

39 - SEX CRIMES
- A - Rape
- B - Molest
- C - Peeper
- D - Expose
- E – Tel. Molest

40 - THEFT
- A - Shoplift
- B - Tilltap
- C - From Premises
- D - From Vehicle
- E - Of Bicycle
- F - Animal

41 - VEHICLE
- A - Stolen
- B - Locate
- C - Recovered
- D - Abandoned

42 - FRAUD
- A - No Pay
- B - Prescription
- C - Credit Card
- D - Check
- K - Other

43 - PROPERTY
- A - Susp. Stolen
- B - Found
- C - Abandoned

44 - MISSING
- A - Child
- B - Aged
- C - Runaway

45 - ANIMAL
- A - Vicious
- B - Wild
- C - Person Bitten
- D - Sick/Injured
- E - Noisy
- F - Stray
- G - Unlicensed
- H - Dead

46 - ASSIST
- A – Incomp. Call
- B - Sick/Injured
- C - Down
- D - Invalid
- E - Stand-By
- F - Lockout
- G - Assist Motorist
- H - Needy Family

47 - PROVIDE INFO
- A - Neighbor Prob
- B - Domestic Prob
- C – Personal Prob

48 - ACCIDENT
- A - Injury
- B – Unk. Injury
- C - No Injury
- D - Hit And Run

49 - TRAFFIC VIO
- A - Attempt Elude
- B - Drunk
- C - Dragging
- D - Parking

50 - DIRECT TRFC.
- A - Accident
- B - Fire
- C - Signal
- D - Sign Down
- E - Congestion
- F - Escort

51 – MARINE
- A - Drowning
- B - Sinking Vessel
- C - Diving
- D - Other

52 - MARINE
- A - Rescue
- B - Escort
- C - Diving

53 - MARINE
- A - Violation
- B - Inspection

54 - PROTECTIVE
- A - Abuse
- B - Neglect
- C - Abandoned
- D - Unsupervised
- E - Child Custody

55 - DEATH
- A - Homicide
- B - Suicide
- C - Accident
- D - Natural
- E - Unknown

56 - VICE
- A - Narcotics
- B - Prostitution
- C - Gambling
- D - Obs. Material
- E - Liquor

All Codes Above: "K" = "Other"

Multnomah County Law Enforcement Callsigns

First Digit (Not Used in Patrol)

1 – Traffic Units
2 – Specialty Units / Detail Units
3 – Sergeants
4 – Detectives and Investigators
5 – Outside Agencies (School Police, Etc.)
6 – Outside Agencies (Parole & Probation, etc.)
7 – Reserve Units & Explorer Scouts
8 – Miscellaneous Police & Admin Units
9 – Command Officers (Lt. and Above)

Second Digit (First Digit for Patrol)

1 – Gresham Police Dept.
2 – Troutdale Police Dept.
3 – Fairview Police Dept.
4 – Port of Portland Police
5 – Portland Police Bureau North
6 – Portland Police Bureau Northeast
7 – Portland Police Bureau Southeast
8 – Portland Police Bureau Central
9 – Portland Police Bureau East

Third Digit (Second Digit for Patrol)

Portland & Gresham – District or Beat
Troutdale & Fairview – Badge Number

Fourth Digit (Third Digit for Patrol)

Portland & Gresham – Shift

0-3	Day Shift
4-6	Afternoon
7-9	Night Shift

Troutdale & Fairview – Badge Number

Multnomah County Sheriff Callsigns

First Digit: Agency

5 – Sheriff's Office

Second Digit: Assignment

A – Administration
C – Civil Process
D – Detectives
E – Search & Rescue
F – Court Services
G – Gang Enforcement
H – Hospital
J – School Police
K – K9 Units
M – River Patrol
P – Patrol
R – Reserve Deputy
S – Patrol Supervisor
T – Traffic Safety
V – SIU & ROCN
W – Inmate Work Crew
Z – Prisoner Transport

Third Digit: District or Unit

1# – West of the River
3# – Maywood Park
5# – East County

Airports

Agency / Use	Location	Frequency	Mode / Output
Aeronautical Radio Incorporated	Portland	128.8250	AM
Airline Company Operations	Portland	128.9250	AM
129.0250 129.1250 129.2500 129.4250 129.6250			
129.8500 130.0250 130.1500 130.2250 130.3750			
131.0500 131.2250 131.4250 131.4500 131.5500			
131.6250 131.8250 131.8500 131.9000 131.9250			
Aerowest Academy Of Aviation	Portland	122.9500	AM
Airport Terminal Services	Portland	460.7750	Simplex
Operations Simplex	Portland	465.7750	Simplex
Air to Air – Portland Area	Portland	122.7500	AM
	Portland	122.8500	AM
	Portland	123.4500	AM
Alaska Airlines [PL 114.8]	Portland	460.8250	465.8250
461.9875 465.8250 466.9875			
Air Operations	Portland	130.5000	AM
Air Operations	Portland	131.2000	AM
America West Airlines			
Ground Operations	Portland	464.6000	469.6000
CactusNet Air Operations	Portland	130.9250	AM
Continental Airlines	Portland	461.2500	466.2500
Operations Repeaters	Portland	464.4500	469.4500
460.8000 460.8500 460.8750 460.7500			
Country Squire Airpark	Sandy	122.9000	AM
Delta Airlines [PL 127.3]	Portland	460.7500	465.7500
DHL Airways Incorporated	Portland	855.0625	810.0625
Estacada Airport	Estacada	122.9000	AM
Flight Watch	Portland	122.0000	AM
Hillsboro Airport: See Portland Hillsboro			
Hillsboro Aviation Incorporated	Portland	153.1550	Simplex
Operations Simplex	Portland	153.2600	Simplex
Aircraft Operations	Portland	122.8500	AM
Hillsboro Helicopters Incorporated	Portland	123.5000	AM
Horizon Airlines Incorporated	Portland	464.3750	469.3750
Ground Support Repeater	Portland	461.8875	466.8875
Air Operations	Portland	132.0000	AM
Air Operations – Maintenance Control	Portland	131.3500	AM
Operations Repeater [PL 77.0]	Portland	460.6750	465.6750
464.3375 469.3375			
JetBlue Airways	Portland	464.4125	469.4125
Northwest Airlines Incorporated	Portland	460.7000	465.7000
Portland Downtown Heliport	Portland	123.0750	AM
Portland Hillsboro Airport			
Approach / Depature	Portland	126.0000	AM
UNICOM	Portland	122.9500	AM
ATIS (Airport Info)	Portland	127.6500	AM
Ground Control	Portland	121.7000	AM
Tower / CTAF	Portland	119.3000	239.3000
Portland International Airport (PDX)			
Approach / Dep. (100-279º / South)	Portland	118.1000	284.6000
Approach / Dep. (280-099º / North)	Portland	133.0000	360.8000
Approach / Dep. (280-099º / North)	Portland	124.3500	299.2000

Final Approach Control	Portland	126.9000	294.7000
ATIS (Airport Info) - Departure	Portland	120.6250	239.2500
ATIS (Airport Info) - Arrivals	Portland	128.3500	269.9000
Clearance Delivery	Portland	120.1250	318.1000
Emergency	Portland	121.5000	243.0000
Radar Control	Portland	127.8500	294.7000
Flight Service Station	Portland	122.6000	AM
Flight Watch	Portland	122.0000	AM
Ground Control	Portland	121.9000	348.6000
Ground Control – Alternate	Portland	132.2750	AM
Tower – Runways 10L/28R	Portland	118.7000	257.8000
Tower – Runways 03/21 & 10R/28L	Portland	123.7750	251.1250
UNICOM	Portland	122.9500	AM
800 MHz Contingency Frequency	Portland	866.2750	Simplex
800 MHz Contingency Frequency	Portland	868.3500	Simplex
Oregon National Guard & Other Military Units at PDX:			
Air Force Reserve Operations		138.4500	252.8000
142nd Fighter Wing Command Post		288.9000	AM
142nd Fighter Wing Operations		280.5000	AM
939th Rescue Wing – Search & Rescue		282.8000	AM
939th Rescue Wing – Search & Rescue Training		251.9000	AM
939th Rescue Wing – Search & Rescue Training		259.0000	AM
939th Rescue Wing – Operations		252.8000	AM
939th Rescue Wing – Command Post		311.0000	AM
046.8500 138.4500 381.0000			
244th and 272nd Combat Comm, Squadrons			
12th Special Forces Group			
Dispatch		238.9000	AM
Have Quick		300.1000	AM
Air Tactical		239.7000	AM
036.7000 040.2000 139.1250			
300.2250 343.5000 351.1000			
Alerting / Paging System		413.4500	Simplex
Avionics		354.2000	AM
National Guard Base Operations		413.0250	Simplex
National Guard Fire & Crash Crews		413.3000	Simplex
National Guard Flightline Operations		413.0000	Simplex
National Guard Maintenance		298.3000	AM
National Guard Security & Law Enforcement		413.2000	Simplex
National Guard Security & Law Enforcement		413.0500	Simplex
Petroleum/Oil/Lubricants/Munitions		413.4000	Simplex

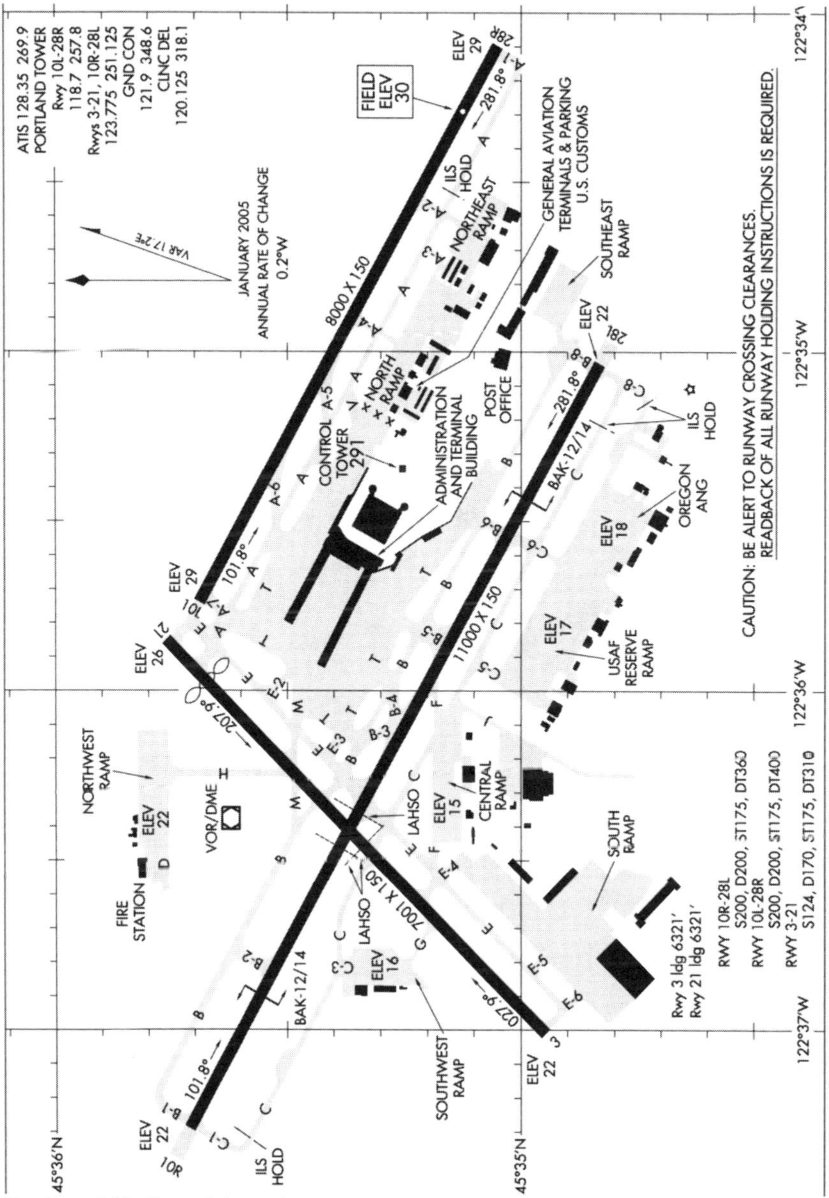

Portland Mulino Airport

CTAF / UNICOM	Portland	123.0500	AM
Clearance Delivery	Portland	119.9500	AM

Portland Troutdale Airport

ATIS – Airport Information	Portland	135.6250	AM
CTAF	Portland	120.9000	AM
UNICOM	Portland	122.9500	AM
Ground Control	Portland	121.8000	AM
Control Tower	Portland	120.9000	254.3000

Approach / Departure – SE / S / SW	Portland	118.1000	AM
Approach / Departure	Portland	124.3500	AM
Sandy River Airport	Sandy	122.8000	AM
Skyport Airport	Cornelius	122.9000	AM
Skywest Airlines	Portland	451.7250	
Southwest Airlines Company [PL 79.7]	Portland	460.6500	465.6500
Air Operations	Portland	130.8500	AM
Operations Simplex	Portland	452.4875	Simplex
Operations Simplex	Portland	457.4875	Simplex
Stark's Twin Oaks Airpark	Hillsboro	123.0500	AM
Clearance Delivery	Hillsboro	119.9500	AM
United Airlines Incorporated			
Ground Support Repeater	Portland	460.7250	465.7250
Ground Support Simplex	Portland	461.3125	Simplex
Operations Repeater	Portland	463.2500	468.2500
Operations Repeater	Portland	463.5250	468.5250
Air Operations	Portland	129.5000	AM
Operations Simplex	Portland		

461.7625 461.7875 461.8125 461.8375
461.8625 461.8875 463.7000 466.7625 466.7875
466.8125 466.8375 466.8625 466.8875

Valley View Airport	Estacada	122.9000	AM

Amateur Radio

Beaverton Area Repeater		Beaverton	145.2700	R -
Beaverton Area Repeater		Beaverton	442.9250	R +
Beaverton Area Repeater		Beaverton	443.1500	R +
Beaverton Area Repeater		Beaverton	448.6500	R -
Mount Hood Repeater		Portland	147.1200	R +
Portland Area Repeater		Portland	1291.5000	R -
Portland Area Repeater		Portland	1291.0000	R -
Portland Area Repeater		Portland	145.1900	R -
Portland Area Repeater		Portland	145.3900	R -
Portland Area Repeater		Portland	146.7000	R -
Portland Area Repeater		Portland	146.8400	R -
Portland Area Repeater		Portland	146.9400	R -
Goat Mountain Repeater		Portland	146.9600	R -
Portland Area Repeater		Portland	147.0400	R +
Portland Area Repeater		Portland	147.2800	R +
Portland Area Repeater	[PL 100.0]	Portland	147.4350	R -
Portland Area Repeater		Portland	223.9400	R -
Portland Area Repeater		Portland	224.9400	R -
Portland Area Repeater		Portland	440.0250	R +
Portland Area Repeater	[PL 167.9]	Portland	440.3000	R +
Portland Area Repeater		Portland	440.4500	R +
Portland Area Repeater		Portland	442.0250	R +
Portland Area Repeater		Portland	443.0250	R +
Portland Area Repeater	[PL 100.0]	Portland	443.5750	R +
Portland Area Repeater		Portland	447.8750	R -
Portland Area Repeater		Portland	449.3250	R -
Portland Area Repeater		Portland	449.7500	R -
Portland Area Repeater		Portland	053.2900	R -
West Linn Area Repeater		West Linn	145.2700	R -

Federal Government

Please note: Federal law enforcement agencies utilize encryption or other devices for sensitive communications. Only "routine" communications can be monitored using the information contained in this book.

Federal Bureau Of Investigation

A1	Operations Repeater	Portland	167.6125	Repeater
A2	Operations Simplex	Portland	167.6125	
B1	Operations Repeater	Portland	167.2125	Repeater
B2	Operations Simplex	Portland	167.2125	
	Portland Primary Repeater	Portland	167.2875	Repeater
	SWAT Team Tactical	Portland	163.8625	
	Operations Simplex	Portland	162.7375	
	Operations Repeater	Portland	163.8375	Repeater
	Operations Repeater	Portland	167.5125	Repeater
	Car to Car Simplex	Portland	167.5625	
	Operations Repeater	Portland	167.7625	Repeater
	TRS Link	Portland	TG: 16528	Portland

General Services Admin.	[Dig-P25]	Portland	417.2000	Repeater

Hotels & Country Clubs

Benson Hotel	Portland	464.4250	
Best Western Inn	Portland	464.2250	
Courtyard by Marriott	Portland	462.8125	Simplex
Charbonneau Golf Course	Portland	466.2125	
Delta Inn	Portland	463.2500	468.2500
Doubletree Hotel Portland-Lloy	Portland	463.2000	468.2000
Operations Simplex	Portland	464.9250	
Embassy Suites	Portland	462.3000	Simplex
	Portland	462.5000	Simplex
Fifth Avenue Suites Hotel	Portland	463.8875	468.8875
Glendover Golf Course	Portland	452.7250	457.7250
Heathman Hotel	Portland	461.0250	
Hilton Hotels	Portland	154.6000	
Hilton Hotel Portland	Portland	464.3750	
	Portland	463.5750	468.5750
Holiday Inn	Portland	464.7750	
Holiday Inn Portland Downtown	Portland	462.0250	467.0250
Homewood Suites Hotel	Portland	461.3750	466.3750
Operations Simplex	Portland	467.9250	
Hotel Deluxe Portland Malory	Portland	462.0875	Simplex
467.7625 467.8125 467.8500	Portland	467.8750	Simplex
467.9000	Portland	467.9250	Simplex
Howard Johnson's Hotel	Portland	461.5500	466.5500
Marriott Hotel	Portland	462.8000	
McMenamin's Edgefield Resort	Troutdale	461.5250	Simplex
463.2750 464.7250	Troutdale	469.9500	Simplex
Menucha Retreat & Conf. Center	Portland	464.9250	
Oregon Golf Club	Portland	462.7875	
Operations Simplex	Portland	467.7875	

Oswego Lake Country Club	Portland	467.8125	
Operations Simplex	Portland	467.8375	
467.8625 467.8875 467.9125			
Paramount Hotel	Portland	461.4625	Simplex
	Portland	461.8375	Simplex
Persimmon Country Club	Portland	151.8650	
462.0875 463.2125 463.9125			
Portland Golf Club	Portland	464.4250	
Portland Hilton	Portland	462.8250	
Portland Marriot Hotel	Portland	151.9550	Simplex
	Portland	462.2000	467.2000
Quality Inn Portland Airport	Portland	154.5400	
Operations Simplex	Portland	154.5700	
Operations Simplex	Portland	154.6000	
Red Lion Columbia River	Portland	463.2000	468.2000
Residence Inn	Lake Oswego	469.2625	Simplex
River Place Hotel	Portland	461.3875	
Riverplace Hotel	Portland	461.1625	
Operations Simplex	Portland	461.5625	
Riverside Golf & Country Club	Portland	452.7250	457.7250
Sheraton Inn Portland Airport	Portland	461.9000	466.9000
Sheraton Portland Airport Hotel	Portland	461.1500	466.1500
Sunset Grove Golf Course	Forest Grove	461.0625	Simplex
	Forest Grove	461.1125	Simplex

Industries & Manufacturing

American Honda Motor Company	Portland	461.3875	Simplex
461.5875 464.9875 466.3875			
Boeing Company	Portland	462.4125	Simplex
Operations Repeaters	Portland		
464.4250 464.6750 464.8250			
854.9125 855.4375 855.6875			
Chevron USA Incorporated	Portland	451.5500	456.5500
Operations Repeater	Portland	451.6750	456.6750
Operations Simplex Frequencies	Portland	451.2625	Simplex
451.2875 451.3625 451.6125			
456.2875 456.3625 456.6125			
Chrysler Corporation	Portland	451.3500	456.3500
Columbia Sportwear Company	Portland	461.4625	Simplex
461.4875 461.5125 461.5375 464.8125 466.4625			
466.4875 466.5125 466.5375 467.1375 468.3625			
469.3375 469.8125			
Fujitsu Microelectronics & Comp	Portland	461.1250	466.1250
Operations Repeater	Portland	461.8250	466.8250
Operations Repeaters	Portland	463.4750	468.4750
464.9750 451.5250 462.5000			
Operations Simplex	Portland	152.4800	Simplex
153.1100 462.2250 462.2500 462.2750 462.3000			
462.3500 467.2500 467.5000			

Gunderson Rail Cars

TRUNKED RADIO SYSTEM (Motorola): TrunkTracker Setting: Type II 900

939.9625 939.9500 936.9750 936.9000
935.9750 935.9125 937.8875

Integrated Device Technology Portland 451.5250 465.5250
Operations Simplex Portland 464.3250 Simplex
464.3750 464.4750 464.5250

Intel Corportation
TRUNKED RADIO SYSTEM (Unknown Type): TrunkTracker Setting: Unk
855.1625 855.8875 856.2875 857.2875 858.2875
859.2875 859.7875 859.8125 859.8625 860.2875
860.7875
Operations Repeaters Portland 461.2250 466.2250
462.4250 462.5000 464.4250
464.5750 464.7750 464.8250
Operations Simplex Portland 462.2500 Simplex
462.3250 464.9250 464.9750

Komatsu Silicon America Inc. Portland 462.2750 467.2750
Operations Repeater Portland 462.3000 467.3000
Operations Repeater Portland 462.4500 467.4500

LSI Logic Corporation
TRUNKED RADIO SYSTEM (Motorola Type II): TrunkTracker Setting: Type II 900
935.3875 935.4000 935.4125 935.4250 935.4375

Nabisco Brands Foods Portland 154.5150 Simulcast
464.0375 469.0375 462.7875
463.4125 464.9875 153.0650

NEC America Incorporated Portland 462.2000 467.2000
Operations Repeater Portland 462.3750 467.3750
Operations Simplex Portland 462.3500 Simplex
Operations Simplex Portland 467.2000 Simplex

Nike Incorporated Portland 462.2250 467.2250
Operations Simplex Portland 154.5700 Simplex
461.2125 461.8625 462.8625 464.2375 464.8625
466.8625 467.8625 469.2375 469.8625

Pennzoil Products Company Portland 173.2500 Simplex
Operations Simplex Portland 173.3500 Simplex

Port Of Portland
Air Operations Portland 123.0000 Simplex
Dredge Operations Portland 453.7500 458.7500
Operations / Maintenance Portland 155.9550 Simplex
Swan Island Shipyard Portland 453.4250 458.4250
Operations Repeater Portland 453.6000 458.6000
Operations Repeater Portland 460.2250 465.2250
Operations Simplex Portland
457.5375 461.5625 464.9750 467.7500 467.8000
467.8750 467.9250 468.2625 468.3625 468.3875
468.4125 468.4625 468.5375 468.5875 468.7125
468.7625 468.8125 468.8875 469.2125
TRUNKED RADIO SYSTEM (Motorola SmartZone) ("Portland" Below)
Part of City of Portland ComNet TRS.
TrunkTracker Setting: Type II 800
Portland Airport Intellirepeater: Portland
868.3500* 868.8500* 866.2750 866.7125 867.7500

Note: Virtually all talkgroups EXCEPT Port of Portland related talkgroups can be heard on the Portland Simulcast site at all times. Port of Portland related talkgroups can generally only be heard on the Portland Airport Intellirepeater, and non-Port related traffic is generally not heard on the Portland Airport Intellirepeater.

General Aviation Portland TG: 22768 Portland

Aviation Maintenance Equipment	Portland	TG: 22800	Portland
PDX Operations – Red Zone	Portland	TG: 25136	Portland
PDX Operations – Bus Dispatch	Portland	TG: 25264	Portland
PDX Operations – Bus TAC 1	Portland	TG: 31280	Portland
PDX Operations – Landside	Portland	TG: 25328	Portland
PDX Operations – TAC 1	Portland	TG: 25456	Portland
PDX Operations – TAC 2	Portland	TG: 25072	Portland
PDX Operations – TAC 3	Portland	TG: 25104	Portland
PDX Operations – Toll Plaza	Portland	TG: 31312	Portland
PDX Operations – Starters	Portland	TG: 31344	Portland
Aviation Emergency Operations Center	Portland	TG: 23216	Portland
Airport EOC	Portland	TG: 24528	Portland
Airport ICS – Incident Command	Portland	TG: 24560	Portland
EOC TAC 1	Portland	TG: 25616	Portland
EOC TAC 2	Portland	TG: 25648	Portland
EOC TAC 3	Portland	TG: 25680	Portland
EOC TAC 4	Portland	TG: 25712	Portland
EOC TAC 5	Portland	TG: 25744	Portland
EOC TAC 6	Portland	TG: 25776	Portland
EOC TAC 7	Portland	TG: 25808	Portland
EOC TAC 8	Portland	TG: 25840	Portland
Aviation FIRE Department Dispatch	Portland	TG: 24720	Portland
Airport Disaster 1	Portland	TG: 24592	Portland
Airport Disaster 2	Portland	TG: 24624	Portland
Aviation Security	Portland	TG: 24656	Portland
Fire TAC 1	Portland	TG: 24752	Portland
Fire TAC 2	Portland	TG: 24784	Portland
Fire TAC 3	Portland	TG: 24816	Portland
Fire TAC 4	Portland	TG: 24848	Portland
Fire TAC 5	Portland	TG: 24880	Portland
Fire TAC 6	Portland	TG: 24912	Portland
Fire TAC 7	Portland	TG: 24944	Portland
Fire TAC 8	Portland	TG: 24976	Portland
EMS TAC 1	Portland	TG: 25904	Portland
EMS TAC 2	Portland	TG: 25936	Portland
EMS TAC 3	Portland	TG: 25968	Portland
EMS TAC 4	Portland	TG: 26000	Portland
Aviation POLICE Department Dispatch	Portland	TG: 26128	Portland
Parking Control	Portland	TG: 25872	Portland
Police TAC 1	Portland	TG: 26192	Portland
Police TAC 2	Portland	TG: 26224	Portland
Police TAC 3	Portland	TG: 26320	Portland
Police TAC 4	Portland	TG: 26352	Portland
Aviation FAA Links – PDX Tower	Portland	TG: 16464	Portland
Hillsboro Airport Tower	Portland	TG: 29488	Portland
Troutdale Airport Tower	Portland	TG: 16496	Portland
Port of Portland Marine Terminals SECURITY	Portland	TG: 24080	Portland
Marine Terminal 4 Operations	Portland	TG: 24112	Portland
Port of Portland GTO – Ground Transport Operations			
GTO – PDX Portland Int'l Airport	Portland	TG: 25488	Portland
GTO – Hillsboro	Portland	TG: 27152	Portland
GTO – Mulino	Portland	TG: 27504	Portland
GTO – Troutdale	Portland	TG: 27280	Portland

Port of Portland & Air National Guard Fire Station Locations

81	NE Marine Drive	810, 811, 815, 816, 817, S881, E890, M880
1	6801 NE Cornfoot Rd, ANG	E3, WT7, Crash 4, Crash 6, R5

Sequent Computer Systems	Portland	461.4750	466.4750
Operations Repeater	Portland	462.2000	467.2000

Operations Simplex	Portland	462.0375	Simplex
464.0375 464.8125 464.8625			
464.8875 467.0375 469.8125			
Weyerhaeuser Company	Portland	451.1875	Simplex
Operations Simplex	Portland	456.1875	Simplex
456.6125 463.2625 466.2125			

Hospitals

American Red Cross	Countywide	047.4200	Simplex
Simplex	Countywide	047.4600	Simplex
Links to Low Band Freqs [Varies]	Countywide	463.6375	468.6375
Adventist Medical Center	Portland	155.3400	Simplex
Hospital Paging System	Portland	152.0075	Simplex
Emanuel Hospital	Portland	152.0075	Simplex
Hospital Paging System	Portland	157.4500	Simplex
Good Samaritan Hospital	Portland	151.8650	Simplex
Operations Simplex	Portland	155.3400	Simplex
Hospital Paging System	Portland	152.0075	Simplex
Legacy Health System	Portland	464.5250	469.5250
Operations Repeater	Portland	464.8750	469.8750
Operations Simplex	Portland	463.6875	Simplex
464.2875 468.6875 469.2875			
Legacy Meridian Park Hospital	Portland	464.8250	Simplex
Meridian Park Hospital	Portland	155.3400	Simplex
Operations Simplex	Portland	467.9250	Simplex
Hospital Paging System	Portland	157.4500	Simplex
Engineering Personnel	Portland	467.8750	Simplex
Mount Hood Medical Center	Portland	155.3400	Simplex
Hospital Paging System	Portland	152.0075	Simplex
Operations / Security Simplex	Portland	154.5400	Simplex
Oregon State Hospital	Portland	151.8050	Simplex
Portland Adventist Medical Center	Portland	463.6250	468.6250
464.3750 464.5250 464.8750			
Providence Medical Center	Portland	463.5750	468.5750
Operations Repeater	Portland	464.9000	469.9000
Providence Milwaukie Hospital	Portland	155.3400	Simplex
Providence Newberg Hospital	Portland	462.8000	Simplex
Providence Saint Vincent's Hospital	Portland	461.8000	466.8000
Operations Repeater	Portland	464.5250	469.5250
Operations Simplex	Portland	155.3400	Simplex
461.8375 464.7375 469.7375			
Hospital Paging System	Portland	157.4500	Simplex
Shriners Hospital For Children	Portland	461.4250	466.4250
Hospital Security	Portland	TG: 20176	Portland
Tuality Community Hospital	Hillsboro	155.3400	Simplex
Operations Simplex	Hillsboro	155.4000	Simplex
Operations Repeater	Hillsboro	462.0750	467.0750
	Hillsboro	461.7000	466.7000
	Hillsboro	464.7000	469.7000
Willamette Falls Comm. Hospital	Portland	155.3400	Simplex
Hospital Paging System	Portland	453.1750	Simplex
Willamette Falls Hospital	Portland	469.9625	Simplex

Veterans Affairs Medical Center		Portland	456.3125	Simplex
Police – Dispatch		Portland	TG: 6480	Portland
Police – TAC 1 Encrypted		Portland	TG: 6512	Portland
Police – TAC 2 Encrypted		Portland	TG: 6544	Portland
Transportation		Portland	TG: 6576	Portland
Facilities Management		Portland	TG: 6608	Portland
GOTO 1		Portland	TG: 6640	Portland
GOTO 2		Portland	TG: 6672	Portland
Police – Phone Patch		Portland	TG: 6704	Portland

Media

KATU Television (Channel 2)				
Ch1 Engineering/Traffic Reports		Portland	450.1125	Simplex
Ch2 Engineering Alternate		Portland	455.1125	Simplex
Ch3 Operations/Dispatch		Portland	450.4125	455.4125
IFB (Producer to Reporter)		Portland	161.7300	Simplex
KEX News (AM 1190)		Portland	450.9250	Simplex
IFB (Producer to Reporter)		Portland	455.9250	Simplex
Traffic Reports	[PL 82.5]	Portland	450.5500	455.5500
KGW Television (Channel 8)		Portland	455.2875	Simplex
Ch1 Operations		Portland	450.5125	
Ch2 Dispatch / Coordination		Portland	450.2875	
Ch3 Multi-Use / Reserve		Portland	450.6125	
IFB (Producer to Reporter)		Portland	450.3875	Simplex
Remotes		Portland	170.1500	Simplex
Remotes		Portland	455.6125	Simplex
		Portland	455.5125	Simplex
KKSN Radio (97.1 FM)		Portland	450.8000	Simplex
"Sky Spy" Helicopter	[PL 114.8]	Portland	450.0625	455.0625
		Portland	450.8500	Simplex
		Portland	455.8500	Simplex
KOIN Television (Channel 6)		Portland	466.1125	Simplex
Ch1 Operations & Chopper 6	[PL 146.2]	Portland	450.0500	455.0500
Ch2 Engineering/Traffic Watch	[PL 146.2]	Portland	450.1500	Simplex
Engineering Simplex	[PL 146.2]	Portland	455.1500	Simplex
KPAM Radio (AM 860)	[PL 103.5]	Portland	450.9500	
KPTV (Oregon) Television (Ch 12)		Portland	461.2500	
Operations Repeater	[DPL 032]	Portland	450.0875	455.0875
		Portland	161.6400	Simplex
		Portland	450.1875	
KFXX (1520 AM)		Portland	450.2500	Simplex
		Portland	455.2500	Simplex
KUFO Traffic Reports	[PL 110.9]	Portland	450.3500	
KXL Traffic Reports	[PL 141.3]	Portland	450.4875	455.4875
Oregonian Newspaper	[DPL 025]	Portland	453.0000	458.0000

Military & Coast Guard

Coast Guard, United States			
Ch6 Marine / Search & Rescue	Portland	156.3000	Simplex
Ch15 Marine / Weather Emergency	Portland	156.7500	Simplex
Ch16 Marine / Calling / Distress	Portland	156.8000	AM
Ch21 Marine / Bend Working Ch	Portland	157.0500	Simplex

Ch	Description	Location	Frequency	Mode
Ch22	Marine / Notices to Mariners	Portland	157.1000	Simplex
Ch23	Marine / Portland Working Ch.	Portland	157.1500	Simplex
Ch82	Marine	Portland	157.1250	Simplex
Ch83	Marine / Restricted to USCG	Portland	157.1750	Simplex
	Air - Emergency - VHF	Portland	121.5000	243.0000
	Air - Search & Rescue Training	Portland	122.9000	AM
	Air - Search & Rescue	Portland	123.1000	282.8000
	Air Support - Primary	Portland	381.8000	AM
	Air Support - Secondary	Portland	381.7000	AM
	Air Operations Secondary	Portland	383.9000	AM
	Air Operations Primary	Portland	381.8000	AM
	Aircraft Search & Rescue	Portland	237.9000	AM
	Portland Law Enforcement	Portland	165.5000	Repeater
	Portland Law Enforcement	Portland	171.1500	Repeater
	Rescue Beacons	Portland	240.6000	AM
	Rescue Beacons	Portland	275.1000	AM
	Search & Rescue	Portland	282.8000	AM
	Search & Rescue	Portland	156.3000	Simplex
	TRS Link	Portland	TG: 16208	Portland
Oregon Air National Guard		Portland	165.1625	Simplex
	Aircraft Maintenance	Portland	165.0125	Simplex
	Security F1	Portland	163.5250	Simplex
	Security F2	Portland	163.4875	Simplex
	Crash/Fire/Rescue	Portland	173.5375	Simplex
	Air Rescue Squad	Portland	149.1750	Simplex
	Base Operations	Portland	280.5000	AM
	Guard Operations	Portland	288.9000	AM
Oregon Army National Guard		Portland	139.1875	
Ch1	Operations Simplex	Portland	139.2125	
Ch2	Operations Simplex	Portland	139.1375	
Ch3	Operations Simplex	Portland	139.3875	
Ch4	Operations Simplex	Portland	139.1875	
Ch5	Operations Simplex	Portland	141.0500	
Ch6	Operations Simplex	Portland	141.1000	
Ch7	Operations Simplex	Portland	139.2375	
Ch8	Operations Simplex	Portland	141.1500	
Ch9	Operations Simplex	Portland	141.2000	
Ch10	Operations Simplex	Portland	142.8750	
Ch11	Operations Simplex	Portland	142.9250	
Ch12	Operations Simplex	Portland	148.5250	
Ch13	Operations Simplex	Portland	150.6250	
	Helicopter Operations	Portland	040.9000	FM
	Helicopter Operations	Portland	241.6000	AM
	Operations Simplex	Portland	135.0000	AM
	Operations Simplex	Portland	139.2625	

Miscellaneous

Name	Location	Frequency	Mode
Albertsons Incorporated	Portland	463.7000	468.7000
Operations Simplex	Portland	151.7150	Simplex
151.9250 461.0125 464.3375			
American Commercial Security	Portland	464.0750	469.0750
American Protective Services	Portland	461.7750	466.7750
American Red Cross	Portland	463.9000	468.9000
Operations Simplex	Portland	047.4200	Simplex
Operations Simplex	Portland	047.4600	Simplex

047.6600 462.6500 462.7000

American Security	Portland	461.5750	466.5750
Benson Hotel	Portland	462.3250	467.3250
B R Cook Investigations	Portland	151.6550	Simplex
Clackamas Town Center	Portland	462.1500	467.1500
Operations Repeater	Portland	464.6000	469.6000
Operations Repeater	Portland	464.6500	469.6500
Operations Simplex	Portland	462.8875	Simplex
Operations Simplex	Portland	463.3125	Simplex
463.3875 467.8875 468.3125			
Computer City Supercenters	Portland	461.4125	Simplex
Operations Simplex	Portland	464.5875	Simplex
Computer Supercenters			
Operations Simplex	Portland	462.0375	Simplex
Operations Simplex	Portland	462.8375	Simplex
463.6125 463.8375 463.9625 464.1125 467.0375			
467.8375 468.6125 468.8375 468.9625 469.1125			
Costco Wholesale Club	Portland	466.8125	Simplex
Operations Simplex	Portland	154.5700	Simplex
Economic Improvement District (E.I.D.)			
Security Dispatch/Operations	Portland	TG: 18032	Portland
Empire Security	Portland	461.9250	Simplex
Operations Simplex	Portland	469.3250	Simplex
Federal Express Corporation	Portland	857.8625	812.8625
Operations Repeater	Portland	854.6125	809.6125
Operations Repeater	Portland	860.9125	815.9125
Operations Simplex	Portland	461.1125	Simplex
Operations Simplex	Portland	466.1125	Simplex
TRUNKED RADIO SYSTEM (Unknown Type): TrunkTracker Setting: Unk			
856.8625 857.8625 858.8625 859.8625 860.8625			
Holladay Park Plaza	Portland	464.4250	469.4250
Home Depot	Portland	461.2125	Simplex
463.2875 464.5625 464.6375			
IKEA Store 461.3875	Portland	463.3125	468.3125
461.7625 461.8375 464.8875	Portland	463.2375	Simplex
J C Penney Co Inc 288-1	Portland	461.6125	Simplex
JJ Security	Portland	461.5000	466.5000
Operations Repeater	Portland	462.0750	467.0750
Lloyd Center Mall	Portland	464.5750	469.5750
463.5250 464.4250 464.8750 464.9750			
468.5250 469.8750 469.9750			
Metropolitan Service District	Portland	453.7250	458.7250
Memorial Coliseum (Security)	Portland	453.3500	458.3500
Low-Power Simplex Frequencies	Portland		
453.2375 453.3125 453.4875 453.5625 453.6875			
453.8375 453.8875 458.2375 458.3125 458.4875			
458.5625 458.6875 458.8375 458.8875			
Nordstrom's Distribution Center	Portland	463.8375	Simplex
464.0875 464.2625 464.6125			
464.9750 468.8375 469.6125			
Northwest Investigations	Portland	463.9750	468.9750
Northwest Protective Service	Portland	461.4750	466.4750
Pinkerton Security	Portland	461.1750	

Pioneer Courthouse Square	Portland	462.8125	Simplex
Pioneer Place Shopping Center	Portland	462.8125	Simplex
461.7375 461.8125 462.1375	Portland	466.7375	Simplex
	Portland	466.8125	Simplex
	Portland	467.1375	Simplex
Portland Business Alliance			
EID Security / Portland Guides	Portland	TG: 18032	Portland
Portland Expo Center	Portland	451.5875	Simplex
451.6125 456.4875 456.6125			
Protective Security & Investigations	Portland	464.7500	469.7500
Rams Specialized Security Service	Portland	463.3000	468.3000
Reed Institute	Portland	463.8750	468.8750
Safeway Stores Incorporated	Portland	464.7000	469.7000
Saks Fifth Avenue	Portland	462.3750	467.3750
Streets of Tanasbourne Mall	Hillsboro	461.3875	Simplex
462.1375 463.5625 463.7625	Hillsboro	464.0875	Simplex
Target Stores	Portland	467.7625	Simplex
469.0625 461.6625 461.8625			
The Home Depot	Portland	461.2125	Simplex
Operations Simplex	Portland	463.2875	Simplex
Operations Simplex	Portland	464.6375	Simplex
United Parcel Service	Portland	157.6800	Simplex
Wackenhut Corporation	Portland	464.0750	469.0750
Waremart	Portland		
457.5375 461.7375 452.3125			
457.5375 467.7750 467.9250			
Washington Square	Portland	151.8350	159.9750
Operations Simplex	Portland	464.4750	Simplex
Maintenance	Portland	452.3625	457.3625
	Portland	464.4375	Simplex
Zupans Food Pavilion	Portland	469.9625	Simplex

Railroad

Burlington Northern & Santa Fe	Portland	160.6650	Repeater
160.5450 160.5900 160.6800 160.7250 160.9650			
161.0100 161.1000 161.1300 161.2200 161.2500			
161.3700 161.4150 161.4600 161.5050			
Railroad Police Personell	Portland	TG: 15696	Portland
Lewis & Clark Railroad	Portland	160.7850	
	Portland	160.4400	
Molalla Western Railway Company	Portland	160.3350	
Operations Repeaters	Portland	160.5750	
161.1900 161.3400 161.4450			
Port Of Tillamook Bay Railroad	Portland	160.3050	160.9050
Operations Repeater	Portland	161.0550	Repeater
Portland & Western Railroad	Portland	160.7700	
Portland Terminal Railroad	Portland	160.6800	
161.1000 161.4900 161.5500			
Union Pacific Railroad	Portland	160.2900	Repeater
Operations Repeater	Portland	160.9500	Repeater
Operations Repeater	Portland	161.3100	Repeater

Operations Repeater	Portland	452.9000	457.9000
Operations Repeater	Portland	464.1000	469.1000
Operations Simplex Frequencies	Portland		

160.2600 160.3200 160.3500 160.3800 160.4100
160.4700 160.5150 160.6050 160.6800 160.7400
160.8000 160.8300 160.8600 160.9800 161.0250
161.1000 161.1450 161.1600 161.2650 161.4000
161.4300 161.5500 160.2300

Willamette & Pacific Railroad Portland
160.7700 160.2150 160.3650

Willamette Valley Railroad Portland
161.4750 160.5300 161.0700

Recreation

Butch Mock Motorsports	Portland	461.9375	Simplex
	Portland	468.9750	Simplex
Clackamas County Fair	Portland	151.8650	Simplex
	Portland	151.9550	Simplex
Coast To Coast Event	Portland	464.7125	Simplex
	Portland	464.7375	Simplex
Event Rental Communications Inc	Portland		

461.1375 461.3125 461.5125 461.6875 462.1625
466.1375 466.3125 466.5125 466.6875 467.1625

Motorsports Management Inc	Portland	469.5000	Simplex
Mount Hood Racing Team	Portland	464.5500	Simplex
Mount Hood Ski Patrol	Parkdale	155.2050	Simplex
Operations Repeater	Parkdale	154.0850	158.9550
Multnomah Athletic Club	Portland	152.4800	Simplex
	Portland	461.5875	466.5875
	Portland	464.7625	469.7625
	Portland	463.9125	468.9125
Maintenance	Portland	452.6250	457.6250
Oregon Historical Society	Portland	457.5375	Simplex
	Portland	457.6375	Simplex
Oregon Museum of Science & Industry		462.0875	467.0875
468.3625 469.3375 464.2125	Portland	464.3375	Simplex
	Portland	464.9875	Simplex
Oregon Zoo (Washington Park)	Portland	462.1000	467.1000
Washington Park & Zoo Railway	Portland	151.6550	Simplex
Simplex Frequencies	Portland	453.5875	Simplex

453.7375 453.7875 458.5875 458.7375 458.7875
464.8750 464.6750 469.8750 469.6750

Portland Art Museum	Portland	463.4125	Simplex
Operations Repeater	Portland	451.8500	456.8500
Operations Simplex	Portland	461.9875	Simplex
Portland Expo Center	Portland	451.5875	Simplex

451.6125 456.5875 456.6125

Portland International Raceway	Portland	461.0875	Simplex

464.3625 466.0875 469.3625

Portland Speedway	Portland	154.5700	Simplex
	Portland	154.6000	Simplex
Portland Trail Blazers	Portland		

Name	Tone	Location	Frequency	Output / Mode
457.5250 457.5750 462.7625				
463.7875 464.1125 469.1125				
Portland Winter Hawks		Portland	464.8625	Simplex
		Portland	469.8625	Simplex
Wireless Microphone		Portland	178.4000	Simplex
Rose Garden Arena		Portland	451.4125	456.4125
		Portland	452.4125	457.4125
		Portland	463.6125	468.6125
461.9875 462.3875 463.2375 463.3375 463.5625				
463.9375 464.4875 464.6875 464.9375 466.9875				
467.3875 468.2375 468.3375 468.5625 468.9375				
469.4875 469.6875 469.9375				
Heliport Operations		Portland	123.0500	AM
Sundance Marina		Portland	463.8125	Simplex
Sunset Hills Memorial Park		Portland	463.6250	468.6250
151.8950 463.2125 468.2125				
Towne Lounge Venue		Portland	463.3750	468.3750
Tualatin Hills Park & Recreation		Portland	151.7150	Simplex
151.7450 154.6000				
Washington County Fairgrounds				
Ch1 Primary Ops/Admin./Maint.	[DPL 114]	Portland	464.7875	Simplex
Ch2 Fair: Security	[DPL 114]	Portland	462.8375	Simplex
Ch3 Fair: Entertainment	[DPL 114]	Portland	457.5375	Simplex
Ch4 Fair: Cashiers / Parking	[DPL 114]	Portland	457.5625	Simplex
Ch5 Fair: Alternate	[DPL 114]	Portland	457.5875	Simplex
Ch6 Fair: Rodeo Operations	[DPL 114]	Portland	464.5000	Simplex

Washington Park Zoo: See Oregon Zoo

Schools

Name	Tone	Location	Frequency	Output / Mode
Beaverton School District # 48		Beaverton	151.6550	Simplex
		Beaverton	154.6000	Simplex
Ch1 Operations		Beaverton	TG: 59472	Washington
Ch2 Operations		Beaverton	TG: 59440	Washington
Ch3 Operations		Beaverton	TG: 59408	Washington
Ch4 Operations		Beaverton	TG: 59376	Washington
Operations Repeaters		Beaverton	461.2500	466.2500
463.8500 464.1250 935.1500 936.1500 937.1500 939.1500				
High School Maintenance		Beaverton	152.9375	160.1775
Westview High School		Beaverton	153.1100	Simplex
Bus Transportation	[PL 88.5]	Beaverton	462.4750	467.4750
Bus Transportation	[PL 88.5]	Beaverton	464.1250	469.1250
		Beaverton	152.2850	157.5450
Canby High School		Canby	464.9000	469.9000
Cascade School District #5		Portland	464.5000	Simplex
Centennial School District		Portland	467.7625	Simplex
Bus 1 – School Busses		Portland	TG: 17840	Portland
Bus 2 – School Busses		Portland	TG: 17872	Portland
Century High School		Portland	151.5200	Simplex
		Portland	154.6000	Simplex
Clackamas County Community College				
Maintenance & Security		Oregon City	152.3525	157.6125
Maintenance & Security		Wilsonville	152.3975	157.6575
Corbett School District No 39		Portland	151.8050	Simplex

Agency		Location	Frequency	Input/Type
David Douglas School District		Portland	464.5000	Simplex
Operations Repeater		Portland	153.2900	158.2650
Bus 1		Portland	TG: 17776	Portland
Bus 2		Portland	TG: 17808	Portland
Administration		Portland	TG: 17904	Portland
Estacada School District 108		Estacada	152.9150	Repeater
Forest Grove School District		Portland	464.5750	469.5750
Bus Operations	[DPL 565]	Forest Grove	463.7500	468.7500
461.0625 462.7625 464.0750		Forest Grove	461.5250	466.5250
466.6625 469.0750 469.6625		Forest Grove	452.2250	457.2250
		Forest Grove	463.9750	468.9750
Fowler Middle School		Portland	151.8950	Simplex
		Portland	154.6000	Simplex
Grant High School		Portland	464.0375	Simplex
		Portland	469.0375	Simplex
Gresham High School		Gresham	159.7850	Simplex
159.8025 160.1025 160.1700				
	[DPL 431]	Gresham	152.3150	157.5750
Hillsboro School District				
Bus Transportation	[DPL 156]	Portland	151.7750	Simplex
Hillsboro High School	[PL 167.9]	Hillsboro	153.0200	159.5850
Glencoe High School	[DPL 223]	Hillsboro	153.3350	Simplex
Century High School	[DPL 023]	Hillsboro	154.6100	Simplex
Lake Oswego Public Schools		Portland	151.8950	Simplex
		Portland	154.6000	Simplex
Lewis & Clark College		Portland	154.5400	Simplex
Ch1 Operations Repeater		Portland	462.5500	467.5500
Ch2 Operations Direct		Portland	462.5500	Simplex
Ch3 Alternate / Talk-around		Portland	464.5500	Simplex
		Portland	461.2750	Simplex
		Portland	462.2250	467.2250
		Portland	463.4750	468.4750
		Portland	464.7250	469.7250
		Portland	462.3750	467.3750
TRUNKED RADIO SYSTEM (LTR): TrunkTracker Setting: LTR				
463.3750 463.8250 464.8250				
Marylhurst College		Portland	464.3250	Simplex
466.1125 466.8125				
Mission College of Evangelism		Gaston	160.0800	Simplex
151.7000 159.5500 159.9900		Gaston	160.0050	Simplex
Molalla High School		Molalla	451.8250	Simplex
452.1625 452.1875		Molalla	461.3000	Simplex
Mount Hood Community College		Gresham	464.8750	469.8750
		Gresham	467.9250	Simplex
Multnomah School of the Bible		Portland	151.8950	Simplex
North Clackamas School District		Portland	452.1750	457.1750
Operations Repeaters		Portland	452.2750	457.2750
452.3500 452.4750 452.6750				
Oregon City School District #62		Portland	047.5400	Simplex
Oregon Graduate Institute		Portland	461.4500	Simplex
Oregon Health Sciences University		Portland	456.2375	Simplex
		Portland	456.3125	Simplex
West Campus		Beaverton	463.4750	468.4750
Oregon Institute Of Technology		Portland	464.5000	Simplex

			Location	Frequency	Input
			Portland	464.5500	Simplex
	Operations Repeaters		Portland	464.6000	469.6000
	464.6250 464.6500 464.7000 464.7250 464.7500				
Pacific University			Portland	451.3625	Simplex
Ch1	Public Safety - Repeater	[DPL 412]	Portland	452.6250	457.6250
Ch2	Public Safety - Simplex	[DPL 412]	Portland	452.6250	Simplex
Ch3	Public Safety - Alternate	[DPL 412]	Portland	452.3250	Simplex
Ch4	Public Safety - Alternate	[DPL 412]	Portland	452.7500	Simplex
Ch5	Public Safety - Alternate	[DPL 412]	Portland	451.4250	Simplex
Ch1/2	Facilities Engineering	[Multi PL]	Portland	151.8050	Simplex
Ch3/4	Facilities Engineering	[Milti PL]	Portland	154.6000	Simplex
Ch1	Information Services	[PL 67.0]	Portland	467.8500	Simplex
Ch2	Information Services	[PL 67.0]	Portland	467.9250	Simplex
	University Center Staff	[PL 71.9]	Portland	462.6250	Simplex
			Portland	451.1875	Simplex
			Portland	451.2875	Simplex
Portland Community College			Portland	453.8500	458.8500
	Security – Dispatch		Portland	TG: 17456	Portland
	Security – TAC 1		Portland	TG: 18448	Portland
	Security – TAC 2		Portland	TG: 18544	Portland
Portland Public School District			Portland	155.2650	Simplex
	151.5500 159.5700 159.5400		Portland	152.3600	157.5750
	159.6150 159.8100 159.8400		Portland	152.3450	157.6050
	159.9300 160.0950 160.1400		Portland	152.3150	157.6200
Portland State University			Portland	461.2000	466.2000
			Portland	462.1500	467.1500
			Portland	462.8875	Simplex
			Portland	467.8875	Simplex
	POLICE		Portland	453.5250	458.5250
	Health Department		Portland	453.6750	458.6750
Reed College			Portland	451.5625	Simplex
			Portland	452.0875	Simplex
			Portland	456.5625	Simplex
Reynolds School District #7					
	School Buses	[PL 85.4]	Portland	155.1750	Simplex
	District-wide Repeater	[DPL 754]	Portland	461.8500	466.8500
	Troutdale Repeater		Troutdale	462.4250	467.4250
	HB Lee Middle School	[PL 103.5]	Portland	461.8500	Simplex
	Reynolds Middle School	[PL 127.3]	Portland	461.8500	Simplex
	Walt Morey Middle School	[PL 162.2]	Portland	461.8500	Simplex
	Reynolds High School	[DPL 047]	Portland	461.8500	Simplex
	Natural Resources Acad.	[PL 100.0]	Portland	461.8500	Simplex
Sandy Elementary Schools District			Sandy	464.9000	469.9000
Sherwood School District #88			Sherwood	045.9200	Simplex
University Of Portland			Portland	461.4875	Simplex
			Portland	464.4750	469.4750
			Portland	466.4875	Simplex
	Security – Dispatch		Portland	TG: 35504	Portland
	Security – Tac-1		Portland	TG: 35536	Portland
	Athletic Department		Portland	452.9750	Simplex
	457.9750 462.3500 464.1000		Portland	467.3500	Simplex

Utilities

Colton Telephone Company		Portland	451.3000	Simplex
Beaver Creek Coop Telephone Co.		Portland	451.4250	456.4250
Canby Telephone Association		Portland	451.4000	456.4000
Cascade Utilities Inc	[PL 91.5]	Portland	451.3250	456.3250
Clear Creek Mutual Telephone		Portland	451.3000	
Damascus Water District		Portland	451.2250	456.2250
GTE Northwest Incorporated		Portland	451.3500	456.3500
		Portland	451.4500	
		Portland	461.1500	466.1500
Molalla Telephone Company		Molalla	855.1375	810.1375
Mount Scott Water District		Portland	451.2250	456.2250
453.2625 458.2625				
Multnomah County Drainage		Portland	461.7250	466.7250
Northwest Natural Gas	[PL 146.2]	Portland	451.2500	456.2500
		Portland	854.9375	809.9375
		Portland	855.9375	810.9375
Simplex Frequencies		Portland	158.1750	Simplex
153.4100 153.4400 158.1900				
Oak Lodge Water District		Portland	451.2000	456.2000
Pacific Power & Light Company		Portland	153.5900	Repeater
Simplex Frequencies		Portland		
153.5150 158.1300 159.5100				
159.5250 159.7500 159.9150				
Portland General Electric	[PL 203.5]	Gresham	153.4700	Repeater
Operations Repeater		Portland	153.6650	Repeater
Washington County Rptr.	[PL 203.5]	Portland	160.0500	153.3800
Paging System		Portland	465.0000	Simplex
Simplex Frequencies		Portland		
152.9750 048.0600 048.4600 153.2750 153.3800				
153.4250 153.4850 153.5750 153.6800				
153.7100 158.1450 159.6300 159.8700				
Repeater Frequencies		Portland		
159.8850 160.0500 160.1400 451.0250 451.1250				
Rockwood Water District		Portland	453.5750	458.5750
Sunrise Water Authority		Happy Valley	451.2250	456.2250
Tualatin Valley Water District		Portland	153.5000	158.4150
		Portland	158.9325	Simplex
Operations		Portland	TG: 62896	Washington
Operations		Portland	TG: 62864	Washington
West Slope Water District		Portland	153.6200	Simplex

Weather Information

National Weather Service – NOAA Weather Radio

Conditions & Forecasts – Portland	162.5500	Simplex

Multnomah County SAME Alert Code: 041051

POLK COUNTY

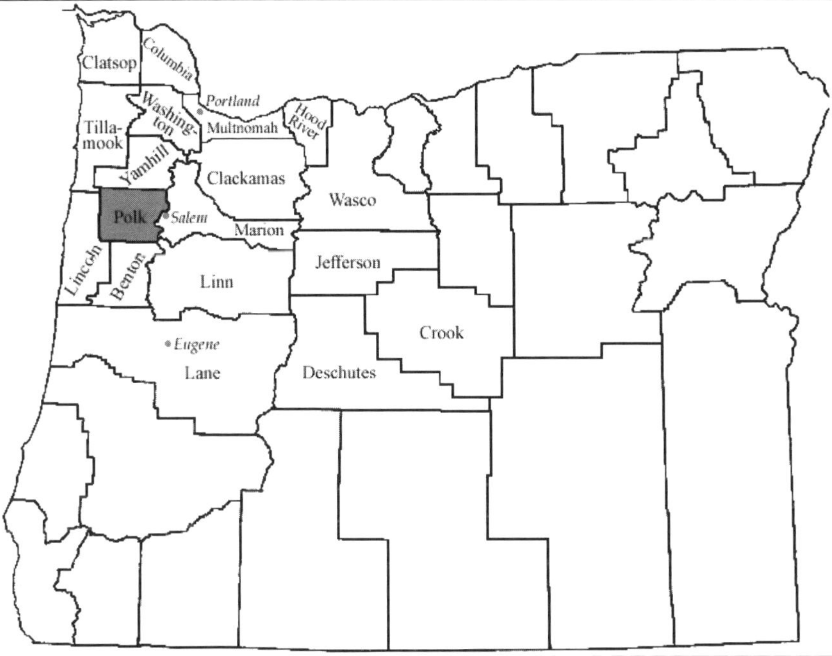

Government & Public Safety

Confederated Tribes of Grand Ronde Fire Department				
		Willamina	155.9550	158.9250
Dallas City Public Works	[PL 110.9]	Dallas	155.1000	Simplex
Dallas Fire Department		Dallas	153.8300	Simplex
Dispatch & Operations	[PL 110.9]	Dallas	154.2350	155.9850
Secondary	[PL 123.0]	Dallas	154.4450	Simplex
Falls City Public Works	[PL 110.9]	Falls City	158.8350	Simplex
Falls City Fire Services: See Dallas Fire Department Frequencies				
Monmouth City Government		Monmouth	154.9500	Simplex
Power & Light		Monmouth	154.8000	Simplex
		Monmouth	155.6700	Simplex
Monmouth Police Department		Monmouth	155.9850	
Oregon State Police				
St 21 North & East Polk	[PL 156.7]	Salem Station	155.9100	159.1500
St 23 Southwest Polk	[PL 156.7]	Albany Station	154.6650	159.0300
Polk County Fire Department: See Dallas Fire Department Frequencies				
Polk County Government		Countywide	453.9250	458.9250
Road Maintenance	[PL 82.5]	Countywide	153.9800	158.8800

Polk County Sheriff's Office	Countywide	154.0850	155.6700
Ch1 Dispatch [PL 110.9]	Countywide	155.2500	156.2100
Ch2 Tactical [PL 110.9]	Countywide	154.8750	158.8500
	Countywide	155.9325	Repeater
	Countywide	155.0325	158.9025
Spring Valley Rural Fire District	Spring Valley	153.7700	

Southwest Polk Rural Fire District: See Dallas Fire Department Frequencies

Polk County Fire Stations

40	Buena Vista	90	Independence
50	Suver	100	Dallas
60	Monmouth	110	Polk Rural
70	Pedee	120	Falls City
80	Airle	130	Rickreall

Polk County Law Enforcement Unit Callsigns

D	Dallas Police	M	Monmouth Police
I	Independence Police	P	Sheriff

Miscellaneous

American Red Cross Disaster Relief	Countywide	453.5750	458.5750
Central School District 13J	Independence	155.1750	Simplex
Independence State Airport	Independence	122.8000	AM
National Weather Service – NOAA Weather Radio			
Conditions & Forecasts – Salem		162.4750	Simplex
Polk County SAME Alert Code: 041053			
Portland General Electric [PL 203.5]	Countywide	153.5750	173.2250
	Countywide	153.3800	160.0500
Spirit Mountain Casino	Grande Ronde	451.1875	Simplex
451.3125 451.6375 456.1875	Grande Ronde	456.3125	Simplex
456.6375 464.8875 464.9125	Grande Ronde	464.9375	Simplex
464.9625 469.4875 469.5125	Grande Ronde	469.5375	Simplex
469.6875 469.7125 469.7375	Grande Ronde	469.7625	Simplex
469.7875 469.8125 469.8375	Grande Ronde	469.8875	Simplex
469.9125 469.9375	Grande Ronde	469.9625	Simplex
Trunked Radio System: Unknown Type			
451.8500 452.3500 452.7250 462.2250 463.7500			
West Valley Hospital	Dallas	159.7575	Simplex
159.7725 159.7875 159.8700	Dallas	160.0200	Simplex
Western Oregon University	Monmouth	151.8050	Simplex
[PL 110.9]	Monmouth	153.9050	Simplex
461.2125 461.5875 462.5125	Monmouth	158.7300	153.8600
	Monmouth	153.1100	160.2000
	Monmouth	462.4750	467.4750

TILLAMOOK COUNTY

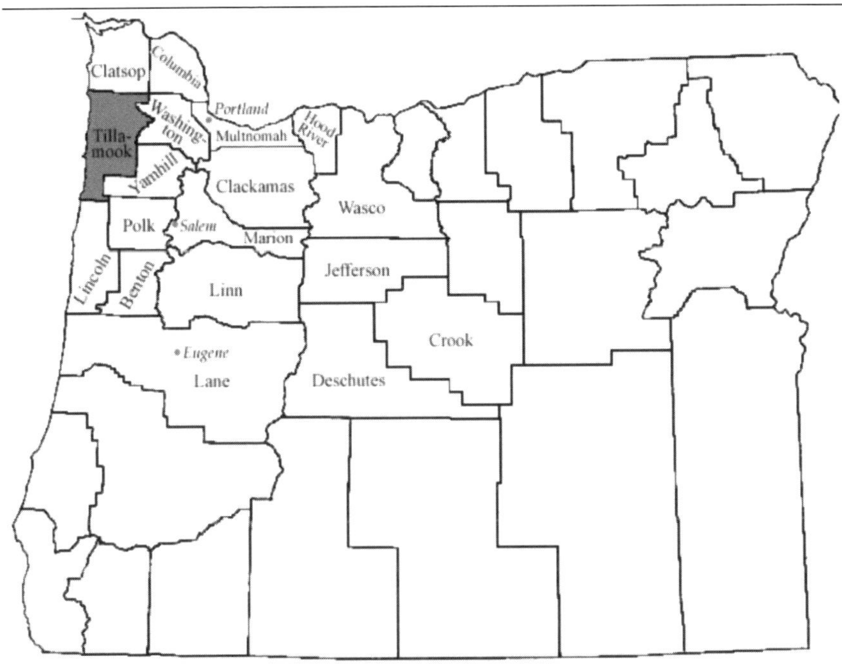

Government & Public Safety

Bay City City Government		Bay City	154.0400	Simplex
Garibaldi City Government		Garibaldi	155.7600	Simplex
Manzanita City Water Department		Manzanita	452.7875	457.7875
Manzanita Dept. of Public Safety		Manzanita	154.0400	Simplex
Nehalem Police Department		Nehalem	154.0400	Simplex

Nestucca Rural Fire Protection District: See Tillamook FD's Radio

Fire Station Locations

81	Cloverdale	83	Beaver	85	Sandlake
82	Pacific City	84	Neskowin	86	Blaine

Oregon Dept. of Forestry	[PL 146.2]	Bald Mtn. W	151.1450	159.2850
	[PL 146.2]	Mount Hebo	151.1675	159.3225
Oregon State Police (Sta 17)	[PL 179.9]	Cape Lookout	154.8600	158.9100
Rockaway Beach Police Department		Rockaway Bch	155.6625	159.0375
Tactical		Rockaway Bch	155.0850	Simplex
Suislaw National Forest	[PL 103.5]	Countywide	162.6125	162.0250
Tillamook County Fire Departments		Countywide	154.0400	Simplex
Ch1 Dispatch		Countywide	154.3550	Repeater
TAC 1 – Tactical – North		Countywide	154.2800	Simplex
TAC 2 – Tactical – Central		Countywide	154.3550	Simplex
Ch5 TAC 3 – Tactical – South		Countywide	154.4450	Repeater
Tactical / Fireground		Countywide	154.4300	Simplex

Agency	Location	Frequency	Type
Tillamook Fire District	Tillamook	154.1450	Repeater
150.7750 153.8300	Tillamook	154.2200	Simplex
155.2050 155.2950	Tillamook	153.8900	Simplex
Tillamook County EMS	Countywide	453.0750	458.0750
Ambulance to T.C. General Hospital	Countywide	462.9500	467.9500
Tillamook County Government	Neahkahnie B.	154.0850	Repeater
	Neahkahnie B.	154.2500	Repeater
	Countywide	154.3100	Repeater
	Tillamook	155.2650	Repeater
	Tillamook	155.0850	Repeater
	Tillamook	155.7150	Repeater
COPW – Public Works	Countywide	153.9650	158.8350
154.1675 155.1075	Countywide	155.8350	Simplex
	Tillamook	155.4900	158.9700
Tillamook County Sheriff's Office	Countywide	155.1300	Simplex
Operations – Mount Hebo [PL 156.7]	Central Cnty	154.7250	155.4150
Operations – South Saddle [PL 114.8]	South Cnty	154.7250	155.4150
Operations – Neahkahnie [PL 179.9]	North Cnty	154.7250	155.4150
Search & Rescue	Countywide	155.8050	Simplex
United States Bureau of Land Management			
Mount Hebo Repeater [PL 100.0]	Countywide	165.9750	166.8500

Miscellaneous

Agency	Location	Frequency	Type
Neahkahnie School District	Rockaway	461.0000	Simplex
School Bus Transportation	Rockaway	461.1000	Simplex
Nehalem Bay State Airport	Manzanita	122.9000	AM
Pacific City State Airport	Pacific City	122.9000	AM
Port of Tillamook Bay Railroad	Tillamook	160.3050	160.9050
Tillamook Airport	Tillamook	122.8000	AM
Seattle Center Approach / Departure	Tillamook	124.2000	317.6000
Tillamook County Creamery Assoc.	Tillamook	461.4375	466.4375
466.3875 467.0875 468.2375	Tillamook	468.6125	Simplex
Tillamook People's Utility District	Countywide	047.8000	Simplex
451.1500 456.1500	Countywide	048.3200	Simplex
United Telephone Company	Tillamook	451.3250	456.3250
	Tillamook	451.4000	456.4000
	Tillamook	451.5000	456.5000

Weather Information

National Weather Service – NOAA Weather Radio

Description	Frequency	Type
Conditions & Forecasts – Mount Hebo	162.5250	Simplex
Conditions & Forecasts – Neahkahnie (North County)	162.4250	Simplex
Conditions & Forecasts – Tillamook	162.4750	Simplex

Tillamook County SAME Alert Code: 041057

WASCO COUNTY

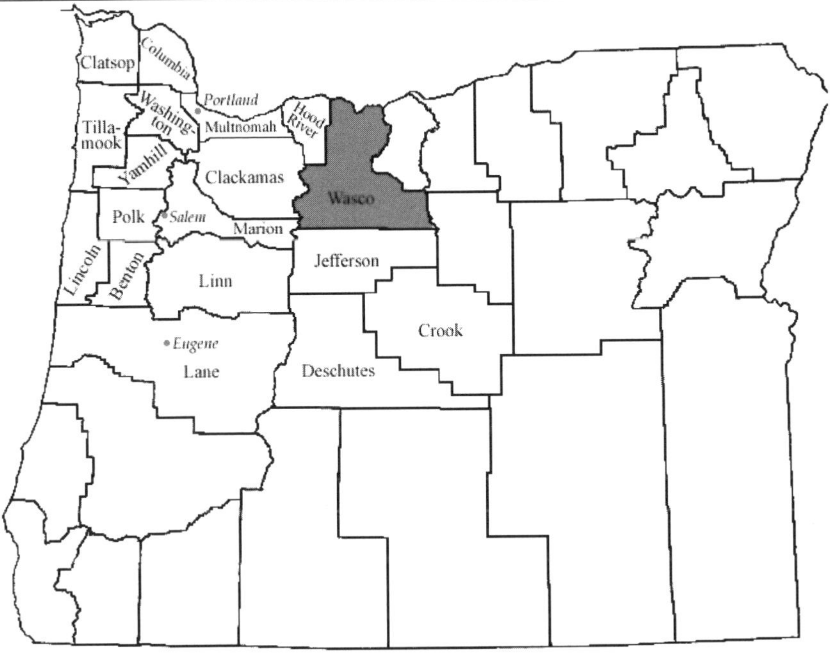

Government & Public Safety

Columbia River Inter-Tribal Fisheries Enforcement			
	The Dalles	156.2250	159.0150
Dufur Fire Department	Dufur	039.8200	Simplex
Operations Simplex	Dufur	155.3400	Simplex
Dufur City Government	Dufur	155.8350	Simplex
Juniper Flat Rural Fire District	Maupin	154.3100	Simplex
	Maupin	158.7450	Simplex
Kah Nee Ta District	Warm Springs	158.7750	153.7550
Maupin Fire Department	Maupin	154.3850	Simplex
Mid Columbia Fire & Rescue	The Dalles	154.3400	Simplex
Ch1 Dispatch [PL 192.8]	The Dalles	154.3700	Simplex
Ch2 Tac 2 – Tactical [PL 192.8]	The Dalles	154.1750	Simplex
Ch3 Tac 3 – Tactical [PL 192.8]	The Dalles	154.2050	Simplex
Repeater	The Dalles	155.2500	158.9700
Repeater	The Dalles	159.0600	153.9650
Mosier Rural Fire District	The Dalles	154.4150	Simplex
	The Dalles	154.4450	Simplex
	The Dalles	156.1650	159.1500
Oregon State Police (Sta 53) [PL 192.8]	The Dalles	155.9100	159.2100
Pine Hollow Fire Department	Tygh Valley	154.1900	158.8500
Shaniko Fire Department	Shaniko	155.7750	Simplex
Southern Wasco County Ambulance	Maupin	155.2200	Simplex

	Maupin	155.2650	Simplex
The Dalles City Government	The Dalles	153.9200	Simplex
The Dalles Police [PL 192.8]	The Dalles	155.0100	Simplex
Tygh Valley Fire Department	Tygh Valley	154.1600	159.1200
Warm Springs Fire & Safety	Simnasho	154.2500	Simplex
Wasco County Highway Maintenance	Countywide	154.9800	159.1950
Wasco County Sheriff [PL 192.8]	Countywide	154.8300	155.6100

Miscellaneous

Columbia Gorge Regional/The Dalles Municipal Airport			
CTAF / UNICOM – Common Traffic	The Dalles	123.0000	AM
Seattle Center Approach / Departure	The Dalles	119.6500	257.6000
Colombia Gorge Tours	The Dalles	151.7750	Simplex
Google	The Dalles	463.9875	468.9875
Security	The Dalles	464.8625	469.8625
Kah Nee Ta Vacation Resort	Warm Springs	152.4800	Simplex
154.5400 157.5600 464.7750 469.7750			
Lyle School District 406	The Dalles	461.2000	466.2000
National Weather Service – NOAA Weather Radio			
Conditions & Forecasts – The Dalles		162.4000	Simplex
Wasco County SAME Alert Code: 041065			
Northern Oregon Correctional Facility	The Dalles	154.8600	Simplex
	The Dalles	156.0150	Simplex
Northern Wasco Electric Power	Countywide	153.5000	158.1600
Wasco County School District	Dallesport	461.2000	466.2000
Wasco Electric Cooperative	Countywide	153.6500	160.0500
Wasco State Airport	Wasco	122.9000	AM
Young Life Washington Family Ranch	Big Muddy	451.3000	456.3000
451.3750 452.2750 452.8250 461.0500 463.2000			

WASHINGTON COUNTY

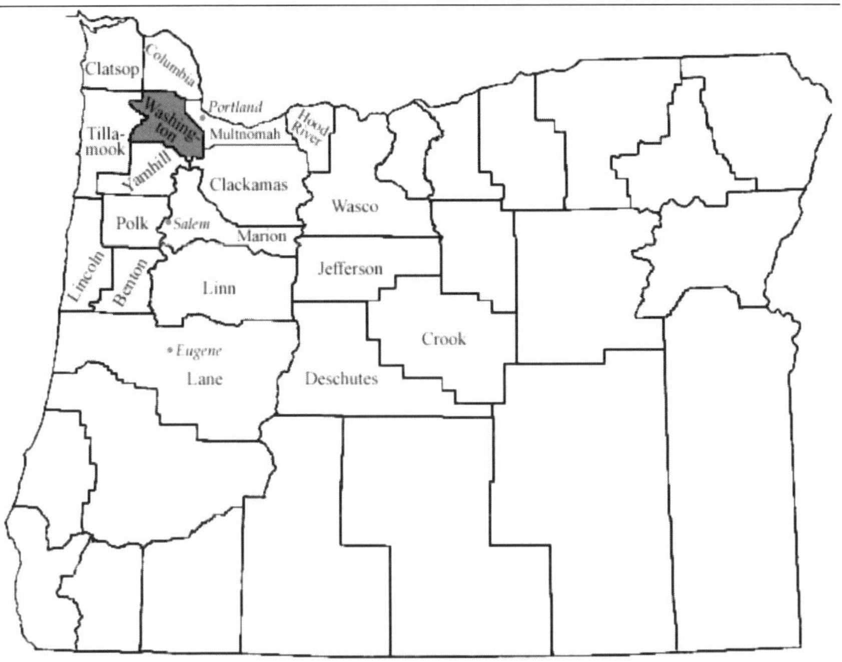

Government & Public Safety

Banks Fire District #13

			Portland	154.4000
13	Banks Station	E13, R13, E16, WT13, WT14, BR13, PU13		
14	Buxton Station	E14		
15	Timber Station	E15		

Banks Police Department

See North Plains Police Department

Beaverton Police Department

Ch1	Dispatch / Operations	Beaverton	TG: 60048	Washington
Ch2	Alternate	Beaverton	TG: 60016	Washington
Ch3	Records Access	Beaverton	TG: 59984	Washington
Ch4	Detectives	Beaverton	TG: 59952	Washington
Ch5	Administration	Beaverton	TG: 59920	Washington
Ch6	Tac 1 (Tactical)	Beaverton	TG: 59888	Washington
Ch7	Tac 2 (Tactical)	Beaverton	TG: 59856	Washington
Ch8	Common Law Channel	Beaverton	TG: 60208	Washington
	Multi-Use / Spare	Beaverton	TG: 59824	Washington

Beaverton City Public Works

Community Development	Beaverton	TG: 59664	Washington
Survey Crew	Beaverton	TG: 59632	Washington
Tactical	Beaverton	TG: 59600	Washington
Special Projects – Operating	Beaverton	TG: 59568	Washington
Public Works 1	Beaverton	TG: 59536	Washington
Public Works 2	Beaverton	TG: 59504	Washington

Beaverton School District

Channel 1	Beaverton	TG: 59472	Washington
Channel 2	Beaverton	TG: 59440	Washington
Channel 3	Beaverton	TG: 59408	Washington
Channel 4	Beaverton	TG: 59376	Washington
Channel 5	Beaverton	TG: 59696	Washington

Cornelius Fire Department

See Washington County Fire Agencies Radio System

Cornelius City Government

Cornelius City Government	Cornelius	155.8800	Simplex

Cornelius Police Department

See Washington County Sheriff's Office for Dispatch Talkgroups

Records Access	Cornelius	TG: 59344	Washington
Detectives	Cornelius	TG: 59312	Washington
Administration	Cornelius	TG: 59280	Washington
Tac 1 (Tactical)	Cornelius	TG: 59248	Washington
Tac 2 (Tactical)	Cornelius	TG: 59216	Washington
Training	Cornelius	TG: 59184	Washington

Forest Grove Fire Department: See Washington County Fire Agencies Radio

Forest Grove Police Department

See Washington County Sheriff's Office for Dispatch Talkgroups

Records Access	Forest Grove	TG: 58640	Washington
Detectives	Forest Grove	TG: 58608	Washington
Administration	Forest Grove	TG: 58576	Washington
Tac 1 (Tactical)	Forest Grove	TG: 58544	Washington
Tac 2 (Tactical)	Forest Grove	TG: 58512	Washington
Training	Forest Grove	TG: 58480	Washington

Forest Grove City Government

Forest Grove City Government	Forest Grove	153.4850
Road Maintenance	Forest Grove	156.0450
Water Department	Forest Grove	155.8800

Gaston Police Department

See Washington County Sheriff's Office for Dispatch Talkgroups

Records Access	Gaston	TG: 58000	Washington
Detectives	Gaston	TG: 57968	Washington
Administration	Gaston	TG: 57936	Washington
Tac 1 (Tactical)	Gaston	TG: 57904	Washington
Tac 2 (Tactical)	Gaston	TG: 57872	Washington
Training	Gaston	TG: 57840	Washington

Hillsboro City Government

Hillsboro City Government	Hillsboro	155.8800	
Engineering Department	Hillsboro	155.7750	
Emergency Operations Center	Hillsboro	TG: 61904	Washington
Administration / Info Technology	Hillsboro	TG: 61872	Washington
Engineering Department	Hillsboro	TG: 61840	Washington
Public Works Dispatch	Hillsboro	TG: 61808	Washington
Water Treatment	Hillsboro	TG: 61776	Washington
Facilities	Hillsboro	TG: 61744	Washington
Parks & Recreation	Hillsboro	TG: 61712	Washington
Building & Codes Division	Hillsboro	TG: 61680	Washington

Hillsboro Police Department

Ch1	Dispatch / Operations	Hillsboro	TG: 60528	Washington
Ch2	Alternate	Hillsboro	TG: 60496	Washington
Ch3	Records Access	Hillsboro	TG: 57360	Washington
Ch4	Detectives	Hillsboro	TG: 57328	Washington
Ch5	Administration	Hillsboro	TG: 57296	Washington
Ch6	Tac 1 (Tactical)	Hillsboro	TG: 57264	Washington
Ch7	Tac 2 (Tactical)	Hillsboro	TG: 57232	Washington

King City Police Department

See South County Police Agencies			
Records Access	King City	TG: 56720	Washington
Detectives	King City	TG: 56688	Washington
Administration	King City	TG: 56656	Washington
Tac 1 (Tactical)	King City	TG: 56624	Washington
Tac 2 (Tactical)	King City	TG: 56592	Washington
Metro West Ambulance	Portland	151.8650	Simplex
Operations Simplex	Portland	155.3550	Simplex
Low-Power Simplex	Portland		
463.2125 463.2375 463.2875			
468.2125 468.2375 468.2875			
North Plains Public Works Dept.	North Plains	153.8150	Simplex
North Plains Police Department			
See Washington County Sheriff's Office for Dispatch Talkgroups			
Records Access (& Car/Car)	North Plains	TG: 56080	Washington
Detectives	North Plains	TG: 56048	Washington
Administration	North Plains	TG: 56016	Washington
Tac 1 (Tactical)	North Plains	TG: 55984	Washington
Tac 2 (Tactical)	North Plains	TG: 55952	Washington
Training	North Plains	TG: 55920	Washington
Oregon Dept. of Forestry [PL 146.2]	Buxton	151.1450	159.2850
[PL 146.2]	South Saddle	151.1900	159.4425
Oregon State Police			
Sta 10 East Washington [PL 179.9]	Portland Sta.	154.9350	159.0900
Sta 14 West Washington [PL 179.9]	McMinnville St.	156.2250	159.1050
Sherwood City Government	Sherwood	453.4250	
Public Works Department	Sherwood	453.9125	Simplex
Public Works Department	Sherwood	458.9125	Simplex
Sherwood Police Department			
See South County Police Agencies			
Records Access	Sherwood	TG: 55440	Washington
Detectives	Sherwood	TG: 55408	Washington
Administration	Sherwood	TG: 55376	Washington
Tac 1 (Tactical)	Sherwood	TG: 55344	Washington
Tac 2 (Tactical)	Sherwood	TG: 55312	Washington
South County Police Agencies			
Includes: Sherwood PD, Tigard PD, Tualatin PD, King City PD			
Dispatch / Operations (Shared)	Countywide	TG: 60848	Washington
Alternate (Shared)	Countywide	TG: 60816	Washington
South Cities Tactical	Countywide	TG: 60720	Washington
Tigard City Government	Tigard	155.0850	Repeater
	Tigard	156.0150	
Tigard Police Department			
See South County Police Agencies			
Records Access	Tigard	TG: 54800	Washington
Detectives	Tigard	TG: 54768	Washington
Administration	Tigard	TG: 54736	Washington
Tac 1 (Tactical)	Tigard	TG: 54704	Washington
Tac 2 (Tactical)	Tigard	TG: 54672	Washington
Common Law Channel (Shared)	Countywide	TG: 60208	Washington
Tri City Rural Fire Protection: See Banks Fire District #13			
Tualatin City Government	Tualatin	155.9250	155.0100
Maintenance Department	Tualatin	151.1225	156.0525
Tualatin Valley Fire & Rescue			
See Washington County Fire Agencies Radio System			

210 square mile area includes the cities of Beaverton, Durham, King City, Rivergrove, Sherwood, Tigard, Tualatin, West Linn, and Wilsonville, as well as unincorporated areas in Clackamas, Multnomah, and Washington counties. The District has 22 fire stations, an administration office, a training facility, and two division offices (North and South).

TVF&R was the first fire department in Oregon to utilize Peak Activity Units (PAU). A PAU works an alternate work schedule designed to maximize their utilization. The crew may staff an Engine or Rescue unit. The PAU is deployed every shift to roam the high volume areas of the District and provide additional emergency response and coverage. These flexible units are also deployed for non-emergency tasks including, but not limited to: fire station coverage due to scheduled training, community events, meetings, emergency move-up, high call volume and targeted work loads. Currently, TVF&R deploys two Advanced Life Support PAU Engines (E90 & E91) six days a week (Monday-Saturday), 12 hours a day. When the PAU shuts down each evening, part of the crew goes off-duty while one member augments the staffing of the home station, becoming the fourth member of their crew.

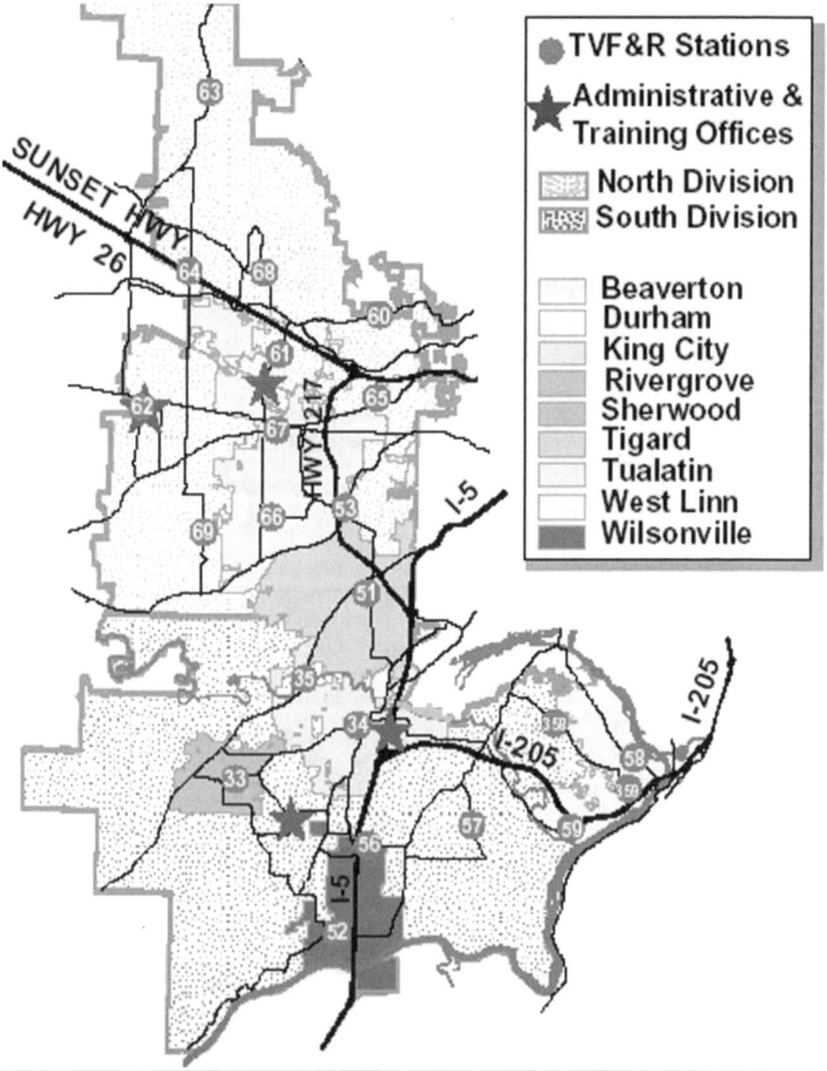

Tualatin Valley Fire & Rescue Station Locations & Apparatus

South Division		7401 SW Washo Ct, Tualatin	Division Chief, 2BC's
33	Sherwood	655 NE Oregon	E33, BR33, WT33, Reserve Engine
34	Tualatin	19365 SW 90th Court	E34, R34, HazMat, Reserve Engine
35	King City	17135 SW Pacific Hwy	E35, R35, BR35, Reserve Engine
51	Tigard	8935 SW Burnham St	E51, Trk51, WT51, TRT, R51
52	Wilsonville	29875 SW Kingsman Rd	E52, BR52, WT52, Reserve Engine
53	Progress	8480 SW Schools Ferry Rd	E53, HM9, PAU E91
56	Wilsonville	8455 SW Elligsen Rd	E56, WT56, PAU E
57	West Linn	24242 SW Mountain Rd	E57, BR57
East Division		6050 Failing Street	East BC
58	Bolton (W Linn)	6050 Failing Street	E58, BR58
59	Willamette	1860 Willamette Falls Dr	E59, R59, WT59, Marine Boat
359	Sunset	2214 Long St, W Linn	S59, E359, WR59, BR59
North Division		14480 SW Jenkins Rd, Beav.	Division Chief, 2BC's
60	Cornell Road	8585 NW Johnson St	E60, BR60
61	Beaverton	13730 SW Butner Rd	SQT61, PAU E90
62	Aloha	3608 SW 209th	SQT62, T62, BR62, E62, E362, C5
63	Skyline (Vol)	11646 NW Skyline Blvd.	E63, BR63, R63, WT63, BR363
64	Somerset	3355 NW 185th	E64, BR64
65	West Slope	8661 SW Canyon Rd	E65, E65R, MED90
66	Beaverton	13900 SW Brockman Rd	E66
67	Farmington	13810 SW Farmington Rd	E67, E67R, T67, C5
68	Oak Hills	3260 NW 147th Place	E68
69	Cooper Mtn.	9940 SW 175th, Aloha	E69, BR69, Urban Interface Engine
358	West Linn (Vol)	1000 Fire Station Lane	

Tualatin Valley Water District

Operations	Countywide	TG: 62896	Washington
Alternate	Countywide	TG: 62864	Washington

Tualatin Police Department

See South County Police Agencies

Dispatch / Operations (Shared)	Countywide	TG: 60848	Washington
Alternate (Shared)	Countywide	TG: 60816	Washington
Records Access	Tualatin	TG: 54160	Washington
Detectives	Tualatin	TG: 54128	Washington
Administration	Tualatin	TG: 54096	Washington
Tac 1 (Tactical)	Tualatin	TG: 54064	Washington
Tac 2 (Tactical)	Tualatin	TG: 54032	Washington
Common Law Channel (Shared)	Countywide	TG: 60208	Washington

Washington County Clean Water Services

SOCOM	Countywide	TG: 62224	Washington
FOFLAG	Countywide	TG: 62256	Washington
FOTV	Countywide	TG: 62288	Washington
FOER	Countywide	TG: 62320	Washington
FOCOM	Countywide	TG: 62352	Washington
ISPTEC	Countywide	TG: 62384	Washington
FGHILL	Countywide	TG: 62416	Washington
BLDCOM	Countywide	TG: 62448	Washington
USACOM	Countywide	TG: 62480	Washington

Washington County Consolidated Communications Agency

WCCCA Dispatches the following public safety agencies:

Banks FD, Banks PD, Cornelius FD, Cornelius PD, Forest Grove FD, Forest Grove PD, Gaston FD, Gaston PD, Hillsboro FD, King City PD, North Plains PD, Sherwood PD, Tigard PD, Tualatin PD, Tualatin Valley Fire & Rescue, Washington County Sheriff.

Washington County Sheriff's Department

WCSO Dispatches All West County city police departments, including:
Forest Grove, Gaston, Banks, North Plains, and Cornelius

Ch1	Dispatch / Operations	Countywide	TG: 64048	Washington
Ch2	Alternate	Countywide	TG: 64016	Washington
Ch3	Records Access	Countywide	TG: 63984	Washington
Ch4	Detectives	Countywide	TG: 63952	Washington
Ch5	Administration	Countywide	TG: 63920	Washington
Ch6	Tac 1 (Tactical)	Countywide	TG: 63888	Washington
Ch7	Tac 2 (Tactical)	Countywide	TG: 63856	Washington
Ch8	Common Law	Countywide	TG: 60208	Washington
	West Cities Tactical	Countywide	TG: 60400	Washington
	Car to Car	Countywide	TG: 61104	Washington
	Court Security	Countywide	TG: 63664	Washington
	Search & Rescue	Countywide	TG: 63792	Washington
	Jail - Operations	Countywide	TG: 63824	Washington

WCSO Unit Identifiers

First Digit:	Second Digit:	Third Digit:	Fourth Digit:
County Sheriff	Shift or Other	District/Beat	Individual unit ID
5	0 – Corrections	# (1-7, see below)	#
5	1 – Day	# (1-7, see below)	#
5	2 – Swing	# (1-7, see below)	#
5	3 – Graveyard		
	5 – Administration		
	7# – Special Units		
	8 – Services		
	9# – Auxiliaries		

Second Digit Specifics are as follows:

5700-5729	Traffic	5770-5779	Transports	5940-5969	Posse
5730-5739	Motor Units	5780-5789	School Rsrc.	5970-5999	SAR/Explorer
5750-5759	Marine	5790-5799	K-9 Units		
5760-5769	Bikes	5910-5939	Reserves		

Districts (3rd callsign digit) are as follows:

1 – County Line (N), 185th Ave (W), Sunset Hwy (S), County Line (E)
2 – Hwy 26 (N), Murray Blvd (W), TV Hwy (S), County Line (E)
3 – Scholls Ferry Road (N), Hwy 219 (W), County Line (S), County Line (E)
4 – Respond anywhere in the county (Wild)
5 – County Line (N), Cornelius Pass Road (W), Hwy 8 (S), Murray Blvd. (E)
6 – Hwy 8 (N), Hwy 219 (W), Scholls Ferry Road (S), Murray Blvd (E)
7 – Cornelius Pass Road (E)

Washington County Fire Agencies

Ch1	Dispatch / Alarms	Countywide	TG: 50640	Washington
Ch2	Tac 2 – West County Minor Incidents	Countywide	TG: 50576	Washington
Ch3	Tac 3 – South County Minor Incidents	Countywide	TG: 50544	Washington
Ch4	Tac 4 – Beaverton/Hillsboro Minor	Countywide	TG: 50512	Washington
Ch5	Tac 5 – Working Fires	Countywide	TG: 50480	Washington
Ch6	Tac 6 – Working Fires / Staging	Countywide	TG: 50448	Washington
Ch7	Tac 7 – Working Fires – Staging	Countywide	TG: 50608	Washington
Ch8	Administration	Countywide	TG: 50128	Washington
Ch11	Training - 1	Countywide	TG: 50160	Washington
Ch12	Training - 2	Countywide	TG: 50032	Washington
	Fire Common	Countywide	TG: 50320	Washington
	MAYDAY Channel	Countywide	868.3875	Simplex
	HEAR – Forest Grove Hospital	Countywide	TG: 51504	Washington
	HEAR – Tualatin Hospital	Countywide	TG: 51536	Washington
	HEAR – Saint Vincent's Hospital	Countywide	TG: 51568	Washington
	HEAR – Meridian Park Hospital	Countywide	TG: 51600	Washington

Washington County Fire Station Locations

Gaston Rural FD	11	E Main St	E1171, E1172, WT11, E1173, R11, BR1120
Tri City Rural FD	13	Main St,Banks	E1361, E1362, WT13, R13, BR13
	14	NW Fisher Rd	E1465, WT14
	15	Cochran Rd	E1568, WT15, R15
County RFPD #2	17	NW Comm'l St	E171, E173, E175, WT17, R17, BR179
	19	SW Midway	E192, E194, E196, R19, WT19, BR19
Forest Grove Fire	42	Ash St	E421, E422, E423, WT4, WT7, L, R4, W429
	43	NW Sargent	E427, E428, BR417
Cornelius FD	81	N Barlow St	E813, E814, E815, R8, WT8, BR8, BA8, BR812
Hillsboro FD	101	NE 25th St	R10, Rsv R10
	102	SE Drake Rd	E12, Rsv E102
	103	NW 299th/Evergreen	E103, WT104
	104	S 1st Ave	E104, E107, L10
Oregon Forestry	43	Gales Creek	E431 432 434 437 438

Washington County Government

Washington County Consolidated Communications Agency Trunked Radio System

IF THESE FREQUENCIES CHANGE, FREE UPDATES ARE AT SCANOREGON.COM

TRUNKED RADIO SYSTEM (Motorola SmartZone): Setting: Type II 800
854.9875 855.2375 855.9625 856.2375 857.2375
858.2375 859.2375*859.7375*860.2375*860.7375*

Buxton 1 (WCSO 1)	[PL 192.8]	Countywide	855.7375	810.7375
Buxton 2 (Fire Tac 2)	[PL 192.8]	Countywide	851.6375	806.6375
Common Law 1		Countywide	TG: 60208	Washington
Common Law 2		Countywide	TG: 60176	Washington
KPDX Tower (BVPD 1)	[PL 192.8]	Council Crest	855.8625	810.8625
Parks Department - 1		Countywide	TG: 62672	Washington
Parks Department - 2		Countywide	TG: 62640	Washington
Petes Mtn. (WCSO 1)	[PL 192.8]	Countywide	851.7125	806.7125
Public Safety 1		Countywide	TG: 53520	Washington
Public Safety 2		Countywide	TG: 53488	Washington
Public Safety 3		Countywide	TG: 53456	Washington
Public Safety 4		Countywide	TG: 53424	Washington
Public Works Department		Countywide	453.2750	458.2750
Road Maintenance		Countywide	151.0550	Repeater
Roundtop 1 (WCSO 1)	[PL 192.8]	Countywide	851.7625	806.7625
Roundtop 2 (Fire Tac 2)	[PL 192.8]	Countywide	854.4375	809.4375
Search & Rescue		Countywide	155.8050	
Service Net / City Net		Countywide	TG: 60272	Washington

(Records Access, Plus a "super group" during low dispatch staffing)

Special Events 1	Countywide	TG: 62736	Washington
Special Events 2	Countywide	TG: 62704	Washington
Talk 1 (Talk-around)	Countywide	866.4375	Simplex
Talk 2 (Talk-around)	Countywide	866.0375	Simplex
Talk 3 (Talk-around)	Countywide	866.9500	Simplex
Talk 4 (Talk-around)	Countywide	866.1750	Simplex
Transportation - 1	Countywide	TG: 62992	Washington
Transportation - 2	Countywide	TG: 62960	Washington
Transportation - 3	Countywide	TG: 62928	Washington
	Countywide	156.1200	
	Countywide	153.9350	

West County Police Agencies

See Washington County Sheriff's Office Talkgroups

Washington County Law Enforcement Radio Codes

10-1	Receiving Poorly	10-29	Check Wanted	12-1	Non-Injury Motor Vehicle Accident
10-2	Receiving Clearly	10-30	Check Person	12-2	Injury Motor Vehicle Accident
10-4	Msg Understood	10-35	Confidential Info	12-5	Assist Motorist
10-7	Off Duty	-D	Driver Suspension	12-6	Tow Truck
10-8	On Duty	-F	Felony Want	12-8	Meet with _
10-9	Repeat Last	-M	Misdemeanor	12-13	DUI
10-10	Meal or Break	-R	Restraining Order	12-14	Death
10-15	In Custody	-W	Warrant	12-34	Behavioral Problems
10-19	Return to Station	10-52	Status Check	19-15	Custody Transport
10-20	Location	10-53	No Status Checks		
10-21	Call by Phone	10-97	Arrived		
10-26	Check Veh. Reg.	10-98	Switch to _		

Washington County Law Enforcement Call Signs

Washington County Law Enforcement agencies are identified by a 4 digit call sign. While each department has different policies for call sign assignment, you can identify which agency a particular unit works for by the first (and sometimes 2nd) digit(s) of their call sign:

1000	Hillsboro PD	3600	Banks PD	7000	Tualatin PD
2000	Tigard PD	4000	Forest Grove PD	8000	Cornelius PD
3300	King City PD	5000	Washington County Sheriff	9000	Sherwood PD
3400	North Plains PD	6000	Beaverton PD		
3500	Gaston PD				

Please see Multnomah County section for Portland area non-public safety information:

Airports, Amateur Radio, Schools, Industries, Recreation, Media, etc.

Weather Information

National Weather Service – NOAA Weather Radio

Conditions & Forecasts – Portland	162.5500	Simplex
Washington County SAME Alert Code: 041067		

YAMHILL COUNTY

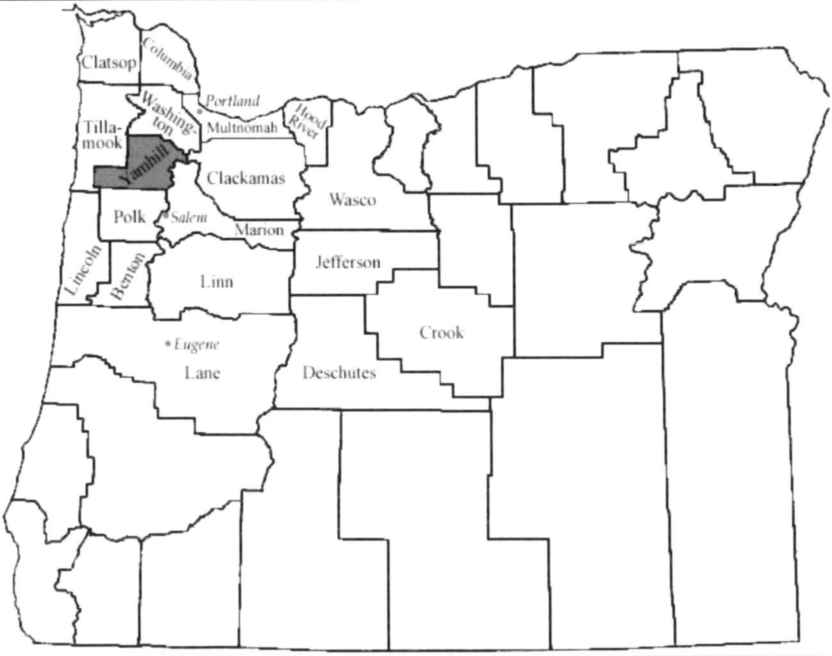

Government & Public Safety

Metro West Ambulance	Newberg	151.8050	Simplex
Newberg Fire Department			
Dispatch / Operations	Newberg	TG: 53200	Newberg
Tac 1 (Tactical)	Newberg	TG: 53392	Newberg
Tac 2 (Tactical)	Newberg	TG: 53360	Newberg
Tac 3 (Tactical)	Newberg	TG: 53328	Newberg
Tac 4 (Tactical)	Newberg	TG: 53296	Newberg
Tac 5 (Tactical)	Newberg	TG: 53264	Newberg
Tac 6 (Tactical)	Newberg	TG: 53232	Newberg
Providence Hospital Link	Newberg	TG: 52976	Newberg
Public Safety Link	Newberg	TG: 53168	Newberg

Newberg City Government

TRUNKED RADIO SYSTEM (Motorola SmartNet): TrunkTracker Setting: Type II 800

867.4125 867.9375* 868.2125* 868.9250*

City Services Talkgroups:	Newberg	TG: 52592	Newberg
52624 52656 52688 52752			
52784 52816 52976 53008			
Public Works - 1 (Electrical)	Newberg	TG: 52912	Newberg
Public Works - 2 (Water/Sewer)	Newberg	TG: 52880	Newberg
Public Works - 3 (Grounds)	Newberg	TG: 52848	Newberg

Newberg Police Department

Dispatch / Law 1	Newberg	TG: 52976	Newberg
Operations	Newberg	TG: 52720	Newberg
Administration / Narcotics	Newberg	TG: 56400	Newberg
Detectives	Newberg	TG: 53040	Newberg
Emergency Ops Center – Dundee	Newberg	TG: 53136	Newberg
Emergency Ops Center – Newberg	Newberg	TG: 53104	Newberg
Public Safety Link	Newberg	TG: 53168	Newberg
Public Service Administration	Newberg	TG: 53072	Newberg

Oregon State Police (Sta 14) [PL 179.9]	McMinnville	156.2250	159.1050
West Valley Fire District	Sheridan	453.3000	458.3000
	McMinnville	453.0250	458.0250

Yamhill County Communications Agency

TRUNKED RADIO SYSTEM (MPT1327): TrunkTracker Setting: None At This Time

460.0375	460.2625	460.3375	460.6375
460.1000	460.1750	460.3250	460.6250
453.1750	453.3625	453.5375	460.5875
453.2125	453.4625	453.7125	453.8625
453.2375	453.3375	453.4375	453.6375
453.8125	453.5125	453.7875	460.1000
460.2750	460.3250	460.6250	

Yamhill County Fire Departments

	Countywide Dispatch		Countywide	163.2500	Simplex
Ch1	Dispatch West	[PL 110.9]	Countywide	046.4200	045.8800
Ch2	Fire West	[PL 110.9]	Countywide	046.1000	046.5000
Ch3	TAC West	[PL 110.9]	Countywide	046.2600	Simplex
Ch4	Dispatch Central	[PL 110.9]	Countywide	046.4200	045.8800
Ch5	Fire Central	[PL 110.9]	Countywide	046.1000	045.8800
Ch6	TAC Central	[PL 110.9]	Countywide	046.3200	Simplex
Ch7	Dispatch East	[PL 110.9]	Countywide	046.4200	045.8800
Ch8	Fire East	[PL 110.9]	Countywide	046.1000	046.5000
Ch9	TAC East	[PL 110.9]	Countywide	046.2800	Simplex
Ch10	TAC County	[PL 110.9]	Countywide	046.2200	Simplex

Fire Station Locations

1 McMinnville: 175 E 1st
2 Newberg: E 2nd St
3 Dundee: Hwy 99 West
4 Yamhill
5 Amity
6 Dayton FD
7 Carlton
8 Willamina
9 Sheridan
10 LaFayette

Yamhill County Sheriffs Department

		Countywide	045.6400	Simplex
Law-1 West	[PL 110.9]	Countywide	045.3200	045.9800
Law-2 West	[PL 110.9]	Countywide	045.4600	046.0200
Law-1 Central	[PL 110.9]	Countywide	045.3200	045.9800
Law-2 Central	[PL 110.9]	Countywide	045.4600	046.0200

Miscellaneous

Albany & Eastern Railroad	Amity	161.0700	160.2450
Amity School District 4J	Amity	464.5750	Simplex
Chehalem Airpark	Newberg	122.9000	AM

Federal Correctional Institution Sheridan

TRUNKED RADIO SYSTEM (Motorola- "FCI Sher"): TrunkTracker Setting: Type II UHF

Base: 408.8125 Spacing: 380 Offset: 12.5

408.8125 409.0125 409.4125 409.9500 410.5250

OPS-1	Operations	Sheridan	TG: 32	FCI Sher
OPS-2	Operations	Sheridan	TG: 48	FCI Sher

George Fox University	Newberg	451.9500	456.9500

452.7500 461.1750 464.3250 464.4250

Linfield College	McMinnville	461.4000	466.4000
	McMinnville	461.7500	466.7500
McMinnville Municipal Airport	McMinnville	123.0000	AM
Approach/Departure South (Portland)	McMinnville	126.0000	AM
Approach/Departure North (Seattle Ctr.)	McMinnville	125.8000	291.7000
Clearance Delivery	McMinnville	118.3500	AM
Evergreen International Aviation	McMinnville	122.7500	AM

Providence Newberg Hospital

Maintenance Department	Newberg	457.7750	Simplex
Sportsman Airpark	Newberg	122.9000	AM
Clearance Delivery	Newberg	126.0000	AM

Weather Information

National Weather Service – NOAA Weather Radio

Conditions & Forecasts – Salem	162.4750	Simplex

Yamhill County SAME Alert Code: 041071

WESTERN WASHINGTON

Conventional Mutual Aid & Coordination

OSCCR – On-Scene Command & Coordination		156.1350	Simplex
Search & Rescue Operations		155.1600	Simplex
REDNET/FIRECOM – State Fire Control		153.8300	Simplex
NLECC – National Law Enforcement Emergency		155.4750	Simplex
HEAR – Hospital Emergency Administration Radio		155.3400	Simplex
HEAR – Hospital Emergency Administration Radio		155.2800	Simplex
LERN – Law Enforcement Radio Network		155.3700	Simplex
I-CALL – Squad Mountain Repeater	[PL 156.7]	866.0125	821.0125
I-TAC 1 – South King County Repeater	[PL 156.7]	866.5125	821.5125
I-TAC 2 – Bellevue Repeater	[PL 156.7]	867.0125	822.0125
I-TAC 3 – Seattle Repeater (Capitol Hill)	[PL 156.7]	867.5125	822.5125
I-TAC 4 – Shoreline Repeater	[PL 156.7]	868.0125	823.0125
State Ops 1 – Simplex On-Scene Comms	[PL 156.7]	867.5375	Simplex
State Ops 2 – Simplex On-Scene Comms	[PL 156.7]	867.5625	Simplex
State Ops 3 – Simplex On-Scene Comms	[PL 156.7]	867.5875	Simplex
State Ops 4 – Simplex On-Scene Comms	[PL 156.7]	867.6125	Simplex
State Ops 5 – Simplex On-Scene Comms	[PL 156.7]	867.6375	Simplex

STATE GOVERNMENT

Washington State Department of Corrections

Agency	Location	Frequency	Input/Mode
Washington Corrections Center	Mason Cnty	155.0550	Simplex
Monroe Correctional Com [PL 110.9]	Monroe	155.6550	Simplex
Monroe Correctional Com [DPL 043]	Monroe	866.5625	Simplex
Monroe Correctional Com [DPL 043]	Monroe	868.1375	Simplex
Correctional Centers	Statewide	155.7000	Simplex
Clallam Bay – External	Clallam Bay	155.7750	Simplex
Clallam Bay – Internal	Clallam Bay	453.8000	Simplex
Larch Corrections Center – Camp Rptr.	Yacolt	866.5625	821.5625
Larch Corrections Center – Camp Rptr.	Yacolt	867.0625	822.0625
Larch Corrections Center – Wide Area	Yacolt	868.1375	823.5125
Stafford Creek Correctional Center	Aberdeen	866.5625	Simplex

867.0625 867.7625 868.1375 868.5000

Western Washington Corrections Centers

Snohomish County	Monroe Correctional Complex	167740 170th Dr SE, Monroe
Clallam County	Clallam Bay Corrections Center	1830 Eagle Crest Way, Clallam Bay
Jefferson County	Olympic Corrections Center	11235 Hoh Mainline, Forks
Pierce County	McNeil Island Corrections Center	1403 Commercial St, Steilacoom
Pierce County	WA Corrections Center for Women	9601 Bujacich Rd NW, Gig Harbor
Grays Harbor Cnty	Stafford Creek Corrections Center	191 Constantine Way, Aberdeen
Mason County	Washington Corrections Center	2321 Dayton Airport Rd, Shelton
Kitsap County	Mission Creek Corrections Center	NE 3420 Sand Hill Rd, Belfair
Thurston County	Cedar Creek Corrections Center	12200 Bordeaux Rd SW, Littlerock
Clark County	Larch Corrections Center	15314 NE Dole Valley Rd, Yacolt
Walla Walla County	Washington State Penitentiary	1313 N. 13th Ave, Walla Walla

Washington State Department of Natural Resources – Regions

TAC1	Simplex Alternate	Statewide	151.4150	Simplex
TAC2	Simplex Alternate	Statewide	151.3400	Simplex
TAC4	Simplex Alternate	Statewide	151.3100	Simplex
	Northwest Region Operations	Countywide	159.2400	151.2350
	South Puget Sound Region Operations	Countywide	159.2550	151.2050
	Olympic Region Operations	Countywide	159.3000	151.2650
	Pacific Cascade Region Ops – Ch1	Countywide	159.3300	151.2500
	Pacific Cascade Region Ops – Ch2	Countywide	151.3850	159.3600
	Pacific Cascade Region Ops – Ch3	Countywide	159.4500	151.2350
	Pacific Cascade Region Ops	Countywide	159.2475	151.1975
	Southeast Regions Operations	Countywide	159.3750	151.3700
	Statewide Parks Operations	Statewide	159.4200	151.2950

Washington State Department of Transportation

TRUNKED RADIO SYSTEM (EFJ/Transcript MultiNet II): Setting: No Scanner Available

Seattle Area	866.0375	866.5625	867.0625	867.1375	868.1375
Vancouver 39th St:	852.0875	852.8375	854.3375		
Vancouver Orchards:	868.1125	868.9875	867.0625	866.5625	
Signal Peak	856.2125	858.2125			
Rainier Hill (Kalama)	851.3375	853.5875	854.3875		
Davis Peak (Ariel)	868.0875	868.9750			
Bawfaw Mtn (Winlock)	852.6625	854.1625	857.2125		
Naselle Ridge	859.2125	860.2125			
Cape Horn (Washougal)	852.0875	853.6375	868.1125	868.9875	
Underwood Mtn	859.2125	860.2125			
Haystack Butte	856.2125	858.2125			
Goldendale	852.1375	853.6375			

WSDOT Regions

Simplex-1	866.5625	Simplex
Simplex-2	866.1125	Simplex
Simplex-3	867.2250	Simplex

Washington State Patrol

Note: Not all WSP units or dispatchers transmit PL tone at all times

District 1: Pierce (Tacoma/Gig Harbor) & Thurston (Olympia) Counties					
Ch4	West Pierce County Disp	[PL 151.4]	Tacoma	155.5200	Simplex
Ch11	Districtwide Car to Car	[PL 151.4]	Countywide	154.9050	Simplex
Ch19	Thurston County Dispatch	[PL 151.4]	Olympia	154.9200	Simplex
Ch31	East Pierce County Disp	[PL 151.4]	East Pierce	154.7400	Simplex
	Thurston County Car-Car	[PL 151.4]	Olympia	155.4600	Simplex
District 2: King County					
Ch3	North King County	[PL 100.0]	Countywide	155.5800	158.7900
Ch7	South King County	[PL 100.0]	Countywide	154.6800	Simplex
Ch12	Districtwide Car to Car	[PL 100.0]	Countywide	155.4600	Simplex
Ch22	King County Mutual Aid		Countywide	155.1900	154.6500
District 3: Yakima, Benton, Franklin, Walla Walla, Columbia, Garfield, Asotin Counties					
Ch4	Benton/Franklin/Walla	[PL 192.8]	Countywide	155.5200	Repeater
Ch19	Yakima/Garfield Counties	[PL 173.8]	Countywide	154.9200	Repeater
District 4: Spokane, Ferry, Stevens, Pen Oreille, Lincoln, Adams, Whitman Counties					
Ch3	Spokane / Whitman Dispatch		Countywide	155.5800	Simplex
Ch15	Stevens / Ferry Dispatch	[PL 179.9]	Countywide	154.6650	Simplex
	West Spokane / Adams		Countywide	154.8450	Simplex
	Metro Spokane Dispatch	[PL 179.9]	Spokane	154.9050	158.8500
District 5: Lewis, Klickitat, Cowlitz, Clark, Skamania, Wahkiakum Counties					
Ch2	Cowlitz County Dispatch		Countywide	154.7700	Simplex
Ch7	Clark / Skamania County	[PL 136.5]	Countywide	154.6800	155.5800
Ch7	Klickitat / Okanogan Counties		Countywide	154.6800	Simplex
Ch14	Lewis County Dispatch		Countywide	155.8500	Simplex
District 6: Chelan, Okanogan, Kittitas, Douglas, Grant Counties					
Ch2	Kittitas County	[PL 123.0]	Ellensburg	154.7700	Simplex
Ch14	Kittitas County		Ellensburg	155.8500	Simplex
Ch7	Klickitat / Okanogan Counties		Countywide	154.6800	Simplex
Ch16	Grant County Car to Car		Ephrata	155.4600	Simplex
Ch17	Chelan / Douglas Counties	[PL 123.0]	Wenatchee	154.9350	Simplex
Ch18	Grant County Dispatch	[PL 123.0]	Ephrata	155.5050	Simplex
District 7: Snohomish, Skagit, Whatcom, San Juan, Island Counties					
Ch13	West Snohomish County	[PL 141.3]	Everett	155.6550	154.8450
Ch17	Districtwide Car to Car	[PL 141.3]	Countywide	154.9350	Simplex
Ch18	Skagit / Island Counties	[PL 141.3]	Countywide	155.5050	159.0450
Ch20	Whatcom County	[PL 141.3]	Bellingham	154.7550	155.2500
	East Snohomish County	[PL 156.7]	Skykomish	154.7550	Simplex
District 8: Clallam, Jefferson, Grays Harbor, Pacific, Mason, Kitsap Counties					
Ch2	Clallam / Jefferson	[PL 110.9]	Port Angeles	154.7700	Simplex
Ch15	Kitsap / Mason Counties	[PL 110.9]	Bremerton	154.6650	158.8950
Ch16	Grays Harbor / Pacific		Hoquiam	154.6950	Simplex
Statewide					
	Liquor Control	[PL 100.0]	Statewide	155.9700	159.0750
	TAC Units		Statewide	155.7000	Simplex
	Commercial Vehicle Enforcement		Statewide	156.0750	
	Aircraft Operations		Statewide	155.4450	
	Aircraft Operations		Statewide	159.0750	

WSP Districts

Western Washington Offices

District 1 HQ:	2502 112th St E, Tacoma	
District 1 Offices:	5524 Soundview Dr, Gig Harbor	222 Airdustrial Way, Tumwater
District 2 HQ:	2803 156th Ave SE, Bellevue	
District 2 Offices:	333 Griffin Ave, Enumclaw	134 Sydney Ave, North Bend
	811 E Roanoke, North Seattle	15666 International Blvd, S Seattle
District 3 HQ:	2715 Rudkin Road, Union Gap	
District 3 Offices:	110 North Avenue, Sunnyside	143302 E Law Lane, Kennewick
	406 Wellington, Walla Walla	
District 5 HQ:	11018 NE 51st Circle, Vancouver	
District 5 Offices:	850 NW Louisiana Ave, Chehalis	Goldendale
	1823 Baker Way, Kelso	342 Morton Rd, Morton
District 7 HQ:	2700 116th St NE, Marysville	
District 7 Offices:	3860 Airport Wy, Bellingham (713)	10945 Chuckanut, Burlington (712)
	909 W. Main Street, Monroe (711)	840 SE 8th Ave., Oak Harbor (714)
	Silverlake Rest Area off I-5 (710)	
District 8 HQ:	4811 Wener Rd, Bremerton	
District 8 Offices:	3111 Pacific Ave, Hoquiam	62 Old Olympic Hwy, Port Angeles
	22065 Viking Way NW, Poulsbo	797 SR4, Naselle
	629 W Dayton Airport Rd, Shelton	

Western Washington WSP Radio Callsigns

150-299	Sergeants
300-1299	Patrol Officers

Washington State Ferry System		Statewide	866.1375	Simplex
	Handheld Radios Ch1	Statewide	151.0400	Simplex
Marine Ch68	Handheld Radios Ch2	Statewide	156.4250	Simplex
Marine Ch79A	Administration to Ferries	Statewide	156.9750	Simplex

CLALLAM COUNTY

Government & Public Safety

Clallam Bay Correctional Facility (Washington Department of Corrections)

			Location	Frequency	
			Clallam Bay	460.1375	465.1375
			Clallam Bay	460.3125	465.3125
			Clallam Bay	155.7000	Simplex

Clallam Bay Fire District #5 (Northwest Clallam County

			Location	Frequency	
	Dispatch & Operations		Clallam Bay	154.2500	153.7700

Clallam County Fire District #1 (West and Southwest Clallam County)

			Location	Frequency	
	Dispatch & Operations		Forks	460.6250	465.6250

Clallam County Fire District #3 (Northeast Clallam County)

			Location	Frequency	
			Countywide	154.2200	Simplex
Ch3	Dispatch	[PL 103.5]	Countywide	154.1750	Simplex
			Countywide	153.7850	Repeater
			Countywide	155.7825	154.0325
			Countywide	154.0100	Simplex
			Countywide	151.7750	Simplex

Fire Station Locations & Apparatus

1	4721 Sequim Dungeness Way, Dungeness	E311, Aid 316
2	256702 Hwy 101, R-Corner	E321, Aid 326
3	70 Carlsborg Road, Carlsborg	E331, E332, WT335, Aid 336, Wildland 333
4	323 N 5th Avenue, Sequim	E341, Ldr, E345, Aid 346, M34, M35, R348, Wildland: E343 E343A E343B
5	121 Sunshine Avenue, Diamond Point	E351, Aid 356
6	40 Texas Valley Road, Lost Mountain	E361, Aid 366
7	269160 Hwy 101, Blyn	E371, Aid 376
#	Wildland Vehicles:	E343A,

Clallam County Government

			Location	Frequency	
	Juvenile Detention Center		Countywide	453.0125	Simplex

Agency / Service	PL	Location	Frequency	Input
Public Works	[PL 118.8]	Countywide	155.9250	158.8350
Transit System		Countywide	453.2000	458.2000
Transit System		Countywide	453.2500	458.2500
Transit System		Countywide	460.2000	Simplex
Clallam County OPSCAN Interoperability Network				
OPSCAN Interoperability		Countywide	154.4525	151.1375
OPSCAN Interoperability		Countywide	453.3750	458.3750

Note: OPSCAN = Olympic Public Safety Communications Alliance Network, with Clallam County the pilot for a major Washington State interoperability project.

Agency / Service	PL	Location	Frequency	Input
Clallam County Sheriff's Office				
Dispatch	[PL 103.5]	Countywide	453.2750	458.2750
Forks Ambulance Service		Forks	463.1000	468.1000
Forks City Government		Forks	453.9750	458.9750
Forks Police Department		Forks	453.3000	458.3000
Lower Elwha Klallam Tribal Police		Port Angeles	453.9500	458.9500
Makah Indian Nation Government		Neah Bay	453.7000	458.7000
Makah Indian Nation Police & Fire		Neah Bay	155.0700	Simplex
Olympic Ambulance Service		Port Angeles	151.8050	Simplex
Port Angeles City Government				
Roads Department		Port Angeles	158.9250	Simplex
Lights Department	[PL 103.5]	Port Angeles	158.2650	Simplex
Public Works	[PL 167.9]	Port Angeles	158.9400	Simplex
Port Angeles Fire Department		Port Angeles	154.2200	Simplex
		Port Angeles	154.1750	Simplex
		Port Angeles	154.3850	Simplex

Apparatus
E1, E2, E3, E5, L4, M10, M11, M12, Air 9

Agency / Service	PL	Location	Frequency	Input
Port Angeles Police Dept.	[PL 127.3]	Port Angeles	460.1000	465.1000
Quileute Indian Tribe Police		Lapush	453.3750	458.3750
		Lapush	453.8750	458.8750
Sequim City Government		Sequim	158.9400	
Sequim Police Department		Sequim	453.3000	458.3000
		Sequim	453.2875	458.2875
		Sequim	453.3125	458.3125

Airports

Airport / Service	Location	Frequency	Mode / Input
Sekiu Airport	Sekiu	122.9000	AM
Sequim Valley Airport	Sequim	122.7000	AM
William R. Fairchild International Airport			
Clearance Delivery	Port Angeles	124.1500	AM
CTAF / UNICOM – Common Traffic	Port Angeles	123.0000	AM
Seattle Radio – Port Angeles RCO	Port Angeles	122.6000	AM
Whidbey Approach / Departure	Port Angeles	118.2000	AM
Aeronautical Radio Inc. Shared TRS	Port Angeles	853.1875	808.1875
Aeronautical Radio Inc. Shared TRS	Port Angeles	853.7875	808.7875

Miscellaneous

Agency	Location	Frequency	Input
Clallam County People's Utility District 1		048.1000	048.4600
Crescent Water Association	Joyce	151.8650	Simplex
	Joyce	463.3125	Simplex
JC Penney Store	Sequim	451.2375	Simplex
451.3375 451.4375 451.5375	Sequim	452.0875	Simplex
Lake Crescent Lodge (Inside Olympic National Park)		151.5200	Simplex
151.5500 151.5800 151.5950	Port Angeles	152.3750	Simplex
National Ski Patrol System	Countywide	155.1600	Simplex
	Countywide	155.1750	Simplex
	Countywide	155.2950	Simplex
Olympic Memorial Hospital	Port Angeles	461.9000	466.9000
	Port Angeles	462.1250	467.1250
National Response – Marine Hazardous Materials Spill Responder			
	Neah Bay	150.9800	159.4800
	Neah Bay	154.5850	Simplex
	Neah Bay	158.4450	Simplex
	Neah Bay	454.0000	459.0000
Seven Cedars Casino	Port Angeles	461.5250	466.5250
Sunland Water District	Sequim	153.5750	Simplex
Westport Shipyard (Yacht Manufacturer)	Port Angeles	452.0000	Simplex
	Port Angeles	462.0625	Simplex
	Port Angeles	464.3750	Simplex

Schools

Agency	Location	Frequency	Input
Crescent School District	Joyce	463.2125	Simplex
Olympic Educational Service District 114			
School Transportation	Port Angeles	155.2650	Simplex
Port Angeles School District	Port Angeles	460.1500	465.1500
School Transportation	Port Angeles	155.2800	Simplex
	Port Angeles	464.9750	Simplex
Quillayute Valley School District 4	Forks	461.8000	466.8000
Sequim School District 323	Sequim	153.7400	Simplex
	Sequim	155.2650	Simplex
	Sequim	460.1500	Simplex

Weather Information

National Weather Service – NOAA Weather Radio

Service	Frequency	Input
Conditions & Forecasts – Neah Bay	162.5500	Simplex
Conditions & Forecasts – Olympic Nat'l Park	162.4250	Simplex
Conditions & Forecasts – Puget Sound Marine	162.4250	Simplex

Clallam County SAME Alert Code: 053009

CLARK COUNTY

Government & Public Safety

Clark = Clark County Trunked Radio System (TRS) in this section
Portland = Portland City Trunked Radio System (TRS) in Multnomah County, OR section

American Medical Response (AMR) Ambulance		855.3625	810.3625
Dispatch-1	Countywide	TG: 44496	Clark
Dispatch-2	Countywide	TG: 44528	Clark
Supervisors	Countywide	TG: 44432	Clark
Administration	Countywide	TG: 44592	Clark
Battle Ground City Police			
Dispatch	Battle Ground	TG: 41456	Clark
Tactical	Battle Ground	TG: 41488	Clark
Battle Ground City Fire – See Clark County Fire Districts			
Battle Ground City Government	Battle Ground	154.0400	
Water Department	Battle Ground	153.0125	158.2875
Camas City Government	Camas	153.9200	
Camas Police Department (Washington)			
Dispatch / Operations	Camas	TG: 41616	Clark
Car to Car	Camas	TG: 41648	Clark
OLD VHF Frequency	Camas	154.8000	
OLD VHF Frequency	Camas	154.8900	
OLD VHF Frequency	Camas	155.0700	

Clark County Government

TRUNKED RADIO SYSTEM ("Clark") (Motorola SmartZone): Setting: Type II 800

Primary 6-Site Simulcast Zone: Countywide

859.7625* 859.9375* 860.7625* 860.9375* 856.7625
856.9375 857.7625 857.9375 858.7625 858.9375

	Yacolt Mountain Intellirepeater	Yacolt Mtn.		
	867.4875* 866.1625 866.4125 866.9625 867.2375			
	Washougal River Intellirepeater	Camas		
	859.9875* 856.9875 857.9875 858.9875 860.9875			
	Detention Center (Jail) Intellirepeater	Vancouver		
	856.9875 857.9875 858.9875			
	Rainier Hill Intellirepeater			
	866.2125 866.9250 867.8250 868.3750 868.9375			
	Cascade Locks Intellirepeater			
	866.1625 866.4125 866.9625 867.2375 867.4875			
	Clark A	Countywide	TG: 48272	Clark
	Clark B	Countywide	TG: 48304	Clark
	Clark C	Countywide	TG: 48336	Clark
TAC-1	Police / Fire Operations	Countywide	TG: 44816	Clark
TAC-2	Police / Fire Operations	Countywide	TG: 44848	Clark
TAC-3	Police / Fire Operations	Countywide	TG: 44880	Clark
TAC-4	Police / Fire Operations	Countywide	TG: 44912	Clark
TAC-5	Police / Fire Operations	Countywide	TG: 44944	Clark
TAC-6	Police / Fire Operations	Countywide	TG: 43472	Clark
TAC-7	Police / Fire Operations	Countywide	TG: 43504	Clark
TAC-8	Police / Fire Operations	Countywide	TG: 43536	Clark
TAC-9	Police / Fire Operations	Countywide	TG: 43568	Clark
TAC-10	Police / Fire Operations	Countywide	TG: 43600	Clark
TAC-11	Police / Fire Operations	Countywide	TG: 43632	Clark
	"Car to Car" [PL 156.7]	Vancouver	867.5625	
	Animal Control – 1	Countywide	TG: 48560	Clark
	Animal Control – 2	Countywide	TG: 48592	Clark
	Health Department	Countywide	TG: 43408	Clark
	Building Inspectors	Countywide	158.8950	Repeater
	Public Works	Countywide	159.1200	Repeater
	Old VHF / UHF Frequencies:			
	453.9000 156.0600 154.8000 158.8650			
	154.8900 155.0700 155.3100 155.6550			
	453.5125 458.0500 460.2875 460.4875			

Clark County Public Transportation Benefit Area Authority (C-Tran)

A1	C-TRAN Tac-1	Countywide	TG: 46416	Clark
A2	C-TRAN Tac-2 Paratransit	Countywide	TG: 46448	Clark
A3	C-TRAN Tac-3	Countywide	TG: 46480	Clark
A4	C-TRAN Tac-4 Bus Maintenance	Countywide	TG: 46512	Clark
A5	C-TRAN Tac-5 Bus Ops	Countywide	TG: 46512	Clark
A6	C-TRAN Tac-6	Countywide	TG: 46576	Clark
A7	C-TRAN Tac-7	Countywide	TG: 46608	Clark
A8	C-TRAN Tac-8	Countywide	TG: 46640	Clark
A9	C-TRAN Connector Bus	Countywide	TG: 46672	Clark
A10	C-TRAN Facility Maintenance	Countywide	TG: 46704	Clark
A11	C-TRAN Security	Countywide	TG: 46736	Clark
A12	C-Van 1 [DPL 503]	Vancouver	856.9375	811.9375

Clark County Sheriff's Office

	Control 1 - Records	Countywide	TG: 44656	Clark
	Control 2 - Dispatch	Countywide	TG: 44688	Clark
	Control 5 - Alternate	Countywide	TG: 44784	Clark
	Eagle-1	Countywide	TG: 44976	Clark
	Eagle-2	Countywide	TG: 45008	Clark
	Eagle-5	Countywide	TG: 45104	Clark
Ch1	Emergency Coordination Center (ECC)	Countywide	TG: 48016	Clark
Ch2	Emergency Coordination Center (ECC)	Countywide	TG: 48048	Clark
Ch3	Emergency Coordination Center (ECC)	Countywide	TG: 48080	Clark
Ch4	Emergency Coordination Center (ECC)	Countywide	TG: 48112	Clark

Ch5	Emergency Coordination Center (ECC)	Countywide	TG: 48144	Clark
Ch6	Emergency Coordination Center (ECC)	Countywide	TG: 48176	Clark
Ch7	Emergency Coordination Center (ECC)	Countywide	TG: 48208	Clark
Ch8	Emergency Coordination Center (ECC)	Countywide	TG: 48240	Clark

Clark County Fire District

	Fire Tap – City/County FD Dispatch	Countywide	TG: 43248	Clark
	Control 8 – North County FD+LifeFlight	Countywide	TG: 43344	Clark
	Fire Com – City/County FD Response	Countywide	TG: 43216	Clark
	Control 10 – County FD Response	Countywide	TG: 43376	Clark
	IC 13 (Incident Command)	Countywide	TG: 43696	Clark
	Skamania County Fire Patch	Countywide	TG: 43280	Clark
	VHF Simulcast of Control 8	Countywide	154.3700	
	VHF Simulcast of Control 10	Countywide	153.9500	
	North County EMS Main Repeater	Countywide	453.3000	458.3000
	North County EMS Backup Repeater	Countywide	453.0750	458.0750

La Center Police Department

	La Center Common	La Center	TG: 41936	Clark
	Dispatch	La Center	TG: 41968	Clark

La Center City Local Government — La Center — 153.9650

North Country EMS

	Main Repeater	Yacolt	453.3000	458.3000
	Backup Repeater	Yacolt	453.0750	458.0750

Ridgefield Police Department

	Ridgefield Common	Ridgefield	TG: 41776	Clark
	Dispatch	Ridgefield	TG: 41808	Clark

Ridgefield City Local Government — Ridgefield — 453.3250

Silverstar Search & Rescue — Washougal — 155.1600 — Simplex

155.2650 155.3400

Vancouver City Services — Vancouver — 453.9500 — 458.9500

	Housing Authority	Vancouver	860.4875	815.4875
	Housing Authority	Vancouver	853.4875	808.4875
	Public Works – Water	Vancouver	460.4000	465.4000
	Public Works – Streets	Vancouver	460.1750	465.1750
1A	Public Works	Vancouver	TG: 46896	Clark
1B	Water Department	Vancouver	TG: 46928	Clark
1C	Public Works	Vancouver	TG: 46960	Clark
1D	Public Works	Vancouver	TG: 46992	Clark
1E	Public Works	Vancouver	TG: 47024	Clark
1F	Public Works	Vancouver	TG: 47056	Clark
1G	Public Works	Vancouver	TG: 47088	Clark
1H	Public Works	Vancouver	TG: 47120	Clark
		Vancouver	159.0300	Repeater
		Vancouver	460.1500	465.1500
		Vancouver	460.4000	465.4000
		Vancouver	462.4750	467.4750

Vancouver Fire Department

	Fire Com	Vancouver	TG: 43216	Clark

Vancouver Police Department

	Control 3 - Operations	Vancouver	TG: 44720	Clark
	Control 4 - Operations	Vancouver	TG: 44752	Clark
	Traffic	Vancouver	TG: 40208	Clark
	Code Enforcement	Vancouver	TG: 40176	Clark
	Eagle 3	Vancouver	TG: 45040	Clark
	Eagle 4	Vancouver	TG: 45072	Clark
	Records Information	Vancouver	TG: 45200	Clark
	Tac 3 (Tactical)	Vancouver	TG: 44880	Clark

Tac 4 (Tactical)	Vancouver	TG: 44912	Clark
Special Events	Vancouver	TG: 40080	Clark

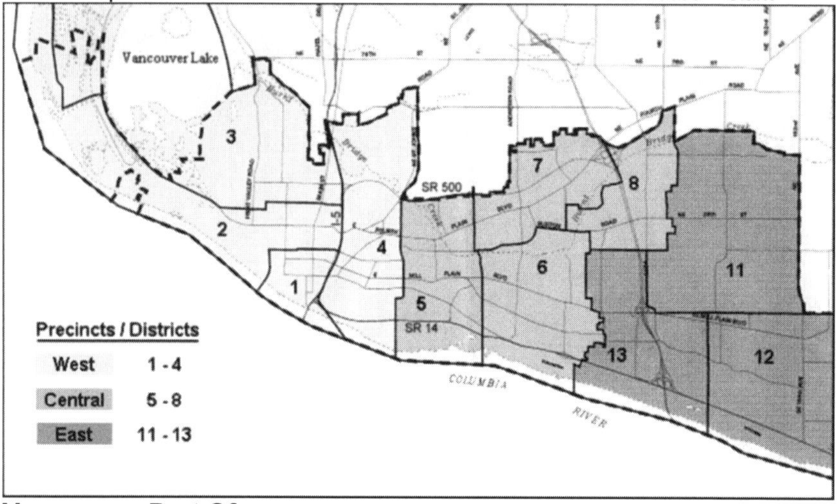

Vancouver, Port Of

Vancouver, Port Of	Vancouver	453.6250	
Port of Vancouver	Vancouver	453.4500	458.4500
Low Power Simplex	Vancouver	458.0875	Simplex
Low Power Simplex	Vancouver	458.1125	Simplex
Low Power Simplex	Vancouver	458.1375	Simplex
Low Power Simplex	Vancouver	458.2875	Simplex

Washington State Department of Natural Resources

Pacific Cascade Region Ops – Ch2	Countywide	151.3850	159.3600
Pacific Cascade Region Ops	Countywide	159.2475	151.1975

Washington State Patrol

Note: Not all WSP units or dispatchers transmit PL tone at all times

Ch7	Clark / Skamania County [PL 136.5]	Countywide	154.6800	155.5800

Washougal Fire & Rescue

Washougal Fire & Rescue	Washougal	154.0700
Washougal City Government	Washougal	155.9400

Washougal Police Department

Dispatch / Operations	Washougal	TG: 41680	Clark
Tactical / Alternate	Washougal	TG: 41712	Clark
Old VHF Frequency	Washougal	155.0100	Simplex

Yacolt City Fire

See Clark County Fire Districts

Yacolt City Government

Yacolt	151.3700	Simplex

Clark County Fire Station Locations & Apparatus

Washougal FD #1

11	39th St	E11, WT11, WT13, S11
12	SE 352nd Av	E12, WT12

Woodland FPD

21	NW 389th St	E21, WT21, BR21

Brush Prairie FPD

31	NE 159th St	E31, E311, WT31, S31
32	NE 227th Ave	E32, WT32
33	NE 259th St	E33 WT33 Amb33 S33
34	NE 182nd Ave	E34, WT34

Camas FD

41	NE 4th Ave	E41, E411, L41, M41
42	Hwy 500/NE 9th St	E42

Hazel Dell/Lakeshore/Felida/ Salmon

Creek/Mt. Vista

61	NE Hazel Hell Av	E61, E611, L61 R61, BC61, B61
62	NE 76th St	E62, E621, BR62
63	NW 134th St	E63, E631, E632
64	Fairgrounds (seas.)	E64, R64, BR64

Washougal FD

71	A St	E71

Vancouver FD

81	7110 NE 63rd St	E81, E811, E821 WT81 HM81 BA81
82	900 W Evergreen	E82, Trk 82, BC82
83	213 NE 120th Av	E83, E831, BR83
84	1110 N Devine	E84, M84
85	3125 NE 112th Av	E85, L85, BR85
86	400 E 37th St	E86
87	11207 NE 70th Av	E87, WT87
88	6701 NE 147th Av	E88, WT88, BC81
89	17408 SE 15th St	E89, WT89
810	NE 58th St	E810

Fern Prairie FD

91	NE 267th Ave	E91, WT91, R91 S91, BC91
92	NE 292nd Ave	E92, WT92
93	NE 312th Ave	E93, WT93

Amboy FPD

101	NE 399th St	E101 WT101 S101
102	NE Yale Bridge Rd	E102
103	NE 119th Ave, View	E103, WT103
104	NE Cedar Creek Rd	E104, S104
105	NE Fargher Lake	E105, R105, S105
106	Columbia Tie Rd, at Lake Merwin Hideaway	E106, S106

Battle Ground FD

111	NE 72nd Ave	E111, WT111 BR111, Amb 111
112	NE 112th Ave	E112
113	West A St	E113, M113, S111 BR113, E1131
114	NE 41st Ave	E114, BR114

Ridgefield FPD

121	NE 10th Ave	E121, WT121
122	(closed)	
123	W 5th St	E123, E125
124	NE 3rd Ave	E124, E1241

Yacolt FD

131	S Parcell Ave	E131, E133, S131 WT131, BR131
132	NE Sunset Falls Rd	E132

La Center FPD

141	NE 41st Ave	E141, WT141, BR141
142	W 5th St	E142, Amb 142

Clark County Law Enforcement Radio Codes

RADIO CODES

Code 0	Officer In Trouble – Emergency
Code 1	At Your Convenience
Code 2	Immediate Response
Code 3	Emergency Response
Code 4	Status OK
Code 5	Stakeout
Code 6M	Possible Misdemeanor Warrant
Code 6F	Possible Felony Warrant
Code 7	Meal Break
Code 20	Officer Needs Help – Two Units
Code 33	Do Not Transmit – Emergency Traffic Only
10-6	Traffic Stop
10-8	In Service / Available For Call
10-7	Out Of Service
10-10	Available at ___ for Calls
10-15	Prisoner In Custody
10-19	Return To Station / Enroute to Station
10-28	(Person) Driver And Wants Check
10-28	(Plate) Registration / Wants Check
10-43	Confidential Information Follows
10-96	Possible Mental Subject
10-97	Arrived At Scene

DISPOSITION CODES

S	**Unable To Locate**
1	Cannot Locate
2	No Such Address
3	Condition Complained Of Does Not Exist
T	**Premise Checked**
1	False Alarm
2	Found Secure
3	Made Secure
4	Alarm Canceled – Prior To Arrival
W	**Call Serviced**
1	Assignment Completed
2	Person Assisted
3	Person Advised or Referred
4	Hazard Corrected or Removed
5	Delivered Person/Message/Package
6	Peace Restored
7	Nuisance Abated
8	Arrest / Investigation Made by Another Officer or Agency
9	Assignment Completed By Phone
X	**Suspicious Activity**
1	Person Checked – Okay
2	Vehicle Checked – Okay
Y	**Action Taken**
1	Field Contact Card Completed

2 Verbal Warning Given
3 Citation
4 Written Warning

Clark County Law Enforcement Unit Callsigns

AGENCY IDENTIFIER		FUNCTIONAL ASSIGNMENT	
1	Clark County Sheriff	A	Command Staff
2	Vancouver Police	B	Patrol – Morning Watch
3	Camas Police	C	Civil
4	Washougal Police	D	Patrol – Day Watch
5	Battle Ground Police	E	Explorer Scouts
6	Ridgefield Police	F	Special Services – Schools
7	LaCenter Police	G	Detention Officers
8	Coroner	H	K9 Unit
9	Clark County Juvenile	K	Detectives / Investigators
10	Clark County Department of Public Works Vehicle Inspection & Weights	L	Personnel Number
		M	Special Details
11	WDOC Larch Mountain Corrections	R	Reserve Personnel
12	Liquor Control	S	Patrol – Evening Watch
13	FBI	T	Traffic
20	Campus Police	W	Warrant Detail
21	Railroad Police	X	Sergeant
22	Department of Emergency Services	Y	Special Details
27	IRS	Z	Tactical Unit
28	State Probation Department		

Airports

Cedars North Airpark	Battle Ground	122.9000	AM
Evergreen Field Airport	Vancouver	122.7000	AM
Clearance Delivery	Vancouver	121.6500	AM
Fly For Fun Airport	Vancouver	122.9000	AM
Goheen Airport	Battle Ground	122.9000	AM
Grove Field Airport	Camas	122.9000	AM
Clearance/Delivery	Camas	121.6500	AM
Pearson Field Airport	Vancouver	123.0000	AM
Portland Approach / Departure	Vancouver	124.3500	AM
Portland Approach – South	Vancouver	118.1000	AM
Clearance Delivery	Vancouver	121.6500	AM
Portland ATC - Advisories	Vancouver	119.0000	AM

Medical & Hospitals

Clark County EMS	Vancouver	155.8050	
North Country EMS	Yacolt	453.0750	
Silverstar Search & Rescue	Washougal	155.1600	Simplex
	Washougal	155.2650	Simplex
Southwest Washington Med Center	Vancouver	856.0375	
Operations Repeater	Vancouver	461.9500	466.9500
	Vancouver	462.8500	Simplex

Miscellaneous

Bonneville Conservation, Restoration & Renewal Team

Operations Repeater	Vancouver	152.9525	159.9675
Alternate Simplex	Vancouver	151.5275	Simplex
Alternate Simplex	Vancouver	151.5575	

Clark County Amphitheater

Security – Repeater	Ridgefield	463.6000	460.6000
Security – Simplex	Ridgefield	461.5000	Simplex
Security – Simplex	Ridgefield	462.1250	Simplex
Venue Operations	Ridgefield	452.6500	Simplex

461.2000 461.4750 461.7250 461.9750 462.0500
463.5500 463.9750 464.0000 464.2000

Clark County Fair Association

	Vancouver	461.1875	Simplex
Low Power VHF Simplex	Vancouver	463.3875	Simplex

463.6125 463.7625 463.8625
463.9625 466.5125 467.8125

Clark County Columbian Newspaper

Vancouver	452.1125	Simplex
Vancouver	452.9875	Simplex

452.1875 452.6625

Club Green Meadows	Vancouver	151.7450	Simplex
Double Down Casino	Vancouver	452.7500	
East County Services	Camas	463.3000	
Heathman Lodge	Vancouver	451.9625	Simplex

452.0125 452.2125 452.2625 452.7375

Hewlett Packard Company

Vancouver	154.5700	Simplex

154.6000 462.2250 463.2000

TRUNKED RADIO SYSTEM ("HP") (Motorola SmartNet): Setting: Type II 900
939.8875 939.9000 939.9125 939.9250 939.9375

Security Services	Vancouver	TC: 112	HP

Holiday Inn Express Hotel	Vancouver	467.8125	Simplex
Lake Merwin Campers Hideaway	Amboy	154.5150	Simplex
	Amboy	462.0500	467.0500
Marine Terminals Corp	Vancouver	460.8875	Simplex

461.0375 461.1125 461.1375

New Phoenix & Last Frontier Casino	La Center	452.2250	Simplex
Nutter Underground Utilities	Battleground	855.5625	
Olson Engineering Inc	Camas	461.8000	466.8000
Pacific Air Helicopters Inc	Camas	464.0500	
Pamplin Broadcasting Inc	Vancouver	161.6700	Simplex
Royal Oaks Country Club	Yacolt	151.8650	Simplex
Sharp Electronic Corporation	Vancouver	461.5125	466.5125
	Vancouver	461.5625	466.5625
	Vancouver	461.9125	466.9125
Sonitrol Of SW Washington	Vancouver	460.9375	465.9375
Target Stores	Vancouver	467.7625	Simplex
	Vancouver	469.0625	Simplex
Tidewater Terminal Company	Vancouver	461.1875	Simplex

461.9375 463.8625 464.0125

Washougal Motorcross Park

Washougal	469.9375	Simplex
Washougal	469.3375	Simplex

469.6750 468.4000 468.8500

Schools

Agency	Location	Frequency	Input / Mode
Battle Ground School District	Battle Ground	462.1750	467.1750
460.7125 464.4125 464.5000	Battle Ground	464.3250	Simplex
	Battle Ground	464.8250	Simplex
Camas School District #117	Camas	151.8950	Simplex
	Camas	464.6000	469.6000
Clark College	Vancouver	464.9250	
Educational Service District #112	Vancouver	151.8050	Simplex
Evergreen School District 114	Vancouver	152.8700	157.6650
152.3450 152.3900 152.4200	Vancouver	152.4050	159.6450
157.5600 157.6450 152.2850	Vancouver	152.3000	Simplex
Hazel Dell School	Vancouver	467.8500	Simplex
	Vancouver	467.9250	Simplex
Hockinson School District	Brush Prairie	461.8500	466.8500
Operations Repeater	Brush Prairie	453.3875	458.3875
453.0000 461.3250 462.1000 463.3250 463.4250			
La Center Intermediate School	La Center	153.3350	
Vancouver School District 37	Camas	464.1500	469.1500
457.5250 461.1125 464.4875	Camas	464.4750	469.4750
464.5000 464.5500 464.6875	Vancouver	464.9125	Simplex
466.1125 467.8500 467.9250	Vancouver	468.4250	Simplex
468.9750 469.5000	Vancouver	469.5500	Simplex
Washington State University Vancouver		462.3875	Simplex
POLICE – Ch1 Dispatch	Vancouver	TG: 42096	Clark
POLICE – Ch2 Tactical	Vancouver	TG: 42128	Clark
Maintenance Department	Vancouver	TG: 42160	Clark
	Vancouver	451.5125	Simplex
Washougal School District	Washougal	151.6850	Simplex

Public Utilities

Agency	Location	Frequency	Input / Mode
Clark County Public Utilities District			
Line Crews	Vancouver	451.0500	456.0500
Service Crews	Vancouver	451.0750	456.0750
Water / Sewer Crews	Vancouver	451.1000	456.1000
Hazel Dell Sewer District	Camas	453.0500	458.0500

Weather Information

Agency	Frequency	Mode
National Weather Service – NOAA Weather Radio		
Conditions & Forecasts – Portland	162.5500	Simplex
Conditions & Forecasts – Davis Peak	162.5250	Simplex
Clark County SAME Alert Code: 053011		

COWLITZ COUNTY

Government & Public Safety

Agency / Channel	Description	Tone	Location	Frequency	Mode / Output
American Medical Response			Countywide	154.5400	Simplex
Ch1	Company Operations	[PL 107.2]	Countywide	155.3250	Simplex
Ch2	County Operations		Countywide	155.2200	Simplex
Castle Rock Ambulance			Castle Rock	155.2200	Simplex
Castle Rock Police			Castle Rock	155.0850	
Cowlitz County Fire Districts			Countywide	155.0850	Simplex
Brown	Dispatch		Countywide	154.2350	Repeater
Yellow	Tactical		Countywide	154.1150	Simplex
Green	Tactical		Countywide	154.4150	Simplex
Red	Tactical		Countywide	154.3700	Simplex
	Ambulances		Countywide	155.2200	Simplex
Cowlitz County Government			Countywide	154.7250	Simplex
	155.3100 155.5350 155.6700		Countywide	158.9250	Simplex
	158.8575 159.0375 153.7475		Countywide	155.2575	Simplex
	155.8875 159.0450 153.8675		Countywide	155.5725	Simplex
	158.9625 153.9425 155.9175		Countywide	159.0825	Simplex
	Sanitation		Countywide	154.0250	Repeater
	Public Works Department		Countywide	151.0850	159.0150
Cowlitz County Sheriff's Office			Countywide	155.9400	Simplex
Ch1	Dispatch	[PL 156.7]	Countywide	154.8150	156.2100
Ch2	Dispatch		Countywide	156.2100	
	Car to Car / Tactical		Countywide	154.9500	Simplex
Black	Search & Rescue / Special Operations		Countywide	159.3900	Repeater
	Search & Rescue		Countywide	155.2800	Simplex
	Search & Rescue – Handhelds		Countywide	154.5400	Simplex
			Countywide	453.5625	458.5625
			Countywide	154.6575	Simplex
	Corrections Department – Jail		Kelso	460.0250	465.0250

Agency	Location	Frequency	Input / Mode
Kalama Police Department	Kalama	155.8350	Simplex
Tactical	Kalama	155.0850	Simplex
Kelso City Government	Kelso	153.5300	Simplex
Kelso Police Department			
Ch1 Dispatch	Kelso	155.6700	154.7250
Ch2 Tactical / Alternate	Kelso	155.9400	Simplex
Longview City Government	Longview	153.8300	
153.9800 155.0550			
Longview Police Department			
Ch1 Dispatch 1	Longview	155.5350	
Ch2 Alternate	Longview	155.9400	
Ch3 Dispatch 2	Longview	156.0900	
Maritime Fire & Safety Association	Countywide	150.9800	154.5850
Operations Repeater	Countywide	159.4800	158.4450
Woodland City Government	Woodland	153.7550	Simplex
Washington State Patrol			
District 5: Lewis, Klickitat, Cowlitz, Clark, Skamania, Wahkiakum Counties			
Ch2 Cowlitz County Dispatch	Countywide	154.7700	Simplex

Miscellaneous

Agency	Location	Frequency	Input / Mode
Castle Rock School District 401	Castle Rock	157.6800	Simplex
	Castle Rock	151.7450	Simplex
Cowlitz County Public Utilities District 1			
Ch1 Operations Repeater [PL 151.4]	Countywide	153.6500	159.7950
Ch2 Alternate Simplex	Countywide	153.7250	Simplex
Daily News	Kelso	453.0000	458.0000
Kelso Longview Airport	Kelso	122.8000	AM
Seattle Center Approach Departure	Kelso	124.2000	317.6000
Kelso School District	Kelso	154.5400	Simplex
151.5350	Kelso	155.2800	Simplex
Longview School District	Longview	151.8350	Simplex
Lower Columbia College	Longview	461.0875	Simplex
464.7625 466.0875 453.7625	Longview	469.7625	Simplex
	Longview	458.7625	Simplex
Riverdale Motorcross Raceway	Toutle	151.5500	Simplex
151.7300 159.7350 159.9900	Toutle	160.0800	Simplex
Saint John Medical Center	Longview	152.3750	157.6350
152.4500 152.8700 157.4500			
Triangle Mall	Longview	461.0625	Simplex
461.7625 463.3125	Longview	464.5875	Simplex
Woodland State Airport	Woodland	122.9000	AM
Woodland School District			
Bus Transportation – Primary	Woodland	464.9500	469.9500
Bus Transportation – Secondary	Woodland	464.7000	Simplex

Weather Information

Agency	Location	Frequency	Input / Mode
National Weather Service – NOAA Weather Radio			
Conditions & Forecasts – Portland		162.5500	Simplex
Conditions & Forecasts – Davis Peak		162.5250	Simplex
Cowlitz Harbor County SAME Alert Code: 053015			

GRAYS HARBOR COUNTY

Government & Public Safety

Aberdeen City Fire Department			
See Grays Harbor County Fire Agencies			
Aberdeen City Government	Aberdeen	159.0375	158.8200
Public Works Department	Aberdeen	158.9400	Simplex
158.8200 155.9850			
Aberdeen City Police Department			
See Grays Harbor County Police/Sheriff WEST			
Conf. Tribes Chehalis Reservation	Oakville	155.8650	
Cosmopolis City Government	Cosmopolis	153.7550	
Cosmopolis Police Department	Cosmopolis	154.7850	Simplex
Elma City Government	Elma	155.0850	
Grays Harbor County Government	Elma	154.9950	
453.5750 460.3750 155.3100	159.2100		
Road Maintenance	Countywide	156.1650	159.1650
Grays Harbor County Fire Agencies	Countywide	154.3850	
Ch3 County Fire Dispatch	Countywide	154.1900	
Ch5 Aberdeen & Cosmopolis Areas	Countywide	154.3100	150.7900
Ch6 Countywide + Hoquiam	Countywide	154.1450	
Ocean Shores/N Beaches [PL 103.5]	Countywide	154.1750	
East – Montesano/Central Park/Elma	Countywide	154.3400	
Copalis Beach (Fire District 7)	Copalis Beach	156.1875	Simplex
Elma (Fire District 5)	Elma	156.0375	151.1000
Grays Harbor County Police/Sheriff	Montesano	155.6100	Simplex
West 1 – West Dispatch [PL 192.8]	Aberdeen	154.7250	158.8500
West: Aberdeen, Hoquiam, Cosmopolis			
West 2 – West Secondary	Aberdeen	155.3100	159.2100

Agency / Channel	Tone	Location	Frequency	Input
West 3 – Hoquiam Tactical		Countywide	155.4300	Simplex
West 4 – Aberdeen Tactical		Countywide	155.7300	Simplex
East 1 – East Dispatch	[PL 203.5]	Elma	155.5650	154.8900
East: Chehalis, Elma, McCleary, Montesano, Ocean Shores, Oakville, Westport				
East 2 – East Secondary	[PL 118.8]	Elma	155.9100	158.9100
Saddle Mountain Repeater	[PL 110.9]	Countywide	155.0100	154.0850
Elma / Montesano TAC		Countywide	155.7900	Simplex
		Elma	156.2400	
LERN		Countywide	155.3700	
Court Security		Aberdeen	158.7600	Simplex
Grays Harbor Transportation Auth.		Countywide	453.3000	458.3000
		Countywide	453.4000	458.4000
Hoquiam City Government		Hoquiam	155.9400	
Road Maintenance – Primary		Hoquiam	155.6400	Simplex
Road Maintenance – Secondary		Hoquiam	155.8650	Simplex
Road Maintenance – Secondary		Hoquiam	156.0750	Simplex
Hoquiam City Fire Department				
See Grays Harbor County Fire Agencies				
Hoquiam Police Department				
See Grays Harbor County Police/Sheriff WEST				
McCleary Fire Department		McCleary	154.2200	Simplex
McCleary Public Works Department		McCleary	154.1150	Simplex
Montesano City Government		Montesano	155.0850	
Montesano City Fire Department				
See Grays Harbor County Fire Agencies				
Ocean Shores City Government		Ocean Shores	158.9550	155.0550
Ocean Shores City Fire				
See Grays Harbor County Fire Agencies				
Ocean Shores City Police Dept.		Ocean Shores	155.4900	159.0450
Dispatch & Operations	[PL 110.9]	Ocean Shores	155.0100	154.0850
Quinault Indian Nation		Taholah	155.5950	
GOVERNMENT Services		Taholah	156.0150	158.9550
FIRE Department		Taholah	154.1000	158.8050
POLICE Department		Taholah	155.0025	158.8725
POLICE Department		Taholah	155.5950	158.8200
POLICE Department		Taholah	159.1500	Simplex
POLICE Department		Taholah	154.8300	Simplex
Washington State Patrol				
Note: Not all WSP units or dispatchers transmit PL tone at all times				
Ch16 Grays Harbor / Pacific		Countywide	154.6950	Simplex
Ch1 Statewide / Aircraft Ops	[PL 100.0]	Statewide	155.9700	159.0750
Westport City Government		Westport	153.8600	
		Westport	155.8725	158.8875
		Westport	156.0975	158.8875
Westport City Fire Department				
Dispatch		Westport	154.0625	159.0150
Tactical		Westport	154.2350	

Westport City Police: See Grays Harbor County Police/Sheriff EAST

Miscellaneous

Agency	Location	Frequency	Mode/Input
Bowerman Airport	Hoquiam	122.7000	AM
Seattle Center Approach Departure	Hoquiam	128.3000	269.0000
Copalis State Airport	Copalis	122.9000	AM
Elma Municipal Airport	Elma	122.8000	AM
Elma School District #68	Elma	154.6000	Simplex
Bus Transportation	Elma	155.1750	Simplex
Grays Harbor Community Hospital	Aberdeen	155.2800	Simplex
155.3400 155.4000 461.2500			
Operations Repeater	Aberdeen	461.2500	466.2500
Grays Harbor College	Aberdeen	453.6750	Simplex
Grays Harbor County	Countywide	155.2350	Simplex
Hoquiam Aberdeen Transportation Cooperative			
School Transportation	Hoquiam	152.9450	159.9450
Mc Cleary School District #65	Mc Cleary	152.4050	
Montesano School District	Montesano	452.1500	Simplex
Security & Maintenance	Montesano	151.5200	Simplex
Security & Maintenance	Montesano	160.2000	Simplex
Bus Transportation – Primary	Montesano	159.5475	Simplex
Bus Transportation – Secondary	Montesano	160.1475	Simplex
North River School District #200	North River	154.5700	Simplex
Ocean Shores Municipal Airport	Ocean Shores	122.9000	AM
Ocean Shores Outdoor Rec Club	Ocean Shores	154.6000	Simplex
Ocean Shores/Grays Harbor Convention Center			
Operations Simplex	Ocean Shores	467.7625	Simplex
Operations Simplex	Ocean Shores	467.8125	Simplex
Ocosta School District #172	Westport	155.2050	Simplex
Pacific Beach Naval MWR Center	Pacific Beach	462.8000	Simplex
Pacific Holiday Incorporated	Ocean Shores	151.8350	Simplex
	Ocean Shores	151.9250	Simplex
Port of Grays Harbor	Aberdeen	153.8150	Simplex
153.9200 153.9800 154.9650			
155.0400 155.8350 155.9550			
Puget Sound & Pacific Railroad	Countywide	160.9350	161.4750
Quinalt Beach Resort and Casino	Ocean Shores	451.2875	Simplex
Westport Airport	Westport	122.9000	AM
Westport Shipyard Incorporated	Westport	451.1750	Simplex
452.0000 464.3750			

Public Utilities

Aberdeen City Utilities	Aberdeen	155.9850	Simplex
158.8200 158.9400			
Cascade Natural Gas	Cosmopolis	048.5000	Simplex
Conf Tribes Chehalis Reservation	Oakville	155.8650	Simplex
Cosmopolis City Utilities	Cosmopolis	153.7550	Simplex
Elma City Utilities	Elma	155.0850	
Grays Harbor Public Utilities	Cosmopolis	452.2500	
Grays Harbor Port Authority	Aberdeen	153.8150	
153.9800 155.8350			
Hoquiam City Utilities	Hoquiam	155.9400	
Water Department	Hoquiam	452.3625	457.3625
Montesano City Utilities	Montesano	155.0850	
Ocean Shores City Utilities	Ocean Shores	155.0550	
	Ocean Shores	158.9550	
Public Utility Dist #1	Grays Harbor	047.7000	
047.8000 048.2600 451.0500			
Quinault Indian Nation	Neilton	156.0150	
	Neilton	158.9550	
Tacoma Public Utilities	Montesano	153.6950	
	Montesano	158.2500	
Washington Public Power System	Elma	451.0750	
451.2500 451.5750			

Weather Information

National Weather Service – NOAA Weather Radio

Conditions & Forecasts – Astoria	162.4000	Simplex
Conditions & Forecasts – Capitol Peak	162.4750	Simplex
Conditions & Forecasts – Olympic Nat'l Park	162.4250	Simplex

Grays Harbor County SAME Alert Code: 053027

ISLAND COUNTY

Government & Public Safety

Island County Fire & EMS Departments (I-COM)				158.9475	Simplex
Ch1	Dispatch	[PL 100.0]	Countywide	154.3400	Simplex
Ch2	Tactical / Responses	[PL 100.0]	Countywide	153.7550	158.7600
Ch3	Tactical Simplex	[PL 100.0]	Countywide	154.1750	Simplex
Ch4	Tactical – Camano	[PL 100.0]	Countywide	154.2950	159.1950
Ch5	Tactical Simplex	[PL 100.0]	Countywide	154.3550	Simplex
Ch6	Ambulance Dispatch	[PL 100.0]	Countywide	154.4300	Simplex
Ch7	WA OSCCR Coordination	[No PL]	Countywide	156.1350	Simplex
Ch8	RedNet – Fire Mutual Aid	[No PL]	Countywide	153.8300	Simplex
Ch9	HEAR – Amb to Hosp.	[PL 156.7]	Countywide	155.3400	Simplex
Ch10	Alternate	[PL 100.0]	Countywide	153.9350	Simplex
Ch11	Alternate	[PL 100.0]	Countywide	154.1300	Simplex
Ch12	Marine Ch16 Emergency		Countywide	156.8000	Simplex
Ch13	Marine Ch21A Coast Guard		Countywide	157.0500	Simplex
Ch14	Marine Ch22A Coast Guard / Ferries		Countywide	157.1000	Simplex
Ch15	LERN – Law Enforcement Interagency		Countywide	155.3700	Simplex
Ch16	Alternate	[PL 100.0]	Countywide	154.4300	Simplex
	Fire District 1		Stanwood	158.7450	153.9350

Whidbey General Hospital EMS Units

Whidbey 1, 2	Oak Harbor Medic Unit
Whidbey 3	Freeland Medic Unit
Whidbey 5	Coupeville Medic Unit
Whidbey 10	Medical Director
Whidbey ##	Volunteer BLS Unit

Island County Fire Districts / Stations / Apparatus

11	Camano Island	E11, WT11
12	Camano Island	E12, WT12, R12
15	Camano Island	E15, WT15, R15

21	North Whidbey Island	E21, R21, T21, Rescue Boat 21
22	North Whidbey Island	E22, Whidbey 22, T22, 200 North (BC)
23	North Whidbey Island	E23, R23
24	North Whidbey Island	E24, R24
25	North Whidbey Island	E25, Whidbey 25, R25, Rescue Boat 25, 200 South (BC)
26	North Whidbey Island	E26, R26
27	North Whidbey Island	E27, WT27, R27
28	North Whidbey Island	E27, Whidbey 28
31	South Whidbey Island	E31, WT31, Aid 31, E312
32	South Whidbey Island	E32, WT32, Aid 32
33	South Whidbey Island	E33, WT33
34	South Whidbey Island	E34
35	South Whidbey Island	E35, WT35, E352
36	South Whidbey Island	E36, WT36, R36, R362, Whidbey 36, Marine 36 (+362)
39	Langley FD	E39, Tower 39, Aid 39
51	Central Whidbey	E51, E512, R51
52	Central Whidbey	E52, WT52, R52
53	Central Whidbey	E53, E532, T53, Whidbey 53, Rescue Boat 53, CMD 500
54	Central Whidbey	E54, WT54, R54
71	Naval Air Station Fire	E70, E71
81	Oak Harbor FD	E81, L81, E812, E813, E814, S81, Medic 81, U81, CMD 8

Island County Government	[PL 127.3]	Countywide	453.7750	458.7750
Engineering – Simplex		Countywide	458.8125	Simplex
Island County Transit		Countywide	867.7125	822.7125
Building Maintenance		Coupeville	453.9375	Simplex

Island County Sheriff's Office

I-COM: Island Public Safety Communications

Dispatch Repeater	[PL 100.0]	Countywide	453.6750	458.6750
Tactical Repeater	[PL 100.0]	Countywide	460.4750	465.4750
LERN – Law Enforcement Interagency		Countywide	155.3700	Simplex

<u>Island County Law Enforcement Unit Numbering / Agency Prefix</u>

Paul:	Coupeville Police Department
Robert:	Langley Police Department
Victor:	Island County Sheriff's Office
Victor 2##:	Island County Sheriff Reserves
Victor 3##:	Island County Jail Deputies

Langley City Government		Langley	155.1000	Simplex
Oak Harbor City Government		Oak Harbor	156.0000	Simplex
Oak Harbor Police	[PL 100.0]	Oak Harbor	453.0500	458.0500
Dispatch Frequency	[PL 94.8]	Oak Harbor	460.5750	465.5750
Old Frequency	[PL 100.0]	Oak Harbor	453.9000	458.9000

Washington State Patrol

<u>Note</u>: Not all WSP units or dispatchers transmit PL tone at all times

<u>District 7</u>: Snohomish, Skagit, Whatcom, San Juan, Island Counties

Ch17	Districtwide Car to Car	[PL 141.3]	Countywide	154.9350	Simplex
Ch18	Skagit / Island Counties	[PL 141.3]	Countywide	155.5050	159.0450

Military

Whidbey Island Naval Air Station

	Description	Location	Frequency	Mode
	ACLS – Automatic Carrier Landing System Approach		300.0000	AM
	Air Tactical		239.7000	AM
	Aircraft Maintenance		140.2250	Simplex
	Approach / Departure Control		118.2000	270.8000
	Approach / Departure Control		120.7000	286.6500
	Approach / Departure Control		299.6000	AM
	ATIS – Automatic Terminal Information System		134.1500	281.5000
	Base Operations		350.0000	
	Clearance Delivery		124.1500	380.8000
	Coast Guad Liason		140.7000	Simplex
	"Dolly" Operations		339.5000	AM
	"DownRigger" Operations		135.3000	337.1000
	Fire Department / Crash Crews		142.7000	Simplex
	Fire Department / Crash Crews		140.1000	Simplex
	Fire Department / Crash Crews – Helipad		161.7000	Simplex
	"Firewood" Base		360.9000	AM
	Flightline Operations		139.5000	Simplex
	Flightline Operations		140.0250	Simplex
	Ground Control		121.7500	336.4000
	GCA – Ground Controlled Approaches		126.0500	226.4000
	GCA – Ground Controlled Approaches		266.8000	AM
	GCA – Ground Controlled Approaches		310.8000	AM
	GCA – Ground Controlled Approaches		322.5000	AM
	GCA – Ground Controlled Approaches		327.0000	AM
	GCA – Ground Controlled Approaches		328.4000	AM
	GCA – Ground Controlled Approaches		343.7500	AM
	Helicopter Search & Rescue Operations		034.1500	Simplex
	Landing Signal Officers (aka "Paddles")		363.6000	AM
	Maintenance Squadron Operations		355.1000	AM
	Metro Operations / Dispatcher		343.4000	AM
	Naval Investigative Service (NIS)	Silverdale	140.0750	Simplex
	Naval Investigative Service (NIS)	Silverdale	140.7750	Simplex
	North American Aerospace Defense Command		282.6000	AM
	Outlying Landing Field (Coupeville) Control Tower		384.4000	AM
	Paging System		149.4000	Simplex
	Petroleum/Oil/Lubricants		140.7500	Simplex
	Pit Fuel		139.5750	Simplex
	Pre-Taxi Clearance		135.1000	AM
	Range Control		344.2000	AM
	Security – Munitions		138.6250	Simplex
Ch1	Security – Primary		139.5250	Simplex
Ch2	Security – Tactical		140.0000	Simplex
	Transportation		142.0750	Simplex
	Totum Operations		386.8000	AM
	Tower		127.9000	340.2000
	Tower		360.2000	AM
	VAQ-129 "Vikings" Electronic Attack Squadron		268.7000	AM
	VAQ-131 "Lancers" Electronic Attack Squadron		385.0000	AM
	VAQ-134 "Garudas" Electronic Attack Squadron		320.4000	AM
	VP-40 Patrol Squadron "Fighting Marlins"		361.9000	AM
	VP-46 Patrol Squadron "Grey Knights"		258.6000	AM
	VR-61 Fleet Logistics Support Squadron		376.8000	AM

Other Frequencies

140.4250 148.3000 281.8000 306.0000

138.6750 140.1750 140.4500 140.5500
140.6000 140.6500 140.8500 142.5250
142.8000 143.7125 143.9250 406.7625

Trunked Radio System – Motorola P25
385.8875* 386.0750* 386.1000* 386.6750
386.3750 386.5500 386.6375 386.7000
388.0000 388.1500

POLICE DEPARTMENT	[Dig-P25]	TG: 23292	Whidbey
FIRE DEPARTMENT	[Dig-P25]	TG: 23322	Whidbey

Miscellaneous

Camano Island Airfield		Stanwood	122.9000	AM
Coupeville School District		Coupeville	464.8625	Simplex
School Bus Operations	[PL 203.5]	Coupeville	155.2350	Simplex
Doss Aviation		Oak Harbor	151.8650	Simplex
Island County Public Utility District 1		Stanwood	153.6550	Simplex
153.5300 153.7250 153.6650		Stanwood	153.7250	Simplex
Oak Harbor School District				
School Bus Operations	[PL 146.2]	Oak Harbor	155.2350	Simplex
Puget Sound Energy		Oak Harbor	153.4250	Simplex
South Whidbey School District		Langley	152.9075	159.6375
School Bus Operations	[PL 85.4]	Whidbey	154.5400	Simplex
Stanwood School District		Stanwood	451.9750	456.9750
		Stanwood	452.0750	457.0750
		Stanwood	452.9750	457.9750
Wes Lupien Airport		Oak Harbor	122.8000	AM
Whidbey Approach / Departure		Oak Harbor	118.2000	AM
Whidbey Air Park		Langley	122.9000	AM
Whidbey General Hospital	[PL 118.8]	Oak Harbor	463.0500	468.0500
Whidbey Golf & Country Club		Oak Harbor	159.7050	Simplex
		Oak Harbor	160.1250	Simplex
Whidbey Island Pony Club		Coupeville	152.9450	Simplex
153.0500 153.0200 153.4550		Coupeville	153.0650	Simplex

Weather Information

National Weather Service – NOAA Weather Radio

Conditions & Forecasts – Puget Sound Marine	162.4250	Simplex

Island County SAME Alert Code: 053029

JEFFERSON COUNTY

Government & Public Safety

Jefferson County Fire District #2, #3, Port Townsend Fire			
Dispatch	East County	153.8600	155.1150
Tactical	East County	154.1600	Simplex
	East County	155.0775	159.0975
	East County	155.3250	Simplex
Jefferson County Fire District #4	Brinnon	154.0925	150.8050
Jefferson County Fire District #6	Pt Townsend	153.9500	Simplex
Jefferson County Government	Countywide	158.8800	Simplex
Public Works	Countywide	154.1150	155.0550
Transit Authority	Countywide	453.2000	458.2000
Transit Authority	Countywide	851.1125	806.1125
Emergency Management	Port Hadlock	154.3025	Simplex
Emergency Management	Port Hadlock	460.6375	Simplex
Jefferson County Sheriff's Office			
East Dispatch (+ Port Townsend Police)	Pt Townsend	453.5750	458.5750
West Dispatch	Kalaloch	453.2750	458.2750
	Countywide	155.1150	Repeater
	Countywide	155.1600	Simplex
Olympic Ambulance	Pt Townsend	462.9500	467.9500
Port Townsend Water & Sewer	Pt Townsend	158.8050	Simplex
Port Townsend Police Department	Pt Townsend	453.5750	458.5750
Quinault Indian Reservation Police	Quinault	460.0375	Simplex
	Quinault	465.0375	Simplex

Jefferson County Sheriff's Office – Patrol Districts

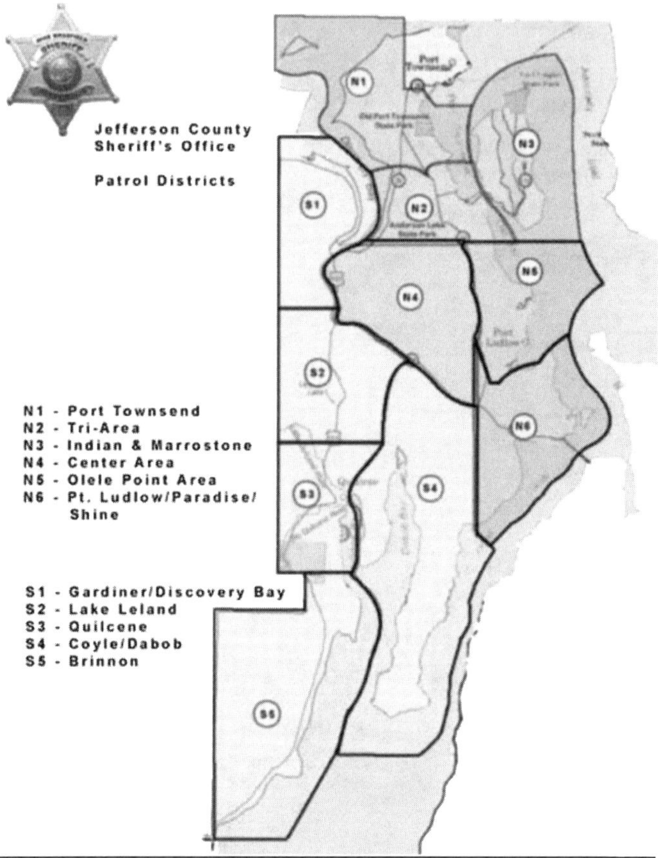

Miscellaneous

Jefferson County Int'l Airport	Pt Townsend	123.0000	AM
Jefferson General Hospital	Pt Townsend	468.4875	Simplex
Security & Maintenance	Pt Townsend	469.4750	
Quilcene School District 48	Quilcine	155.2650	Simplex

Weather Information

National Weather Service – NOAA Weather Radio

Conditions & Forecasts – Neah Bay	162.5500	Simplex
Conditions & Forecasts – Olympic Nat'l Park	162.4250	Simplex
Conditions & Forecasts – Puget Sound Marine	162.4250	Simplex
Conditions & Forecasts – Seattle	162.5500	Simplex
Jefferson County SAME Alert Code: 053031		

KING COUNTY

Government & Public Safety

King County Law Enforcement Radio Callsign Prefixes

A	Auburn PD	D	Des Moines PD	L	Algona PD	T	Tukwila PD
B	Black Diamond	D	Duvall PD	M	Medina PD		
C	Clyde Hill PD	F	Lk Forest Park	P	Pacific PD		

Seattle/King County Regional 800 MHz Trunked Radio System

Scanner Setting: Motorola Type II 800 Standard (Analog)

IF THESE FREQUENCIES CHANGE, FREE UPDATES ARE AT SCANSEATTLE.COM

Note: The Seattle/King County Regional 800 MHz Trunked Radio System is comprised of interconnected subregional analog trunked radio simulcast cells and intellirepeaters owned the City of Seattle, Valley Communications Center, the East Side Public Safety Communications Agency and King County. The system supports approximately 14,000 radio users throughout King County, including police, fire, EMS, general government services, school districts and utilities, with the ability of radio users to communicate with their counterparts even while "roaming" outside their home area by connecting to other subregional simulcast cells and intellirepeaters participating in the regional system.

Subregional Simulcast Cells

Note: Each talkgroup in the lists in this section indicates which subregional simulcast cell (King, EPSCA or Seattle) that particular talkgroup is most active on. To avoid missed conversations, where possible use the indicated Simulcast Cell frequencies to monitor the respective talkgroup. Note, however, that most talkgroups can "roam" between cells and onto the intellirepeaters listed below when needed to provide clear communications with one or more radios that are not receiving that talkgroup's primary cell well. System control channels are noted below by a * and bold. If your scanner is capable of Control Channel Only Mode (see your manual), only control channels need to be entered for each cell. The other (voice) frequencies will be tuned automatically by your scanner using data provided on the control channel.

King: **King County / Valley Communications Simulcast Cell**
This Cell Simulcasts from Cambridge; Crista; Federal Way; Rattlesnake Mtn; Ring Hill; Skyway; Squak Mtn; View Park;

868.4250*	**867.8625***	**851.8125***	851.3625	851.1625	851.9625
852.1125	852.6125	852.8625	853.3125	853.4875	853.6125
854.0875	854.2875	866.3625	866.3875	866.9375	866.9625
867.3375	867.4375	867.8375	868.4500	868.7250	

EPSCA: Eastside Public Safety Communications Agency Simulcast Cell
This Cell Simulcasts from Capitol Hill; Horizon Heights; Norway Hill; W. Tiger Mtn

868.7750*	**868.2000***	851.1375	851.8875	852.6375	853.3875
854.2375	866.2125	866.4625	866.9125	866.9875	867.3125
867.4875	867.8125	867.8875	868.5250	868.8250	

Seattle Seattle City Simulcast Cell
This Cell Simulcasts from: Capitol Hill; Columbia; Northeast; West Seattle

868.1750*	**851.4125***	851.1875	851.9375	851.9875	852.6875
852.9125	853.4375	854.1875	854.3625	866.2875	866.3125
866.3375	866.4375	866.6875	866.7125	866.7375	866.8875
867.2875	867.7625	867.7875	868.4750	868.6750	868.8750

Subregional Intellirepeaters

Apple Cove Intellirepeater (Far Northwest)
868.2250* **868.6500*** 866.1625 866.4125 866.6625 868.9000

Education Hill Intellirepeater (Northeast)
855.9625* 855.4625 855.7125

Grass Mountain Intellirepeater (Far Southeast)
858.4875* **858.2375*** 855.7125 856.2375 856.4875 856.9875
857.2375 857.4875 857.9625

King County Courthouse Intellirepeater (Seattle)
851.0625* **867.9125*** 868.3000 868.7000 868.8500

McDonald Mountain (Kangley) Intellirepeater (Southeast)
860.7125* **860.2625*** 855.2375 855.3125 855.5625 855.8125
856.4375 856.7125 856.9375 857.2625 857.7125 858.2625
858.7125 858.9375 859.2625 859.7125

Metro Tunnel North Intellirepeater (Heard in Tunnel Only)
868.2250* 853.6125 854.0875

Metro Tunnel South Intellirepeater (Heard in Tunnel Only)
868.6750* 852.8625 853.4875

North Seattle (Kenmore) Intellirepeater
868.9500* **868.9750*** 855.4875 867.2225 867.2500 867.3625
867.3875

Snoqualmie Pass (Dodge Ridge) Intellirepeater (Far East)
857.4625* 855.2375 856.4625 856.9625 858.4625 859.7625

Sobieski Mountain Intellirepeater (Far Northeast)
860.9625* **860.4625*** 855.9875 858.9625 859.4625 859.9625
860.9125

Airlift Northwest Air Ambulance Countywide TG: 61776 King

	Helicopter Dispatch	[PL 192.8]	Countywide	155.2950	Simplex
	Operations Repeater	[PL 179.9]	Seattle	463.0000	468.0000
	Fixed Wing Aircraft		Seattle	129.8250	AM

Algona City Government Algona 154.0250

	City Services VHF Simplex		Algona	155.9250	Simplex
	City Services - Trunked		Algona	TG: 42928	King
	Public Safety Talkgroup		Algona	TG: 40048	King

Algona Police

MAIN	Dispatch / Operations		Algona	TG: 36848	King
TAC	Tactical / Alternate		Algona	TG: 36880	King
ADMIN	Administration		Algona	TG: 36912	King
	155.2500 155.5350 155.6850				

American Medical Response (AMR Ambulance) 851.4625 806.4625

Ch1	Seattle Dispatch	[PL 103.5]	Seattle	154.2200	Simplex
Ch2	HEARS	[PL Varies]	Seattle	155.3400	Simplex
Ch3	East King County Dispatch	[PL 103.5]	Countywide	155.2800	Simplex
Ch4	South King County	[PL 103.5]	Countywide	155.1600	Simplex
Ch5	Tactical / Special Events	[PL 103.5]	Countywide	155.4000	Simplex

Auburn Fire: See ValleyCom

	Fire Administration		Auburn	TG: 38768	King
	Fire Prevention		Auburn	TG: 38800	King
	Old VHF Frequencies		Auburn	154.0700	Simplex
	154.2500 154.3100 154.4450				

Auburn City Government Auburn 155.0850 Simplex

			Auburn	155.9250	Simplex
	Operations Repeater		Auburn	460.3625	465.3625
	Operations Simplex		Auburn	158.8050	Simplex
	Public Safety Talkgroup		Auburn	TG: 40080	King
	City Services Talkgroup		Auburn	TG: 42960	King

Auburn Police

MAIN	Dispatch / Operations		Auburn	TG: 36944	King
TAC	Tactical / Alternate		Auburn	TG: 36976	King
TAC 2	Tactical / Alternate		Auburn	TG: 28400	King
ADMIN	Administration		Auburn	TG: 37008	King
	Old VHF Frequencies		Auburn	154.6500	Simplex
	155.2500 155.5350 155.6850				

Beaux Arts Village Law Enforcement: Contract to King County Sheriff Precinct 3

	Sheriff Patrol - Southeast		Countywide	TG: 23216	King
TAC 03	Tactical Southeast Precinct		Countywide	TG: 23248	King

Bellevue City Government Bellevue 037.1000 Simplex

	045.2000 045.3600 045.4800				
	153.8450 158.9400				
	Building Maintenance		Bellevue	TG: 20240	EPSCA
	Code Compliance Inspectors		Bellevue	TG: 20496	EPSCA
	Communications Center		Bellevue	TG: 21072	EPSCA
	Communications Maintenance		Bellevue	TG: 21616	EPSCA
	Engineering / Inspectors		Bellevue	TG: 20336	EPSCA
	Fleet Electronics Services		Bellevue	TG: 20592	EPSCA
	Golf Course Maintenance		Bellevue	TG: 20272	EPSCA
	Parks Department		Bellevue	TG: 20304	EPSCA
	Services Repeater		Bellevue	453.4500	458.4500
	Services Repeater		Bellevue	453.7250	458.7250
	Special Purpose Talkgroup		Bellevue	TG: 20848	EPSCA
	Street Maintenance		Bellevue	159.0000	
	Streets Department - Primary		Bellevue	TG: 20400	EPSCA

Streets Department – Storm Water		Bellevue	TG: 20368	EPSCA
Survey / Inspectors		Bellevue	TG: 20560	EPSCA
Transportation Department		Bellevue	TG: 20528	EPSCA
Transportation Department		Bellevue	TG: 20624	EPSCA
Transportation Department		Bellevue	TG: 20688	EPSCA
Utilities		Bellevue	TG: 20656	EPSCA
Water / Sewers - Dispatch		Bellevue	TG: 20432	EPSCA
Water / Sewers - Working		Bellevue	TG: 20464	EPSCA

Bellevue Fire: See ValleyCom

Fire Administration		Bellevue	TG: 21680	EPSCA
		Bellevue	460.5750	465.5750
		Bellevue	460.6250	465.6250

Bellevue Police

		Bellevue	453.0500	Simplex
DISPATCH 1	Dispatch / Operations	Bellevue	TG: 21424	EPSCA
DISPATCH 2	Alternate	Bellevue	TG: 21392	EPSCA
RECORDS	Records Checks	Bellevue	TG: 21360	EPSCA
TAC-1	Tactical / Alternate	Bellevue	TG: 21328	EPSCA
TAC-2	Tactical / Alternate	Bellevue	TG: 21296	EPSCA
TRAINING	Training	Bellevue	TG: 21264	EPSCA
CAR-CAR	Car to Car	Bellevue	TG: 21232	EPSCA
EVENTS	Special Events	Bellevue	TG: 21200	EPSCA
DETEC	Detectives / Investigators	Bellevue	TG: 21168	EPSCA
SDU		Bellevue	TG: 21136	EPSCA
TRAFFIC	Traffic Enforcement	Bellevue	TG: 21104	EPSCA
Citywide-1	Citywide Alternate	Bellevue	TG: 20912	EPSCA
Citywide-2	Citywide Alternate	Bellevue	TG: 20880	EPSCA

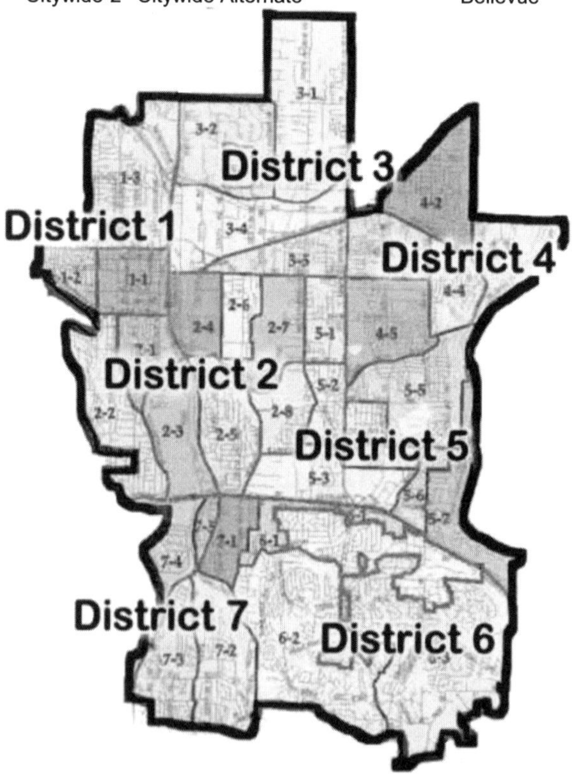

Black Diamond Fire		Blk Diamond	154.2800	Simplex
Dispatch & Operations		Blk Diamond	154.2500	Simplex
Black Diamond City Government		Blk Diamond	155.8350	Simplex
Administration		Blk Diamond	TG: 33360	King
City Services		Blk Diamond	TG: 33328	King
Public Safety		Blk Diamond	TG: 33296	King
Black Diamond/Lake Sawyer Fire: See ValleyCom				
Black Diamond Law Enforcement: See Auburn Police				
Boeing Company Fire		Countywide	154.2500	Simplex
		Countywide	154.3400	Simplex
Boeing Field Police	[PL 100.0]	Seattle	155.8050	Simplex
Bothell Fire: See ValleyCom				
154.1900 154.2050 154.4000				
Bothell City Government		Bothell	045.2400	
Services Repeater		Bothell	453.9750	458.9750
Citywide Talkgroup		Bothell	TG: 20080	EPSCA
Public Works Department		Bothell	TG: 20112	EPSCA
Bothell Police Department				
MAIN	Dispatch / Operations	Bothell	TG: 20208	EPSCA
INVEST	Investigations	Bothell	TG: 17808	EPSCA
RECORDS	Records Checks	Bothell	TG: 17840	EPSCA
TAC	Tactical / Alternate	Bothell	TG: 20176	EPSCA
Burien City Government		Burien	TG: 33456	King
Burien Fire: See ValleyCom				
Burien Law Enforcement: Contract to King County Sheriff Precinct 4				
Sheriff Patrol - Southwest		Countywide	TG: 23280	King
TAC 04 - Tactical Southwest Precinct		Countywide	TG: 23312	King
Carnation City Government		Carnation	155.1150	
City Services		Carnation	TG: 28528	King
Public Safety		Carnation	TG: 28496	King
Carnation Fire Services: See Eastside Fire Rescue				
Carnation Law Enforcement: Contract to King County Sheriff Precinct 2				
Sheriff Patrol - Northeast		Countywide	TG: 23152	King
TAC 02 - Tactical Northeast Precinct		Countywide	TG: 23184	King
Clyde Hill City Public Works		Clyde Hill	TG: 20016	EPSCA
Clyde Hill Police Department: Dispatched by Bellevue Police				
Car to Car		Clyde Hill	TG: 20048	EPSCA
Covington Law Enforcement: Contract to King County Sheriff Precinct 3				
Sheriff Patrol - Southeast		Countywide	TG: 23216	King
TAC 03 - Tactical Southeast Precinct		Countywide	TG: 23248	King
Des Moines Fire: See ValleyCom				
Des Moines City Gov't	[PL 141.3]	Des Moines	155.7600	Simplex
City Services		Des Moines	TG: 42992	King
Public Safety		Des Moines	TG: 40112	King
		Des Moines	158.7300	Simplex
Des Moines City Police				
MAIN	Dispatch / Operations	Des Moines	TG: 37040	King
ADMIN	Administration	Des Moines	TG: 37104	King
TAC	Tactical / Alternate	Des Moines	TG: 37072	King
	OLD VHF Frequencies	Des Moines		
	154.8150 155.0700 155.3100			
Duvall City Government		Duvall	155.1150	Simplex
City Services		Duvall	TG: 32784	King

	Public Safety Common Channel	Duvall	TG: 32816	King

Duvall Police Department: Dispatched by Redmond Police

Eastside Fire Rescue [PL 107.2] Countywide 154.2050 Simplex

Eastside Fire protects Carnation, Issaquah, North Bend, and Sammamish

DISP-1	Dispatch	Countywide	TG: 22384	EPSCA
DISP-2	Dispatch	Countywide	TG: 22352	EPSCA
DISP-3	Dispatch - Issaquah	Countywide	TG: 22320	EPSCA
DISP-4	Dispatch - Mercer Island	Countywide	TG: 22288	EPSCA
DISP-5	Dispatch	Countywide	TG: 22256	EPSCA
DISP-10	Dispatch	Countywide	TG: 31408	EPSCA
TAC-1	Tactical	Countywide	TG: 22224	EPSCA
TAC-2	Tactical	Countywide	TG: 22192	EPSCA
TAC-3	Tactical	Countywide	TG: 22160	EPSCA
TAC-4	Tactical	Countywide	TG: 22128	EPSCA
TAC-5	Tactical	Countywide	TG: 22096	EPSCA
TAC-6	Tactical - Issaquah	Countywide	TG: 22064	EPSCA
TAC-7	Tactical - Mercer Island	Countywide	TG: 22032	EPSCA
TAC-11	Tactical	Countywide	TG: 31440	EPSCA
TAC-12	Tactical	Countywide	TG: 31472	EPSCA
Training-1	Training	Countywide	TG: 22000	EPSCA
Training-2	Training	Countywide	TG: 21968	EPSCA
Training-3	Training	Countywide	TG: 21936	EPSCA
	INTEROP-1	Countywide	TG: 21872	EPSCA
	INTEROP-2	Countywide	TG: 21840	EPSCA
	INTEROP-3	Countywide	TG: 21808	EPSCA
	INTEROP-4	Countywide	TG: 21776	EPSCA
	Mutual Aid Zone - 1	Countywide	TG: 62416	EPSCA
	Mutual Aid Zone - 2	Countywide	TG: 62448	EPSCA
	Mutual Aid Zone - 3	Countywide	TG: 62480	EPSCA
	Mutual Aid Zone - 4	Countywide	TG: 62512	EPSCA
	Mutual Aid Zone - 5	Countywide	TG: 62544	EPSCA
	Mutual Aid - Pool 1	Countywide	TG: 62576	EPSCA
	Mutual Aid - Pool 2	Countywide	TG: 62608	EPSCA

Fire Station Locations & Apparatus

71	190 E Sunset Way, Issaquah	E71, Trk71, WT71, A71
72	1770 NW Maple St, Issaquah	E72, A72
73	1280 NE Park Dr, Issaquah	E73, Trk 73, A73, R73, M73
74	8641 Preston Fall City Rd SE	E74, A74, WT74
75	4225 180th Ave SE, Sunset	E75
76	15132 Tiger Mtn Rd SE	E76, A76
78	16135 SE 113th Pl, Renton	E78, A78
79	20505 SE 152nd St, Renton	E79, A79
81	2030 212th SE, Sammamish	E81, A81, Air Support
82	1851 228th Av, Sammamish	E82, Trk 82, A82
83	3425 Issaquah Pine Lake Rd	E83, A83, BR83, Wildland
85	3600 Told Av, Carnation	E85, A85, WT85, TRT/Rescue
86	10644 E Lake Joy Dr NE	E86
87	122 W 2nd St, North Bend	E87, A87, Medic 3
88	42804 SE 170th, North Bend	E88, A88

Enumclaw City Government Enumclaw 155.0550 Simplex

Enumclaw Fire Department: Self Dispatched, but see ValleyCom for unified list

Enumclaw Police Department Enumclaw 154.8750

155.0100 158.8500 159.0900

Trunked: Use King County Grass Mountain Intellirepeater

Primary Operations	Enumclaw	TG: 29808	King
TAC-1 Tactical	Enumclaw	TG: 29840	King
TAC-2 Link to Buckley PD	Enumclaw	TG: 29904	King

	Citywide Multigroup		Enumclaw	TG: 29776	King
EOC	Operations Center		Enumclaw	TG: 29872	King

Everett: See Snohomish County

Federal Way Public Works [PL 103.5] Federal Way 155.7150 Simplex

	City Services Talkgroup		Federal Way	TG: 32912	King

Federal Way Fire District: See ValleyCom

	Fire 8	[PL 131.8]	Federal Way	154.2200	
	Fire 9	[PL 131.8]	Federal Way	153.9500	
	Fire Administration		Federal Way	TG: 30992	King
	Fire Prevention		Federal Way	TG: 31024	King

Federal Way Police Department

	Primary Operations / Dispatch		Federal Way	TG: 29136	King
	Tactical / Alternate		Federal Way	TG: 29168	King
	Administration		Federal Way	TG: 29200	King
	Public Safety Talkgroup		Federal Way	TG: 32880	King

Fife Police Department

Ch1	Operations Simplex		Fife	155.4900	
Ch4	Operations Simplex		Fife	155.0100	

Hunts Point Law Enforcement: Contracted to Medina Police

Issaquah City Fire: See Eastside Fire Rescue

	Fire Administration		Issaquah	TG: 19664	EPSCA

Issaquah City Government Issaquah 155.7450

	Building Department		Issaquah	TG: 19568	EPSCA
	Citywide Talkgroup		Issaquah	TG: 19536	EPSCA
	Parks Department		Issaquah	TG: 19600	EPSCA
	Public Works Department		Issaquah	TG: 19632	EPSCA
	Unknown Use		Issaquah	155.0625	158.9175
	158.8650 155.0850				

Issaquah City Police Issaquah 154.7850

	Primary Operations / Dispatch		Issaquah	TG: 19888	EPSCA
	Record Checks		Issaquah	TG: 19856	EPSCA
	TAC-1 Tactical / Alternate		Issaquah	TG: 19824	EPSCA
	TAC-2 Tactical / Alternate		Issaquah	TG: 19792	EPSCA

Kenmore Fire: See ValleyCom

Kenmore Law Enforcement: Contract to King County Sheriff Precinct 2

	Sheriff Patrol - Northeast		Countywide	TG: 23152	King
TAC 02 - Tactical Northeast Precinct			Countywide	TG: 23184	King

Kent City Government Kent TG: 43024 King

		[PL 114.8]	Kent	156.0000	153.9050
	Public Works		Kent	154.8600	Simplex

Kent Fire Department: See ValleyCom

	Fire Administration		Kent	TG: 38832	King
	Fire Prevention		Kent	TG: 38864	King

Kent Police Department

	Primary Operations / Dispatch		Kent	TG: 37136	King
	Tactical / Alternate		Kent	TG: 37168	King
	Administration		Kent	TG: 37200	King
	Public Safety		Kent	TG: 40144	King

King County Mutual Aid Radio Countywide 155.1900 154.6500

King County / Valley Communications

	Common Fire Merge		Countywide	TG: 38416	King
	Common Police Merge		Countywide	TG: 36816	King
	King Comm Administration		Countywide	TG: 38384	King
	School Districts Merge		Countywide	TG: 43536	King
	City Governments Merge		Countywide	TG: 42896	King

Common	Countywide	TG: 43120	King
Common Mutual Aid	Countywide	TG: 39696	King
ERT-1	Countywide	TG: 37488	King
ERT-2	Countywide	TG: 37520	King
ERT-3	Countywide	TG: 37552	King
ERT-4	Countywide	TG: 37584	King
Local Governments Merge	Countywide	TG: 40976	King
NARC-1 Narcotics	Countywide	TG: 37424	King
NARC-2 Narcotics	Countywide	TG: 37456	King
PD TRAIN-1 Police Training	Countywide	TG: 37616	King
PD TRAIN-2 Police Training	Countywide	TG: 37648	King
Public Safety	Countywide	TG: 40240	King
Public Safety Merge	Countywide	TG: 40016	King
Special Operations	Countywide	TG: 37680	King

King County EOC Support Team Redmond 155.2350 Simplex

King County Fire Protection Districts (Now mostly trunked)

District - 2	[PL 131.8]	Seattle	154.2500	Simplex
District - 2		Seattle	154.4150	Simplex
District - 2 - Administration		Seattle	TG: 30544	King
District - 2 - Administration		Seattle	TG: 30480	King
District - 4		Seattle	153.7700	Simplex
District - 4	[PL 107.2]	Seattle	154.1900	Simplex
District - 10		Issaquah	046.0800	Simplex
046.4800 154.0850 154.1900 154.2500				
District - 10 - Zone 2 - Battalion 1		Issaquah	TG: 31504	EPSCA
District - 10 - Zone 2 - Battalion 2		Issaquah	TG: 31536	EPSCA
District - 10 - Zone 2 - Battalion 3		Issaquah	TG: 31586	EPSCA
District - 10 - Zone 2 - Battalion 4		Issaquah	TG: 31600	EPSCA
District - 10 - Zone 2 - Battalion 5		Issaquah	TG: 31632	EPSCA
District - 13		Vashon	151.0850	Simplex
District - 16		Bothell	154.1900	Simplex
154.2050 154.4000				
District - 17 - Administration		Countywide	TG: 39056	King
District - 20	[PL 131.8]	Seattle	154.2500	Simplex
District - 20		Seattle	154.4450	Simplex
District - 20 - Administration		Seattle	TG: 39088	King
District - 25	[PL 107.2]	Renton	154.1900	Simplex
District - 26	[PL 131.8]	Des Moines	154.2500	Simplex
District - 26 - Administration		Des Moines	TG: 30512	King
District - 27	[PL 107.2]	Fall City	154.1900	Simplex
District - 27 - Administration		Fall City	TG: 31696	EPSCA
District - 39		Federal Way	153.9500	Simplex
District - 39		Federal Way	154.2200	Simplex
District - 39	[PL 131.8]	Federal Way	154.2500	Simplex
District - 40		Renton	154.0700	Simplex
154.2500 154.3100 154.4450				
District - 40 - Administration		Renton	TG: 39120	King
District - 43 - Administration		Renton	TG: 39152	King
District - 44		Auburn	153.9500	Simplex
154.2200 154.2500 154.2800				
District - 44 - Administration		Auburn	TG: 39184	King
District - 45	[PL 107.2]	Duvall	154.1900	Simplex
District - 45 - Administration		Duvall	TG: 31664	EPSCA
District - 46 - Administration		Countywide	TG: 39216	King
District - 47		Ravensdale	154.0700	Simplex
154.2500 154.3100 154.4000 154.4450				
District - 47 - Administration		Ravensdale	TG: 39248	King
District - 49	[PL 107.2]	Snoqualmie	154.1900	Simplex

District - 50	Skykomish	154.4300	
District - 50 - Administration	Skykomish	TG: 31760	EPSCA
District - 51 - Administration	Countywide	TG: 31728	EPSCA
Zone - 2 - Multigroup	Countywide	TG: 31376	King
Zone - 4 - Administration	Countywide	TG: 31216	King
Zone - 4 - Multigroup	Countywide	TG: 30576	King

King County East Fire Protection Districts

	Bellevue	1035 116th Ave NE	Medic 1
1	Bellevue	766 Bellevue Way SE	E1, E1A, BC1, A1
2	Bellevue	2802 148th Ave SE	E2, A2, Medic 2, M95
3	Bellevue	16100 NE 8th St	E3, L3 A3, R3
4	Bellevue	4216 Factoria Blvd SE	E4, A4
5	Bellevue	9621 NE 24th St	E5, A5
6	Bellevue	1850 132nd Ave NE	E6, HM1, A6
7	Bellevue	11900 SE 8th St	E7, L7
8	Bellevue	5701 Lakemont Blvd SE	E8, A8, Air Unit 1
9	Bellevue	12412 SE 69th Way	E9, A9
11	Redmond	8450 161st Ave NE	E11, A11, BC11, R11, M19
12	Redmond	4211 148th Ave NE	E12, A12, BC12
13	Redmond	8201 208th Ave NE, #34	E13, A13
14	Redmond	5021 264th Ave NE, #34	E14, A14
15	Redmond	4200 228th Ave NE, #34	E15, A15
16	Redmond	6502 185th Ave NE	E16, A16, Trk16
21	Kirkland	9816 Forbes Creek Dr	E21, A21, BC21
22	Kirkland	6602 108th Ave NE	E22, A22
23	Kirkland	Evergreen Hospital	M23
24	Kirkland	8411 NE 141st St, Bothell	A24, Air 21
25	Kirkland	12033 76 Place NE	E25, A25
26	Kirkland	9930 124th Ave NE	Trk 26, A26
27	Kirkland	11210 NE 132nd St	E27, A27, E20
31	Woodinville	144th Ave NE	E31, Trk31, A31, A32, BC31
33	Woodinville	NE 133rd	E33, A33
34	Woodinville	144th St NE	E34, E34A, A34
35	Woodinville	Avondale & Woodinville	E35, A35, M35
42	Bothell	Beardslee Blvd	E42, E43, A42, A43, BC42, U42
44	Bothell	228th St SW	E44, A44
45	Bothell	Bothell Everett Hwy	E45, Trk 45, A45
47	Bothell	Mountain View Rd, Cherry G.	E47
51	Northshore	73rd Ave NE, Kenmore	E51, E52, A51, A52
57	Northshore	Brookside Blvd & 44th	E57, E58
61	Shoreline	Aurora Ave N & N 175th St	E62, Trk 61, BC61, A64 A62 S61
63	Shoreline	NE 180th St & 15th NE	E63, M63, A63
64	Shoreline	N 185th St	E64, A64
65	Shoreline	NE 155th St & I-5	E65, M65
66	Duvall	Duvall	E66, A66, BC66, U66, WT66
67	Duvall	Big Rock Rd	E67
68	Duvall	Lake Margaret	E68, A68
69	Duvall	Lake Marcel	E69, A69
91	Mercer Island	78th SE	E91, E93, L91, R91, M91 A91 BC91
92	Mercer Island	68th SE	E92, E94, M92, A92
271	Fall City	Fall City	E271, A271, R271, WT271, BC271
281	Snoqualmie	Snoqualmie City	E281, A281, R281, BC281
113	White Center	SW 112th St	E113, L163, A133, C100
114	White Center	S 128th St	E114, A134, Air 174, E124

King County Water District

	Seattle	453.9875	Simplex
	Seattle	045.5200	Simplex

King County Sheriff's Office

	Sheriff Patrol – Northeast (Precinct 2)		Countywide	TG: 23152	King
	Sheriff Patrol – Southeast (Precinct 3)		Countywide	TG: 23216	King
	Sheriff Patrol – Southwest (Precinct 4)		Countywide	TG: 23280	King
	Sheriff Patrol – Northwest (Precinct 5)		Countywide	TG: 23344	King
	Sheriff Special Operations		Countywide	TG: 23408	King
	Sheriff STATS		Countywide	TG: 23504	King
	Records Information		Countywide	TG: 23472	King
TAC 02 - Tactical – Northeast (Precinct 2)			Countywide	TG: 23184	King
TAC 03 - Tactical – Southeast (Precinct 3)			Countywide	TG: 23248	King
TAC 04 - Tactical – Southwest (Precinct 4)			Countywide	TG: 23312	King
TAC 05 - Tactical – Northwest (Precinct 5)			Countywide	TG: 23376	King
TAC 06 - Tactical Special Operations			Countywide	TG: 23440	King
TAC 07 - Tactical Countywide			Countywide	TG: 23536	King
TAC 08 - Tactical / Alternate			Countywide	TG: 23568	King
TAC 10 - Tactical / Alternate			Countywide	TG: 25424	King
TAC 12 - Tactical / Alternate			Countywide	TG: 23632	King
TAC 14 - Tactical / Alternate			Countywide	TG: 24976	King
TAC 16 - Tactical / Alternate			Countywide	TG: 26608	King
TAC 18 - Surveillance			Countywide	TG: 27088	King
TAC 20 - Tactical / Alternate			Countywide	TG: 30128	King
TAC 22 - Surveillance			Countywide	TG: 35824	King
TAC 30 - Tactical / Alternate			Countywide	TG: 27344	King
	Court Security - 1 - Courthouse		Countywide	TG: 23120	King
	Court Security - 2 - Juvenile		Countywide	TG: 23088	King
	Event 1 - Alternate/Events		Countywide	TG: 23664	King
	Event 2 - Alternate/Events		Countywide	TG: 23696	King
	Event 3 - Alternate/Events		Countywide	TG: 23728	King
	Event 4 - Alternate/Events		Countywide	TG: 23760	King
	Event 5 - Alternate/Events		Countywide	TG: 23796	King
	Fire Investigations Unit		Countywide	TG: 25456	King
	Multigroup Talkgroup		Countywide	TG: 23056	King
	Search & Rescue – F2		Countywide	TG: 24112	King
	Search & Rescue – F4	[PL 97.4]	Countywide	155.1600	Simplex
	Search & Rescue – Ham	[PL 103.5]	Countywide	145.1100	
F1	M.A.R.S.	[PL 100.0]	Countywide	155.1900	154.6500
F2	S.A.R.	[PL 123.0]	Countywide	154.9650	153.9950
F3	S.A.R.	[PL 123.0]	Countywide	153.7550	153.7550
F10	TAC-10 (SIMPLEX)	[DPL143]	Countywide	867.7000	867.7000

King County Sheriff's Office Callsigns (see maps on next few pages)

A	Shoreline	H2#	Newcastle	W	Woodinville
B	North County	H3#	Covington		
C	North County	K	Southwest		
D1	North Bend	L	Sea-Tac		
D8#	Carnation	N	Burien		
E	Kenmore	O	Sammamish		
F	Southeast	P	Other Units		
G	Southeast	R	Reserves		
H1#	Maple Valley	V	Vashon Islnd		

Precinct 2 Map: NE Unincorporated Areas + Kenmore, Woodinville, Sammamish, Carnation, and North Bend city contracts

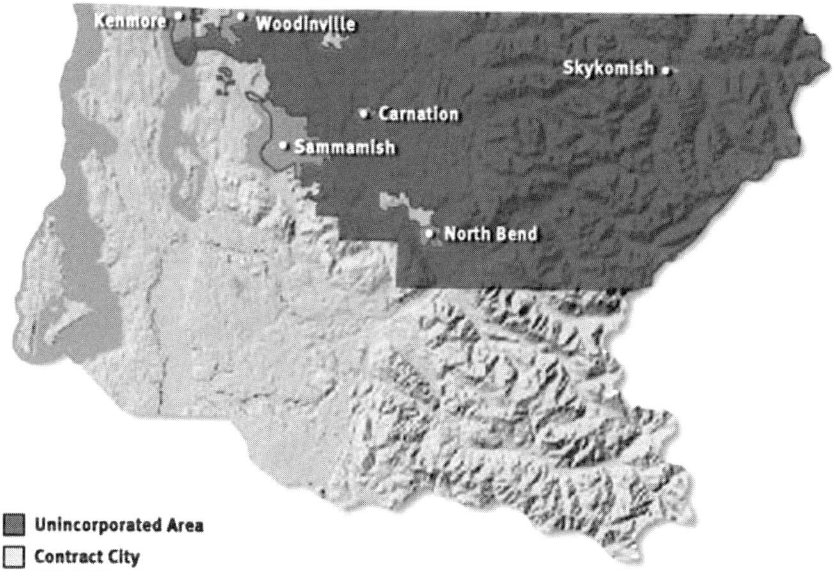

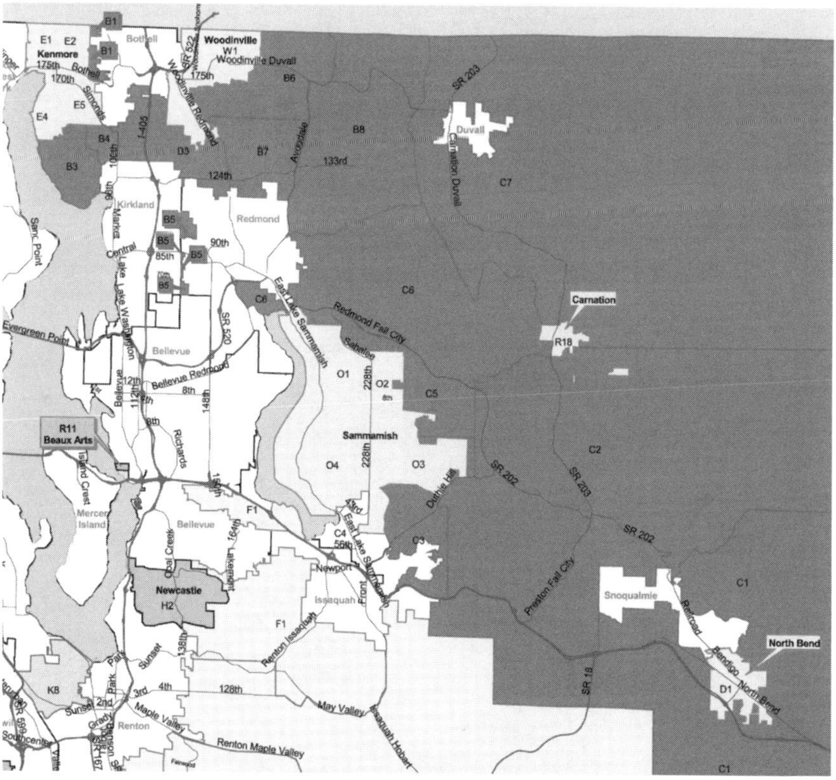

Precinct 3 Map: SE Unincorporated Areas + Newcastle, Covington, Maple Valley, and Muckleshoot Indian Tribe city contracts

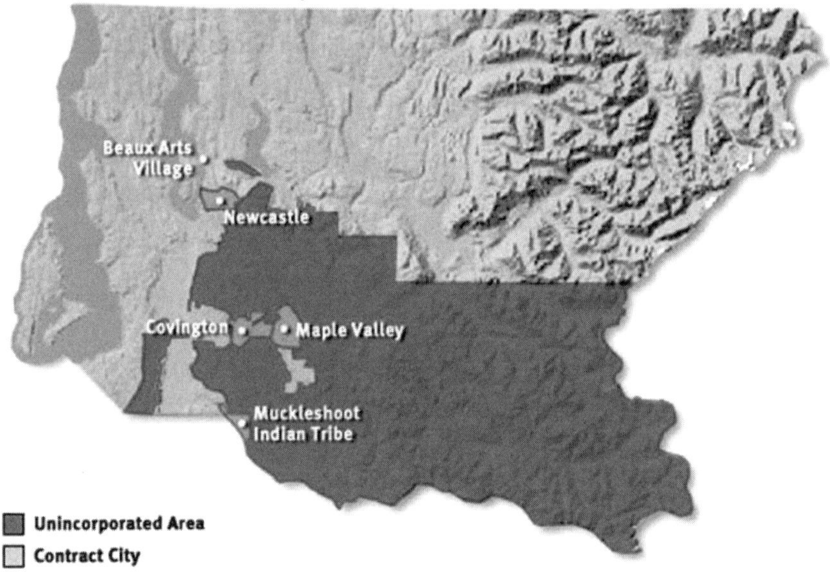

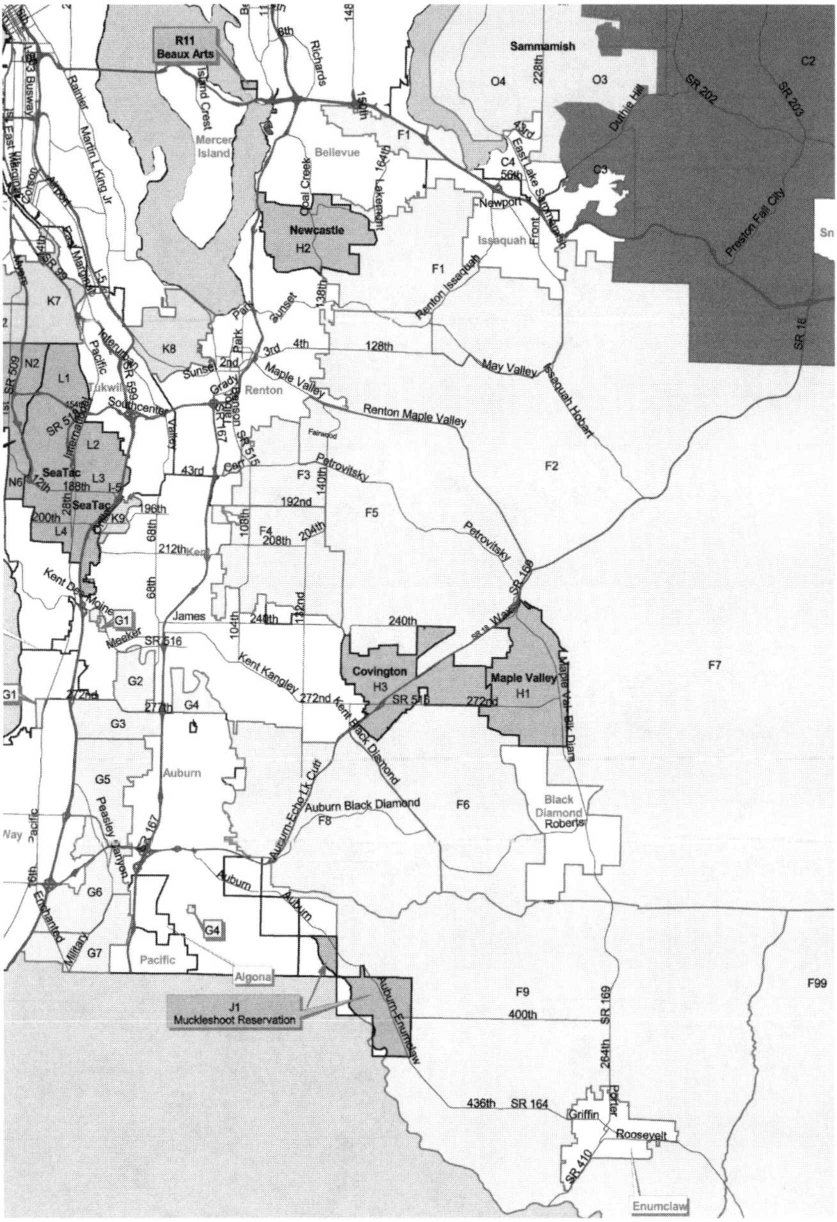
R11
Beaux Arts
Sammamish
C2
O4
O3
SR 202
SR 203
Duthie Hill
Island Crest
Mercer Island
Bellevue
F1
C4
56th
C3
Newport
East Lake Sammamish
Preston Fall City
Richards
Rainier
Martin L King Jr
Opal Creek
164th
Lakemont
Newcastle
H2
Issaquah
Front
F1
Renton Issaquah
138th
Sunset
Park
K7
K8
SR 99
3rd
4th
128th
May Valley
Issaquah Hobart
SR 18
N2
L1
SR 509
Interurban
Pacific
Tukwila
SR 599
Southcenter
Grady
Renton
Maple Valley
Renton Maple Valley
Fairwood
SR 518
International
L2
Valley
SR 167
SR 515
SeaTac
L3
N6
188th
I-5
43rd
Petrovitsky
F3
140th
F2
200th
28th
SeaTac
K9
196th
192nd
F5
L4
68th
108th
F4
208th
204th
Petrovitsky
212th
Kent
SR 169
Kent Des Moines
G1
James
104th
240th
132nd
240th
Meeker
SR 516
Kent Kangley
Covington
H3
Maple Valley
H1
F7
G2
272nd
277th
G4
SR 516
272nd
Kent Black Diamond
G1
G3
Auburn
G5
Peasley Canyon
Auburn-Echo Lk Cutoff
Auburn Black Diamond
F8
F6
Black Diamond
Roberts
Pacific
16th
Enchanted
G6
Auburn
G4
G7
Military
Pacific
Algona
J1
Muckleshoot Reservation
Auburn-Enumclaw
F9
400th
SR 169
F99
264th
Porter
436th
SR 164
Griffin
Roosevelt
SR 410
Enumclaw

Precinct 4 Map: SW Unincorporated Areas + Burien and SeaTac city contracts

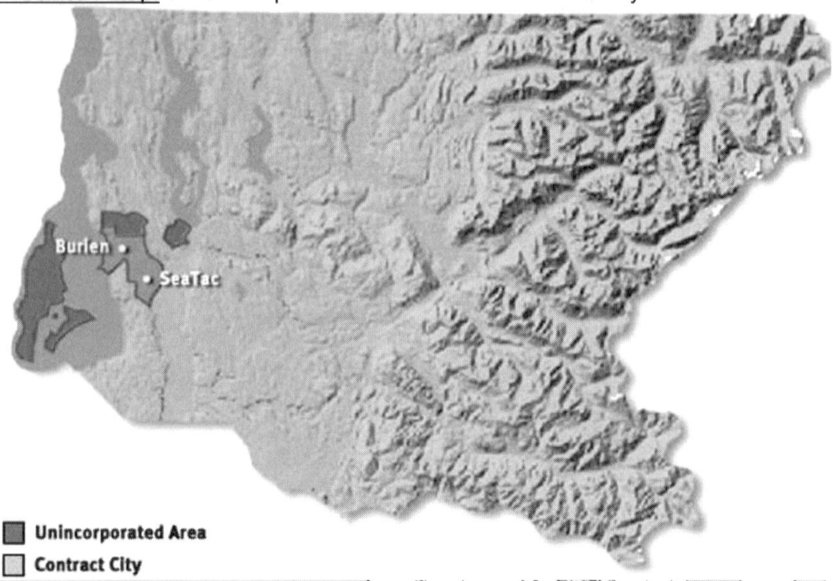

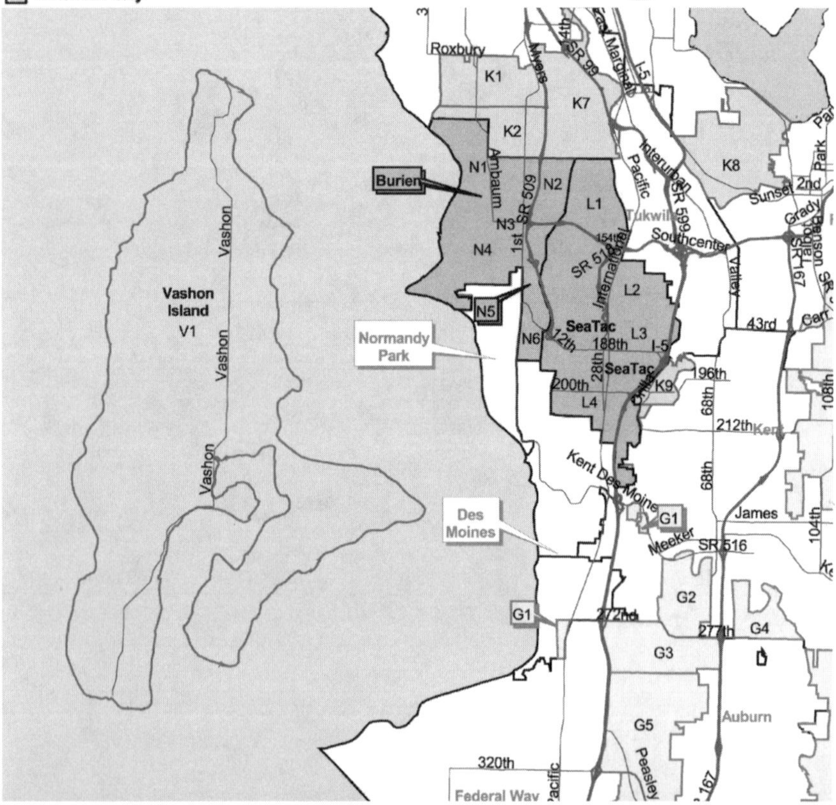

Precinct 5 Map: Shoreline city contract

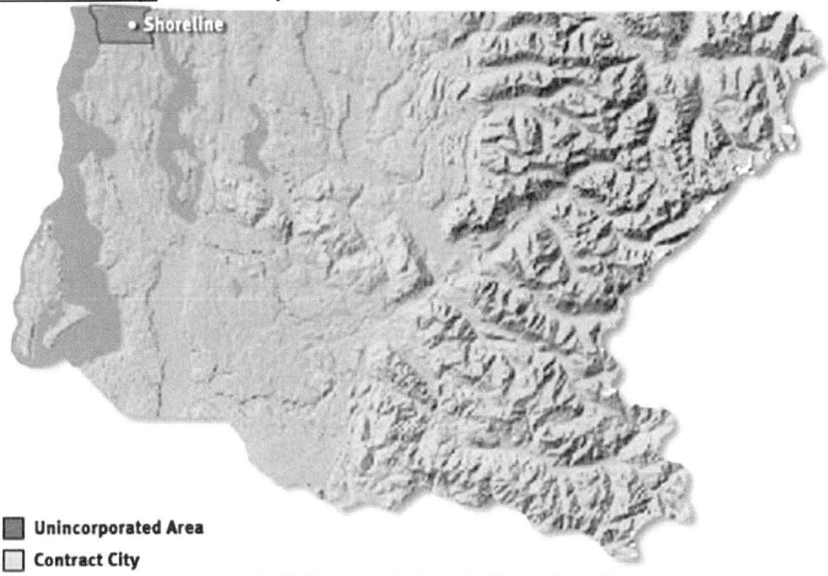

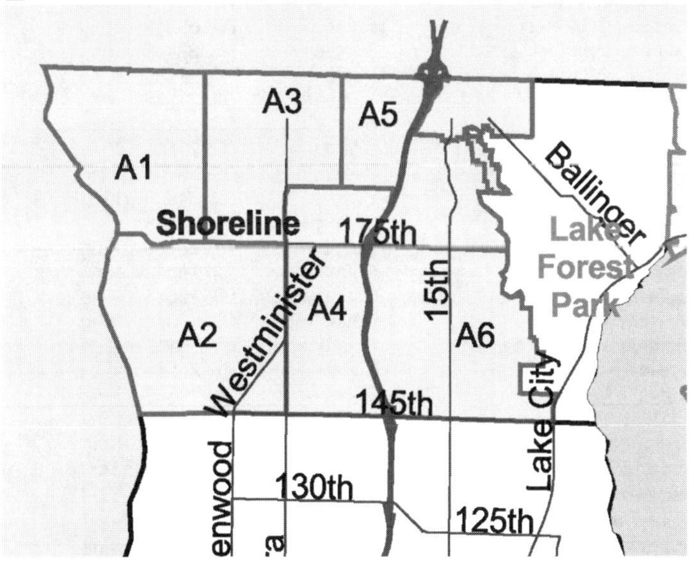

King County Transit Police

Metro - 1 - Transit Police	Countywide	TG: 23952	King
Metro - 2 - Transit Police	Countywide	TG: 23984	King
Metro - 7 - Transit Police	Countywide	TG: 24016	King
Metro - 8 - Transit Police	Countywide	TG: 24048	King

King County Government

Countywide 453.1000 458.1000

154.0400	154.0850	155.8050	453.2875	458.2875
453.8375	458.8375	453.9500	458.9500	458.2625

Animal Control - Primary	Countywide	TG: 26704	King
Animal Control - Tactical	Countywide	TG: 26672	King

Emergency OPS Center - Common	Countywide	TG: 22768	King
Emergency OPS Center - EMD	Countywide	TG: 22832	King
Emergency OPS Center - EMZ-1	Countywide	TG: 22864	King
Emergency OPS Center - EMZ-3	Countywide	TG: 22896	King
Emergency OPS Center - EMZ-4	Countywide	TG: 22928	King
Emergency OPS Center - EMZ-5	Countywide	TG: 22960	King
Emergency OPS Center - Multigroup	Countywide	TG: 22704	King
Emergency OPS Center - OPS	Countywide	TG: 22736	King
Emergency OPS Center - Seismic	Countywide	TG: 22800	King
Executive Cabinet	Countywide	TG: 27216	King
FAIR – 1 King County Fair	Countywide	TG: 27056	King
Fleet Services - Primary	Countywide	TG: 26736	King
Highway Maintenance	Seattle	453.1500	458.1500
Metro - Power & Facilities – Ch1	Issaquah	851.0125	806.0125
Metro - Power & Facilities – Ch2	Seattle	851.7625	806.7625
Metro - Power & Facilities	Countywide	TG: 27664	King
Medical Examiners Office	Countywide	TG: 25552	King
Natural Resources - Administration	Countywide	TG: 26768	King
Natural Resources - Multigroup	Countywide	TG: 26640	King
Parks - 1 - Parks & Recreation	Countywide	TG: 26960	King
Public Health Department - COMM	Countywide	TG: 62064	King
Public Health Department - OPS-1	Countywide	TG: 62032	King
Public Health Department - OPS-2	Countywide	TG: 62000	King
Public Health Department - OPS-3	Countywide	TG: 61968	King
Radio Shops - COMM-1	Seattle	TG: 63184	King
Radio Shops - COMM-2	Seattle	TG: 63216	King
Radio Shops - DEMO-1	Seattle	TG: 22448	King
Radio Shops - DEMO-2	Seattle	TG: 22480	King
Radio Shops - DEMO-3	Seattle	TG: 22512	King
Radio Shops - Project	Seattle	TG: 22416	King
Radio Shops - R.C.S.	Seattle	TG: 22544	King
Radio Shops - Trunking Alert	Seattle	TG: 36784	King
Radio Shops - WNC-1	Seattle	TG: 34800	King
Road Maintenance - 1	Countywide	TG: 26448	King
Road Maintenance - 2	Countywide	TG: 26480	King
Road Maintenance - 3	Countywide	TG: 26512	King
Road Maintenance - 4	Countywide	TG: 26544	King
Road Maintenance - 5	Countywide	TG: 26576	King
Road Maintenance - Multigroup	Countywide	TG: 26416	King
Solid Waste - Administration	Countywide	TG: 26864	King
Solid Waste - Operations	Countywide	TG: 26832	King
Waste Treatment - WTD1	Countywide	TG: 26896	King
Waste Treatment - WTD2	Countywide	TG: 26928	King
Water & Land	Countywide	TG: 26800	King
Youth Services Department	Countywide	TG: 27024	King

King County Department of Adult Detention

COURT-1	Court House Escorts - Primary	Countywide	TG: 24208	King
COURT-2	Court House Escorts - Alternate	Countywide	TG: 24240	King
COURT-3	Juvenile Escorts - Primary	Countywide	TG: 24272	King
COURT-4	Juvenile Escorts - Alternate	Countywide	TG: 24304	King
COURT-5	Reserved Use	Countywide	TG: 24336	King
COURT-6	Reserved Use	Countywide	TG: 24368	King
	Administration	Countywide	TG: 24656	King
COM-1	Common Use	Countywide	TG: 24592	King
COM-2	Common Use	Countywide	TG: 24624	King
MG-1	King Couty Courthouse	Countywide	TG: 24176	King
MG-2	Juvenile Center	Countywide	TG: 24784	King
JAILOPS-1	Courthouse Jail Primary	Countywide	TG: 24400	King

JAILOPS-2 Courthouse Jail Secondary		Countywide	TG: 24432	King
JAILOPS-3 Juvenile Jail Primary		Countywide	TG: 24464	King
JAILOPS-4 Juvenile Jail Secondary		Countywide	TG: 24496	King
King County Local Police Systems			453.0500	458.0500
MARS	[None]	Countywide	155.1900	154.6500
LERN		Countywide	155.3700	
Kirkland City Government				
Building Department		Kirkland	TG: 19120	EPSCA
Citywide Talkgroup		Kirkland	TG: 19056	EPSCA
Engineering Department		Kirkland	TG: 19184	EPSCA
Parks Department		Kirkland	TG: 19088	EPSCA
Public Works Department		Kirkland	TG: 19152	EPSCA
Kirkland Fire Department: See ValleyCom				
Fire Administration		Kirkland	TG: 19504	EPSCA
Kirkland Police Department				
DISP-1	Dispatch / Operations	Kirkland	TG: 19472	EPSCA
DISP-2	Car to Car	Kirkland	TG: 19440	EPSCA
TAC-1	Tactical / Alternate	Kirkland	TG: 19376	EPSCA
TAC-2	Tactical / Alternate	Kirkland	TG: 19344	EPSCA
TAC-3	Tactical / Alternate	Kirkland	TG: 19312	EPSCA
	Records Checks	Kirkland	TG: 19280	EPSCA
	Detectives	Kirkland	TG: 19248	EPSCA
	Traffic Enforcement	Kirkland	TG: 19408	EPSCA
Koll Center Bellevue Juvenile		Bellevue	462.0125	467.0125
		Bellevue	467.1625	Simplex
Lake Forest Park City Services		Lk Forest Pk	TG: 18960	EPSCA
Lake Forest Park Police Department: Dispatched by Bothell Police				
OLD Dispatch / Operations		Lk Forest Pk	TG: 19024	EPSCA
Tactical / Alternate		Lk Forest Pk	TG: 18992	EPSCA
Tactical / Alternate		Lk Forest Pk	TG: 36592	EPSCA
Maple Valley Fire: See ValleyCom				
Maple Valley Law Enforcement: Contract to King County Sheriff Precinct 3				
Sheriff Patrol - Southeast		Countywide	TG: 23216	King
TAC 03 - Tactical Southeast Precinct		Countywide	TG: 23248	King
Medina Police: Dispatched by Kirkland Police on Kirkland Police Channels				
VHF Repeater		Medina	156.0825	153.8225
Mercer Island City Government		Mercer Island	TG: 18640	EPSCA
Parks Department		Mercer Island	TG: 18672	EPSCA
Public Works Department		Mercer Island	TG: 18704	EPSCA
Public Works Department		Mercer Island	453.2500	458.2500
		Mercer Island	453.3250	458.3250
Mercer Island Fire Department		Mercer Island	154.1900	Simplex
See Eastside (King County) Fire Shared Talkgroups				
Fire Tactical		Mercer Island	TG: 18864	EPSCA
Mercer Island Police Department				
Primary Operations / Dispatch		Mercer Island	TG: 18928	EPSCA
Records Checks		Mercer Island	TG: 18896	EPSCA
Alternate / Tactical		Mercer Island	TG: 18864	EPSCA
Milton Law Enforcement: See Pierce County section				
Mountain View Fire: See ValleyCom				
Muckleshoot Indian Tribe Law Enforcement: Contract to KCSO Precinct 3				
Sheriff Patrol - Southeast		Countywide	TG: 23216	King
TAC 03 - Tactical Southeast Precinct		Countywide	TG: 23248	King
Newcastle City Government		Newcastle	TG: 15984	EPSCA

Newcastle Law Enforcement: Contract to King County Sheriff Precinct 3

	Sheriff Patrol - Southeast		Countywide	TG: 23216	King
TAC 03 - Tactical Southeast Precinct			Countywide	TG: 23248	King

Normandy Park Police Department Normandy Pk 155.8650

	Dispatch	[PL 141.3]	Normandy Pk	154.8150	
Ch1	Patched from VHF		Normandy Pk	TG: 33104	King
Ch2	Patched from VHF		Normandy Pk	TG: 33072	King

North Bend Fire: See Eastside Fire Rescue

North Bend Law Enforcement: Contract to King County Sheriff Precinct 2

	Sheriff Patrol - Northeast		Countywide	TG: 23152	King
TAC 02 - Tactical Northeast Precinct			Countywide	TG: 23184	King

North Highline Fire District Seattle 154.2500

See ValleyCom – King County Shared South County Fire Talkgroups

Ch5	Seatac F5	[PL 131.8]	Seatac	154.4150	
	Fire Administration		Seatac	TG: 30928	King
	Fire Prevention		Seatac	TG: 30960	King

Northshore Fire Department: See ValleyCom

	Administration	Northshore	TG: 19984	EPSCA

Pacific Fire Department: See ValleyCom

	Administration	Pacific	TG: 38896	King

Pacific Police Department: Dispatched by Algona Police

Port of Seattle	Seattle	155.7150	Repeater
Unknown Use	Seattle	453.2000	458.2000

TRUNKED RADIO SYSTEM (Motorola): TrunkTracker Setting: Type II 800
853.5125* 853.5625* 854.2625* 854.3125* 851.2625
851.3125 852.0125 852.0625 852.7625 852.8125

ACCESS - 1	Port	TG: 2448	Port
ACCESS - 2	Port	TG: 2480	Port
ACCESS - Announcements	Port	TG: 2416	Port
Airport Underground Trains	Port	TG: 8016	Port
Airport Underground Trains	Port	TG: 8080	Port
Emergency Command – Announce.	Port	TG: 6416	Port
Emergency Command	Port	TG: 6448	Port
Engineering - Announcements	Port	TG: 5616	Port
Engineering - Foxtrot Talkgroup	Port	TG: 5680	Port
Engineering - Port Construction	Port	TG: 5712	Port
Engineering - Survey	Port	TG: 5648	Port
Executive Command	Port	TG: 6480	Port
Facilities Management	Port	TG: 12368	Port
GENO Talkgroup	Port	TG: 4080	Port
Maintenance - Administration	Port	TG: 1648	Port
Maintenance - Conveyor	Port	TG: 1744	Port
Maintenance - DMR-ET	Port	TG: 1680	Port
Maintenance - Snow Removal	Port	TG: 1776	Port
Maintenance - Works	Port	TG: 1712	Port
Marine - Announcements	Port	TG: 12016	Port
Marine - Emergency All	Port	TG: 12560	Port
Marine - Emergency Common	Port	TG: 12528	Port
Marine - Maintenance All	Port	TG: 12048	Port
Marine - Special Projects	Port	TG: 12080	Port
Marine - Terminal - All	Port	TG: 12208	Port
Marine - Terminal - Special Projects	Port	TG: 12240	Port
Operations - Field	Port	TG: 3280	Port
Operations - Supervisor	Port	TG: 3248	Port
PAPA Talkgroup	Port	TG: 4048	Port
Parking & Garage	Port	TG: 4016	Port
Shuttle Operations	Port	TG: 4112	Port
Technical Communications - Private	Port	TG: 7248	Port
Technical Communications - Shared	Port	TG: 7280	Port
Technical Communications	Port	TG: 7216	Port
Port of Seattle Fire			
FIRE - Administration	Port	TG: 944	Port
FIRE - All Talk Group	Port	TG: 816	Port
FIRE - Mutual Aid	Port	TG: 976	Port
FIRE - Prevention	Port	TG: 912	Port
FIRE - TAC 1 - Tactical	Port	TG: 848	Port
FIRE - TAC 2 - Tactical	Port	TG: 880	Port
Port of Seattle Police			
Airport Operations	Port	TG: 48	Port
Police - Announcements	Port	TG: 16	Port
Mutual Aid	Port	TG: 240	Port
TAC 1 - Tactical	Port	TG: 112	Port
TAC 2 - Tactical	Port	TG: 144	Port
TAC 3 - Tactical	Port	TG: 176	Port
TAC 4 - Tactical	Port	TG: 208	Port
Waterfront	Port	TG: 80	Port

Redmond City Government

Administration	Redmond	TG: 18352	EPSCA
Building Inspectors	Redmond	TG: 17936	EPSCA
Building Maintenance	Redmond	TG: 18096	EPSCA
Citywide Talkgroup	Redmond	TG: 18352	EPSCA
Parks Department	Redmond	TG: 18032	EPSCA
Parks Department - Maintenance	Redmond	TG: 17968	EPSCA
Public Works - Operations	Redmond	TG: 18288	EPSCA
Public Works - Supervisors	Redmond	TG: 18256	EPSCA
Recreation Department	Redmond	TG: 18000	EPSCA
Sewer Maintenance	Redmond	TG: 18192	EPSCA
Traffic Signal Maintenance	Redmond	TG: 18064	EPSCA
Flood Control / Storm Water Drainage	Redmond	TG: 18160	EPSCA
Streets Department	Redmond	TG: 18128	EPSCA
Water Department	Redmond	TG: 18224	EPSCA

Redmond Fire Department: See ValleyCom

Fire Prevention	Redmond	TG: 18608	EPSCA
Fire Command Talkgroup	Redmond	TG: 18576	EPSCA

Redmond Fire & Fire District 34 Station Response Areas

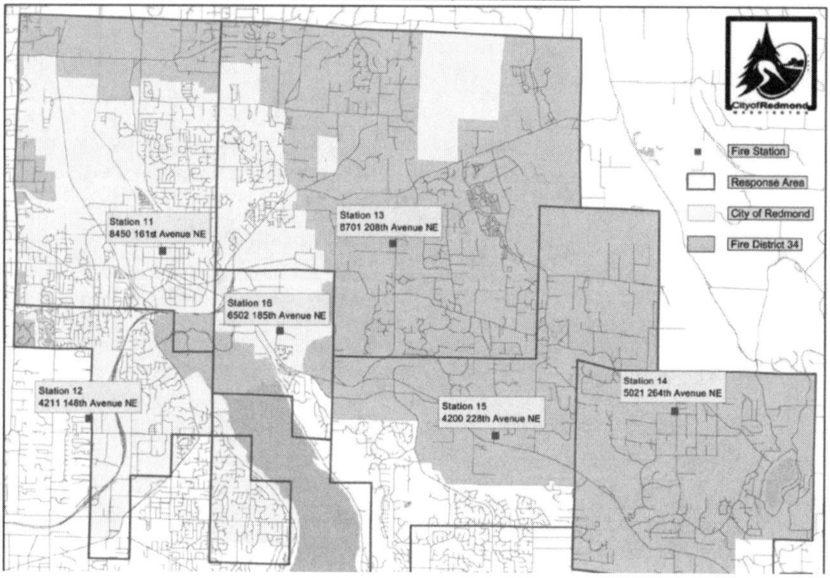

Redmond Police Department

Primary Operations / Dispatch	Redmond	TG: 18544	EPSCA
Administration	Redmond	TG: 18448	EPSCA
Investigations	Redmond	TG: 18480	EPSCA
TAC-1 - Tactical	Redmond	TG: 18416	EPSCA
TAC-2 - Tactical	Redmond	TG: 18384	EPSCA
TAC-4 - Traffic Enforcement	Redmond	TG: 18512	EPSCA

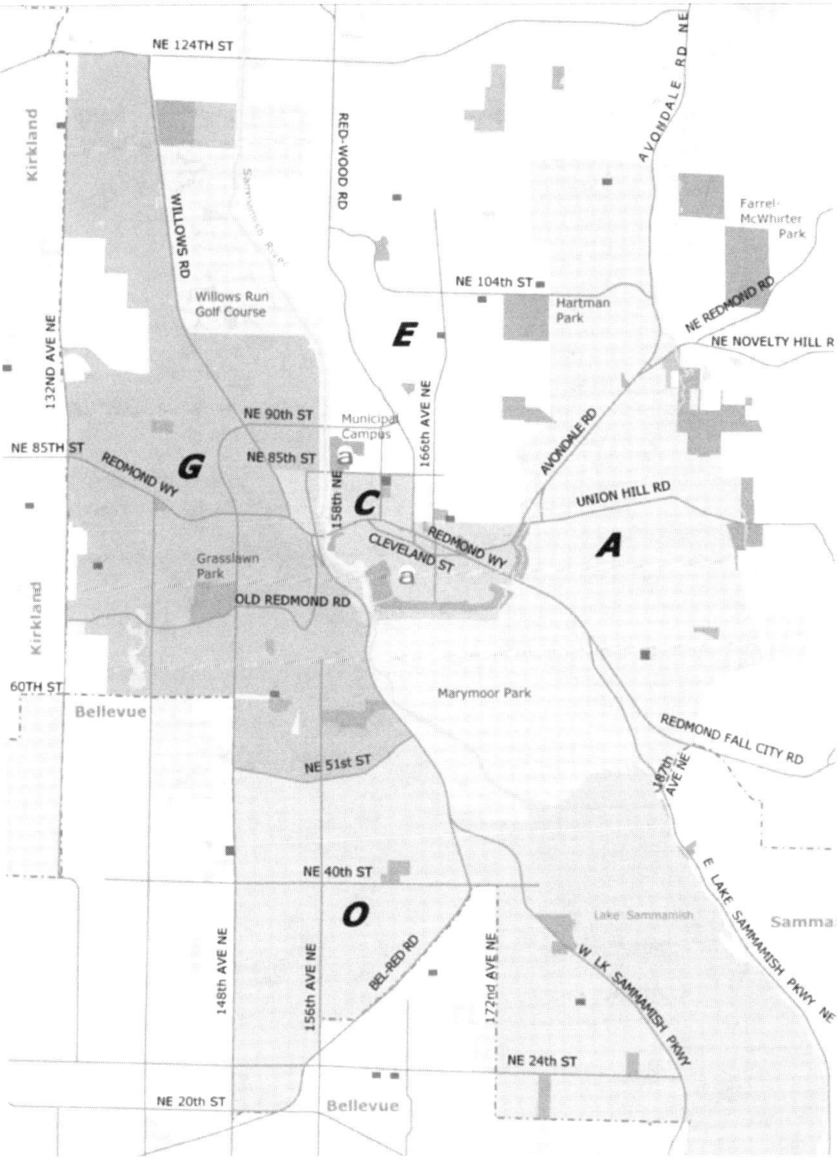

Renton Fire Department: See ValleyCom

Fire Administration	Renton	TG: 38928	King
Fire Prevention	Renton	TG: 38960	King
Renton City Government	Renton	154.1000	
155.9250 453.4375 458.4375			
Renton Police Department	Renton	TG: 40176	King
Main - Dispatch & Operations	Renton	TG: 37232	King
Administration	Renton	TG: 37296	King
Tactical / Alternate	Renton	TG: 37264	King
155.2500 155.5350 155.6850			

Sammamish Fire Services: See Eastside Fire Rescue

Sammamish Law Enforcement: Contract to King County Sheriff Precinct 2

Sheriff Patrol - Northeast	Countywide	TG: 23152	King
TAC 02 - Tactical Northeast Precinct	Countywide	TG: 23184	King
SeaTac City Government	Sea Tac	TG: 33488	King

SeaTac Fire Department: See ValleyCom

Fire Administration	Sea Tac	TG: 31056	King
Fire Prevention	Sea Tac	TG: 31088	King

SeaTac Law Enforcement: Contract to King County Sheriff Precinct 4

Sheriff Patrol - Southwest	Countywide	TG: 23280	King
TAC 04 - Tactical Southwest Precinct	Countywide	TG: 23312	King

Seattle City Government

Emergency Management [PL 141.3]	Seattle	453.6000	458.6000
Housing Authority	Seattle	462.0000	467.0000
Housing Authority	Seattle	463.2250	468.2250
Wastewater Treatment Facilities	Seattle	451.1750	456.1750
Communications Shop - 1	Seattle	TG: 5296	Seattle
Communications Shop - 2	Seattle	TG: 5328	Seattle
Communications Shop - 3	Seattle	TG: 5360	Seattle
Engineering - Announcements	Seattle	TG: 7056	Seattle
Engineering - Bridge	Seattle	TG: 7152	Seattle
Engineering - Drainage	Seattle	TG: 7088	Seattle
Engineering - Emerg. Ops. Center Link	Seattle	TG: 7664	Seattle
Engineering - Landfill	Seattle	TG: 7632	Seattle
Engineering - OPS-6	Seattle	TG: 7248	Seattle
Engineering - OPS-7	Seattle	TG: 7280	Seattle
Engineering - Solid Waste	Seattle	TG: 7184	Seattle
Engineering - Street Maintenance	Seattle	TG: 7120	Seattle
Engineering - Traffic	Seattle	TG: 7216	Seattle
EOC - Citizen Information	Seattle	TG: 9136	Seattle
EOC - Health Department	Seattle	TG: 7824	Seattle
EOC - Health Net - 1	Seattle	TG: 9072	Seattle
EOC - Health Net - 2	Seattle	TG: 9104	Seattle
EOC - Housing	Seattle	TG: 9008	Seattle
EOC - Human Services - 1	Seattle	TG: 8944	Seattle
EOC - Human Services - 2	Seattle	TG: 9104	Seattle
EOC - Human Services	Seattle	TG: 7760	Seattle
EOC - Logistics	Seattle	TG: 6192	Seattle
EOC - Parks	Seattle	TG: 7728	Seattle
EOC - Personnel TAC	Seattle	TG: 9040	Seattle
EOC - PIO Primary	Seattle	TG: 7856	Seattle
EOC - PIO TAC	Seattle	TG: 8912	Seattle
ESD - Emergency Management Staff	Seattle	TG: 6256	Seattle
ESD - Operations Center	Seattle	TG: 6224	Seattle
EXEC-1	Seattle	TG: 5712	Seattle
EXEC-2	Seattle	TG: 5744	Seattle

Fleet Services	Seattle	TG: 6288	Seattle
Freeway Parking Garage	Seattle	TG: 5680	Seattle
Grounds	Seattle	TG: 5776	Seattle
Lights - Emergency OPS Center	Seattle	TG: 8656	Seattle
Lights - Talkgroup 1	Seattle	TG: 8048	Seattle
Lights - Talkgroup 2	Seattle	TG: 8080	Seattle
Lights - Talkgroup 3	Seattle	TG: 8112	Seattle
Lights - Talkgroup 4	Seattle	TG: 8208	Seattle
Lights - Talkgroup 5	Seattle	TG: 8240	Seattle
Lights - Talkgroup 6	Seattle	TG: 8272	Seattle
Lights	Seattle	451.1500	456.1500
451.2125 451.3875 451.6875	Seattle	456.2125	Simplex
Municipal Court Security	Seattle	TG: 5584	Seattle
OEM-1	Seattle	TG: 5968	Seattle
OEM-2	Seattle	TG: 6000	Seattle
Operations Calling - CALL	Seattle	TG: 6032	Seattle
Operations Calling - COM1	Seattle	TG: 6064	Seattle
Operations Calling - COM2	Seattle	TG: 6096	Seattle
Operations Calling - COM3	Seattle	TG: 6128	Seattle
Parks Department	Seattle	TG: 5488	Seattle
Recreations	Seattle	TG: 5808	Seattle
Seattle Center - 1	Seattle	TG: 4848	Seattle
Seattle Center - 2	Seattle	TG: 4880	Seattle
Seattle Center - 3	Seattle	TG: 4912	Seattle
Seattle Center - 4	Seattle	TG: 4944	Seattle
Seattle Center - 5	Seattle	TG: 4976	Seattle
Seattle Center - 6	Seattle	TG: 5008	Seattle
Treasurers Office	Seattle	TG: 5648	Seattle
Water Department - Control 1	Seattle	TG: 6448	Seattle
Water Department - Control 2	Seattle	TG: 6480	Seattle
Water Department - Control 3	Seattle	TG: 6512	Seattle
Water Department - Control 4	Seattle	TG: 6544	Seattle
Water Department - Control 5	Seattle	TG: 6576	Seattle
Water Department - Control 6	Seattle	TG: 6608	Seattle
Water Department - To EOC	Seattle	TG: 6704	Seattle

Seattle Fire Department

		Seattle	TG: 2064	Seattle
A01	Fire Responses – Primary	Seattle	TG: 1648	Seattle
A02	Fire Responses – Secondary	Seattle	TG: 1680	Seattle
A03	Fire Responses – Tertiary	Seattle	TG: 1712	Seattle
A04	Dispatch	Seattle	TG: 1744	Seattle
A05	Aid/Medic Responses – Primary	Seattle	TG: 1776	Seattle
A06	Aid/Medic Responses – Secondary	Seattle	TG: 1808	Seattle
A07	Trauma Doctor	Seattle	TG: 2032	Seattle
A08	Medic One Doctor	Seattle	TG: 2000	Seattle
A09	AMR Ambulance requests	Seattle	TG: 1968	Seattle
A10	Car-to-Car	Seattle	TG: 2096	Seattle
A11	Command	Seattle	TG: 1840	Seattle
A12	Command	Seattle	TG: 1872	Seattle
A13	Command	Seattle	TG: 2064	Seattle
A14	Site Trunking	Seattle	TG: 1904	Seattle
A15	Emergency Zone 5	Seattle	TG: 2128	Seattle
A16	Dynamic Regroup Channel			
B1	State OPS 1 + Fireground + LZ's	Seattle	867.5375	Simplex
	Announcements / Emergency	Seattle	TG: 1616	Seattle
C06	Fire Investigations	Seattle	TG: 2160	Seattle
C07	Administration	Seattle	TG: 2192	Seattle
C08	Airlift Northwest	Seattle	TG: 2224	Seattle
	Emergency Operations Center	Seattle	TG: 3184	

Fire Station Locations & Apparatus

2	2334 4th Ave N & Bell	E2, A2, L4
3	1735 W Thurman	FireBoat 3
5	925 Alaska Way & Marion	E5, A5, FireBoat 4
6	101 23rd Ave S & Yesler	E6, L3
8	110 W Lee St & 1st Ave N	E8, L6
9	3829 Linden Ave N & 38th St	E9, Air 9
10	301 2nd Ave S & Main	E10, RS10, L1, BC1, HM77, Air 10
11	1514 SW Holden St/16th Av	E11
13	3601 Beacon Av S/Spokane	E13, BC5
14	3224 4th Ave S / Horton	L7, A14
16	6846 Oswego Plc NE/69th St	E16, M1
17	1050 NE 50th St/Roosevelt	E17, L9, BC6, A17
18	1521 NW Market St/Russell	E18, L8, BC4, A18
20	3205 13th Ave W/Dravus	E20
21	7304 Greenwood Ave N/73rd	E21
22	901 E Roanoke St/Broadway	E22
24	401 N 130th St/Phinney	E24
25	1300 E Pine St/Harvard E	E25, A25, L10, BC2
26	800 S Cloverdale St/8th Av	E26
27	1000 S Myrtle St/Ellis S	E27, F1
28	5968 Rainier Ave S/Juneau	E28, M28, L12
29	2139 Ferry Ave SW/Walker	E29, BC7
30	2931 Mt. Baker Blv S/Rainier	E30
31	1319 N Northgate Way	E31, M31, L5
32	3715 SW Alaska St/37th Av	E32, M32, L11, Air32
33	9645 Renton Av/Roxbury S	E33
34	633 32nd Av/Mercer NE	E34
35	8729 15th Av NW/87th St	E35
36	3600 23rd Av SW/Spokane	E36, U99
37	7300 35th Av/Webster SW	E37
38	5503 33rd Av/55th St NE	E38
39	12705 30th Av/127th St NE	E39
40	9401 35th Av/94th St NE	E40
41	2416 34th Av/McGraw W	E41

Seattle Fire Station Locations

Seattle Police Department

EAST-DISP	East Dispatch	Seattle	TG: 3344	Seattle
EAST-TAC	East Tactical	Seattle	TG: 3632	Seattle
NORTH-DISP	North Dispatch	Seattle	TG: 3280	Seattle
NORTH-TAC	North Tactical	Seattle	TG: 3696	Seattle
SOUTH-DISP	South Dispatch	Seattle	TG: 3312	Seattle
SOUTH-TAC	South Tactical	Seattle	TG: 3664	Seattle
WEST-DISP	West Dispatch	Seattle	TG: 3248	Seattle
WEST-TAC	West Tactical	Seattle	TG: 3728	Seattle
CIS-1	Criminal Intelligence Section	Seattle	TG: 4272	Seattle
CIS-1-D	Criminal Intelligence Digital	Seattle	TG: 44880	Seattle
CIS-2	Criminal Intelligence Section	Seattle	TG: 4304	Seattle
COMMAND-1	Command	Seattle	TG: 3888	Seattle
COMMAND-2	Command	Seattle	TG: 3920	Seattle
COMMAND-3	Command	Seattle	TG: 3952	Seattle
COMMAND-4	Command	Seattle	TG: 3984	Seattle
COMMAND-5	Command	Seattle	TG: 4016	Seattle
CTR-1		Seattle	TG: 3824	Seattle
CTR-2		Seattle	TG: 3856	Seattle
DEMO-1		Seattle	TG: 4688	Seattle
EOC-SPD	Emergency Operations	Seattle	TG: 3184	Seattle
ERT-1	ERT / SWAT	Seattle	TG: 4336	Seattle
ERT-1-D	ERT / SWAT Digital	Seattle	TG: 44912	Seattle
ERT-2	ERT / SWAT	Seattle	TG: 4368	Seattle
EVENT-1	Events	Seattle	TG: 6768	Seattle
EVENT-2	Events	Seattle	TG: 6800	Seattle
EVENT-3	Events	Seattle	TG: 6832	Seattle
EVENT-4	Events	Seattle	TG: 6864	Seattle
EVENT-5	Events	Seattle	TG: 6896	Seattle
HARBOR		Seattle	TG: 4048	Seattle
HNT-1	Hostage Negotiations	Seattle	TG: 4400	Seattle
HNT-1-D	Hostage Negotiations Digital	Seattle	TG: 44944	Seattle
HNT-2	Hostage Negotiations	Seattle	TG: 4432	Seattle
INT INVEST	Internal Investigations	Seattle	TG: 4464	Seattle
INVEST T1	Investigation w/ Patrol	Seattle	TG: 3760	Seattle
INVEST T2	Investigation w/ Patrol	Seattle	TG: 3792	Seattle
NARC-1	Narcotics Enforcement	Seattle	TG: 4080	Seattle
NARC-1-D	Narcotics Enforcement Digital	Seattle	TG: 44816	Seattle
NARC-2	Narcotics Enforcement	Seattle	TG: 4112	Seattle
NARC-3	Narcotics Enforcement	Seattle	TG: 4144	Seattle
NTF	Narcotics Task Force	Seattle	TG: 4176	Seattle
REGROUP		Seattle	TG: 3568	Seattle
TAC-1	Tactical	Seattle	TG: 3440	Seattle
TAC-1-D	Tactical Digital	Seattle	TG: 44976	Seattle
TAC-2	Tactical	Seattle	TG: 3472	Seattle
TAC-2-D	Tactical Digital	Seattle	TG: 45008	Seattle
TAC-3	Tactical	Seattle	TG: 3504	Seattle
TAC-4	Tactical	Seattle	TG: 3536	Seattle
VICE-1	Vice Squad	Seattle	TG: 4208	Seattle
VICE-1-D	Vice Squad Digital	Seattle	TG: 44848	Seattle
VICE-2	Vice Squad	Seattle	TG: 4240	Seattle

Seattle Police - Precincts

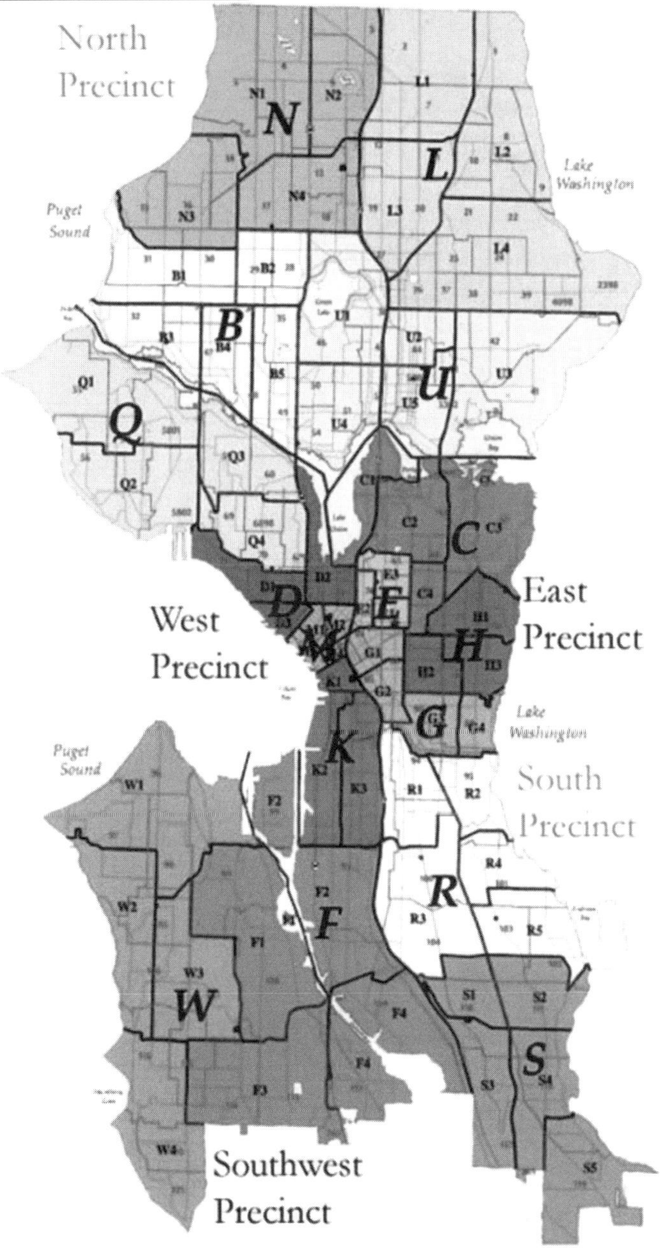

Seattle Police Unit Radio Callsigns
Patrol Units: Shift Number + Area Letter + Beat Number or Special Designator
Special Designators: 10 = Jail Wagon; #1 (i.e. "1F31") = Second Unit in F3 beat
Lack of a Beat or Special (i.e "1F") = Area Sergeant for 1st shift
"CSO" = Non-commissioned community service officers
"DOC" = Department of Corrections / Probation & Parole
P = Parking Enforcement Officers
T = Traffic Enforcement (no assigned areas)

Seattle Police Radio Codes

10	Murder	181	Narcotics Call
20	Sexual Assault	182	Narcotics Warrant
21	Sexual Assault	183	Narcotics Found
30	Robbery - Armed	184	Narcotics Arrest
31	Robbery - Mugging	185	Narcotics Overdose
40	Assault	191	Warrant - Felony
41	Threats	192	Warrant - Misdem.
42	Threats - Phone	211	Alarm - Bank
43	Assault with Gun	212	Alarm - Commercial
49	Assault - Gang Rel.	213	Alarm - Fire
50	Burglary - No Occup.	214	Alarm - Residential
51	Burglary - Occupant	220	Mental Subject
52	Burglary - Commercial	242	Fight
54	Burglary - Vacant	243	Juveniles Disturbing
61	Theft - from car	244	Noise Complaint
62	Theft - Bicycle	245	Misc Disturbance
63	Theft - Prowler	250	Disturbance
64	Theft - Shoplifting	280	Suspicious Person
65	Theft	281	Suspicious Vehicle
80	Domestic - Threat	282	Suspicious Situation
81	Domestic - Disturbing	291	Man with a gun
82	Domestic - Violent	292	Man w/ other weapon
83	Domestic - Arrest	342	Water Emergency
84	Keep the Peace	361	Person Found
85	Court Order Service	362	Person Lost
86	Court Order Assist	363	Person Missing
87	Court Order Enforc.	364	Runaway Juvenile
90	Explosives	390	Public Assist
91	Arson	415	Traffic Blockage
92	Reckless Fire	440	Motorist Assist
100	Fraud	450	DUI
101	Fraud - Forgery	482	Vehicle Pursuit
110	Kidnapping	510	Officer needs help!
111	Custodial Dispute	520	Officer assist
121	Gambling	926	Hostage Situation
122	Liquor Violation	930	Break - At Station
125	Prostitution	931	Break - Meal
130	Vandalism	932	Break - Coffee
131	Vandalism	933	Vehicle Maintenance
139	Graffiti Problem	950	Out of Service
141	Indecent Exposure		
142	Molestation		
150	Abandoned Child		
160	Prowler		
161	Trespassing		
174	Drunk - Minor		
175	Drunk - Disturbing		
176	Drunk - Adult		
179	Shots Fired from car		

Selleck/Palmer Fire: See ValleyCom

Shoreline City Utilities		Shoreline	TG: 16656	King
Shoreline Fire Department: See ValleyCom				
	Administration	Shoreline	TG: 19920	King
Shoreline Law Enforcement: Contract to King County Sheriff Precinct 5				
	Sheriff Patrol - Northwest	Countywide	TG: 23344	King
TAC 05	Tactical Northwest Precinct	Countywide	TG: 23376	King
Skyway Fire: See ValleyCom				
Snoqualmie City Government		Snoqualmie	TG: 32848	King
Snoqualmie Police Department				
MAIN	Dispatch / Operations	Snoqualmie	TG: 28336	King
TAC1	Tactical	Snoqualmie	TG: 28368	King
TAC2	Tactical	Snoqualmie	TG: 28400	King
Spring Glen Fire: See ValleyCom				
Tukwila Fire Department: See ValleyCom				
	Explorer Post 51 Operations	Tukwila	151.7450	Simplex
	Fire Administration	Tukwila	TG: 38992	King
	Fire Prevention	Tukwila	TG: 39024	King
Tukwila City Government		Tukwila	TG: 43088	King
		Tukwila	154.1150	155.0550
	155.9250 453.6375	Tukwila	458.6375	Simplex
Tukwila Police Department		Tukwila	155.2500	Simplex
	155.5350 155.6850			
MAIN	Dispatch / Operations	Tukwila	TG: 37328	King
TAC	Tactical	Tukwila	TG: 37360	King
ADMIN	Administration	Tukwila	TG: 37392	King
University of Washington Police				
MAIN	Dispatch / Operations	Countywide	TG: 9232	King
TAC	Tactical	Countywide	TG: 9264	King
ADMIN	Administration + Parking	Countywide	TG: 9296	King
ValleyComm – King County Shared South East County Fire Dispatch				
FIRE-1	ValleyComm – Fire & EMS (Aid) DISPATCH		TG: 38448	King
FIRE-2	ValleyComm – Primary EMS (Aid) Response		TG: 38480	King
FIRE-3	ValleyComm – Primary Fire Response		TG: 38512	King
FIRE-4	ValleyComm – Secondary Fire & EMS Response		TG: 38544	King
FIRE-5	ValleyComm – Secondary Fire & EMS Response		TG: 30672	King
FIRE-6	ValleyComm – Secondary Fire & EMS Response		TG: 30704	King
FIRE-7	ValleyComm – Secondary Fire & EMS Response		TG: 30736	King
FIRE-10	Enumclaw Fire – Dispatch		TG: 32048	King
FIRE-11	Enumclaw Fire – Response		TG: 32080	King
FIRE-12	Vashon Island Fire – Operations		TG: 31280	King
FIRE-13	Vashon Island Fire – Alternate		TG: 31248	King
OPS 1	ValleyComm – Training & Secondary Response		TG: 38576	King
OPS 2	ValleyComm – Training & Secondary Response		TG: 38608	King
OPS 3	ValleyComm – Training & Secondary Response		TG: 38640	King
OPS 4	ValleyComm – Training & Secondary Response		TG: 38672	King
OPS 5	ValleyComm – Training & Secondary Response		TG: 38704	King
OPS 6	ValleyComm – Training & Secondary Response		TG: 38736	King

ValleyCom Fire Departments / Stations / Apparatus

11	Renton	211 Mill Ave S	E11, L11, HM11, A11, Air 11, E111
12	Renton	1209 Kirkland Ave NE	E12, A12, Boat 12, BC12, E121
13	Renton	17040 108th Ave SE	E13, A13, E131
14	Renton	1900 Lind Av SW	E14, M5, BC14, E141
16	Renton	12923 156th Ave SE	E16, A16
18	North Highline	SW 112th St	E18, A18, L18, E181
19	North Highline		E19, A19, Air 19
21	Skyway	84th Ave S, Bryn Mawr	E24, E25, A23, Air 20
22	Skyway	76th Ave S, Skyway	E22, A22
26	Des Moines	S 223rd St	E26, L26, A26, E261, Marine 26
28	Burien	8th Ave SW	E28, A28, S28, E281
29	Burien	S Normandy Rd	E29, L29, A29, E291
31	Auburn	D St	E31, A31, L31, M6, U31, E311
32	Auburn	R St & 22nd SE	E32, A32, E311
33	Auburn	C Street SW	E33, E331, A33, L33, R33, HM33
34	Auburn	17th & West Valley Hwy	E34
39	Pacific	3rd St SE	E39, A39, E38, Trk39, A38
41	Spring Glen	Petrovitsky Rd SE, SpgGln	E41, M41, A41, E411, BC41
42	Spring Glen	176th St SE, Renton	E42, M42, A42, BC42, E421
45	SeaTac	2929 S 200th St, Angle Lk	E45, A45, BC45, S45, L441, E451
46	SeaTac	3521 S 170th St, McMicken	E46, A46, E461
47	SeaTac	3215 S 152nd St & 32nd Av	E47, M4
51	Tukwila	Andover Park E	E51, E511, S51
52	Tukwila	147th St S	E52, HM52
53	Tukwila	S 115th St	E53, R53, E531
54	Tukwila	144th St S	E54, L54, A54, A541
55	Vashon Island	SW Bank Rd	E55, E12, M55
56	Vashon Island	SW Burton Rd, Burton	E56, WT56
57	Vashon Island	131st Ave SW, Tahlequah	E57
58	Vashon Island	99th Ave SW, Dockton	E58, WT58, M58
59	Vashon Island	Vashon Hwy SW, Vashon Ht	M59
61	Federal Way	S 360th St, Five Mile Lake	E61, E611, HM61, BC61
62	Federal Way	1st Ave S, Woodmont	E62, L62, A62, U62
63	Federal Way	21st Ave SW, Twin Lakes	E63, R63, E631
64	Federal Way	S 320th St	E64, HM64, M8, A64
65	Federal Way	S 298th St	E65, E651
66	Federal Way	15th Ave S, Star Lake	E66, A66, M13
71	Kent	504 W Crow	E71, A71, Boat, E711, BC71
72	Kent	25620 140th Av SE	E72, A72, WT72
73	Kent	26512 Military Rd S	E73, A73, Mobile Command, BC73
74	Kent	24611 116th Ave SE	L74, E74, A74, Air 74, BR74, Boat
75	Kent	15635 SE 272nd St	E75, A75, MCI 75, M11, BC75
76	Kent	20676 72nd Ave S	E76, HM76, M7, E761, BC76
77	Kent	20717 132nd Ave SE	E77, E771, BC77
80	Maple Valley	23775 SE 264th St	E80, U80
81	Maple Valley	22225 231st St SE	E81, A81, BC81 BR81 Air 81 WT81
82	Maple Valley	27519 Kent/Kangley Rd SE	E82, A82, R82, BR82
83	Maple Valley	27251 216th Ave SE	E83, A83, Air 83
84	Maple Valley	16855 194th Ave SE	E84, A84, E87
85	Maple Valley	27605 SE 208th St/276th Av	E85, R85
88	Palmer Selleck	268th St SE, Selleck	E88, A88, WT88
89	Palmer Selleck	309th St SE, Palmer	E89, A89, R89, BC89
91	Mountain View	124th Ave SE, Lea Hill	E91, A91, E911, S91, BC91
92	Mountain View	Kent Black Diamond Rd	E92, A92, E921, BR92
93	Mountain View	Covington Sawyer Rd SE	E93, A93, BC93
94	Mountain View	Mountain View	E94, A94
95	Mountain View	148th Ave SE	E95, A95, E951, WT95, BR95, R95
96	Mountain View	Mountain View	E95, A95, WT95, BR95, R95, E951

97	Mountain View	Mountain View	E97, A97
98	Black Diamond	SE 296th St	E98, A98, BR98, S98, M12
99	Black Diamond	Baker St	E99, A99, E991, BR99, S99

Vashon Island Fire: Self Dispatched, but see ValleyCom for unified list

Washington State Patrol

Note: Not all WSP units or dispatchers transmit PL tone at all times

District 2: King County

Ch3	North King County	[PL 100.0]	Countywide	155.5800	158.7900
Ch7	South King County	[PL 100.0]	Countywide	154.6800	Simplex
Ch12	Districtwide Car to Car	[PL 100.0]	Countywide	155.4600	Simplex
Ch22	King County Mutual Aid		Countywide	155.1900	154.6500

Weyerhauser Emergency Alert Countywide 158.2800

Weyerhauser Fire Services Countywide 153.7700

White Center Fire: See ValleyCom

Woodinville Law Enforcement: Contract to King County Sheriff Precinct 2

Sheriff Patrol - Northeast	Countywide	TG: 23152	King
TAC 02 - Tactical Northeast Precinct	Countywide	TG: 23184	King

Woodinville Fire & Life Safety: See King Count East Fire Dispatch

153.7700 154.1900 154.2050 154.2500 154.4000 154.4300

Yarrow Point Law Enforcement: Contracted to Clyde Hill Police

Airports

Not all of the following airlines operate at airports in and around Seattle, but the following air operations frequencies can be heard in and around Seattle.

Action Aviation Incorporated		Renton	151.8050	Simplex
Aeroflight Executive Air Service		Seattle	131.7500	AM
Aeroflot Russian Int'l Airlines		Seattle	467.1375	Simplex
Operations Simplex		Seattle	469.2125	Simplex
Air Operations		Seattle	128.8250	AM
Aeronautical Radio Inc		Seattle	460.7750	Simplex
Operations Simplex		Seattle	460.8500	Simplex
San Francisco Radio		Seattle	129.4000	AM
San Francisco Radio		Seattle	129.8500	AM
Airborne Express / DHL Cargo		Seattle	131.0000	AM
Aircraft Service International		Seattle	452.3000	Simplex
Operations Repeater		Seattle	452.7500	457.7500
Air Canada		Seattle	129.9000	
Air Operations – United Contract		Seattle	129.0000	AM
Airlift Northwest		Seattle	129.8250	AM
Airpac Airlines		Seattle	128.9000	AM
Alaska Airlines Incorporated		Seattle	461.7625	466.7625
Ground Operations	[PL 100.0]	Seattle	460.7000	465.7000
Ground Operations	[PL 100.0]	Seattle	460.7750	465.7750
461.3125 466.3125				
Air Operations – Departing Flights		Seattle	130.8750	AM
Air Operations – Arriving Flight		Seattle	131.2000	AM
Ground Operations		Seattle	452.6750	457.6750
America West Airlines		Seattle	460.6750	AM
CactusNet Air Operations		Seattle	130.9250	AM
Air Operations		Seattle	129.4500	AM
American Airlines		Seattle	460.6500	
Ground Operations		Seattle	460.6750	465.6750
Ch1 Ground Operations	[PL 114.8]	Seattle	461.6375	Simplex
Aircraft Maintenance Channel		Seattle	462.8625	Simplex
Air Operations		Seattle	129.2250	AM
Low-Power Simplex Frequencies		Seattle	466.6375	Simplex
461.9125 466.9125 467.8625 463.7125 468.7125				
463.8375 468.8375 464.2375 469.2375				
American Trans Air (ATA)		Seattle	129.6000	AM
Asiana Airlines		Seattle	131.6500	AM
Auburn Municipal Airport		Auburn	122.8000	AM
Aviation West Corporation		Seattle	461.7375	Simplex
British Airways		Seattle	464.4500	469.4500
Air Operations		Seattle	130.0500	AM
Air Operations (Vancouver)		Seattle	130.2000	AM
Canada 3000 Airlines		Seattle	129.0000	AM
Cargolux Air Cargo		Seattle	129.7250	AM
Cathay Pacific Airways		Seattle	128.9000	AM
Civil Air Patrol	[PL 127.3]	Seattle	148.1500	143.9000
Continental Airlines Inc		Seattle	460.7000	465.7000
Operations Repeater		Seattle	460.8500	465.8500
Operations Repeater		Seattle	461.2500	466.2500

Name	PL	Location	Freq	Mode/Input
Air Operations		Seattle	129.9250	AM
Crest Airpark		Kent	123.0000	AM
Delta Air Lines Inc	[PL 136.5]	Seattle	460.7000	
Ground Operations	[PL 114.8]	Seattle	460.7500	465.7500
Air Operations		Seattle	130.1000	AM
DHL Airways Air Cargo		Seattle	131.6250	AM
EVA Air		Seattle	131.4250	AM
Evergreen Air Freight		Seattle	131.7250	AM
Federal Express		Seattle	131.9250	AM
Frontier Airlines		Seattle	130.8250	AM
Ground Operations		Seattle	460.6750	465.6750
Kenmore Seaplane Base		Seattle	122.7000	AM
King County International Airport / Boeing Field			122.9500	AM
Airport Administration		Seattle	TG: 25712	King
Airport Command		Seattle	TG: 25744	King
Airport Information (ATIS)		Seattle	127.7500	AM
Boeing Operations		Seattle	123.5500	AM
Boeing Flight Testing		Seattle	227.8000	AM
291.8000 341.6000 349.6000				
Ground Control		Seattle	121.9000	AM
Tower – North & East		Seattle	118.3000	257.8000
Tower – South & West		Seattle	120.6000	257.8000
Approach / Departure		Seattle	119.2000	284.7000
Approach / Departure		Seattle	120.1000	290.9000
Approach / Departure		Seattle	120.4000	395.0000
Approach / Departure		Seattle	125.9000	290.9000
Approach / Departure		Seattle	126.5000	391.9000
Clearance Delivery		Seattle	132.4000	AM
Kitty Hawk Air Cargo		Seattle	130.2500	AM
Lester State Ultralight Flight Park		Lester	122.9000	AM
Hawaiian Airlines Inc		Seattle	460.8000	
Air Operations – Northwest Contract		Seattle	131.7000	AM
Horizon Airlines Inc	[PL 192.8]	Seattle	452.1000	457.1000
Air Operations		Seattle	129.3250	AM
Air Operations		Seattle	130.7250	AM
Ground Operations		Seattle	460.8000	Simplex
Seattle Maintenance		Seattle	132.0000	AM
Ground Operations		Seattle	451.4500	456.4500
Northwest Airlines Inc		Seattle	460.7500	
Operations Repeater	[PL 192.8]	Seattle	460.8750	465.8750
Air Operations		Seattle	131.7000	AM
Operations Repeater		Seattle	452.5250	457.5250
Operations Repeater		Seattle	463.8250	468.8250
Renton Municipal Airport		Renton	124.7000	256.9000
ATIS – Terminal Information		Renton	126.9500	AM
Ground Control		Renton	121.6000	256.9000
UNICOM		Renton	122.9500	AM
Scandinavian Airlines		Seattle	130.4500	AM

Seattle Tacoma Int'l Airport (Sea Tac)

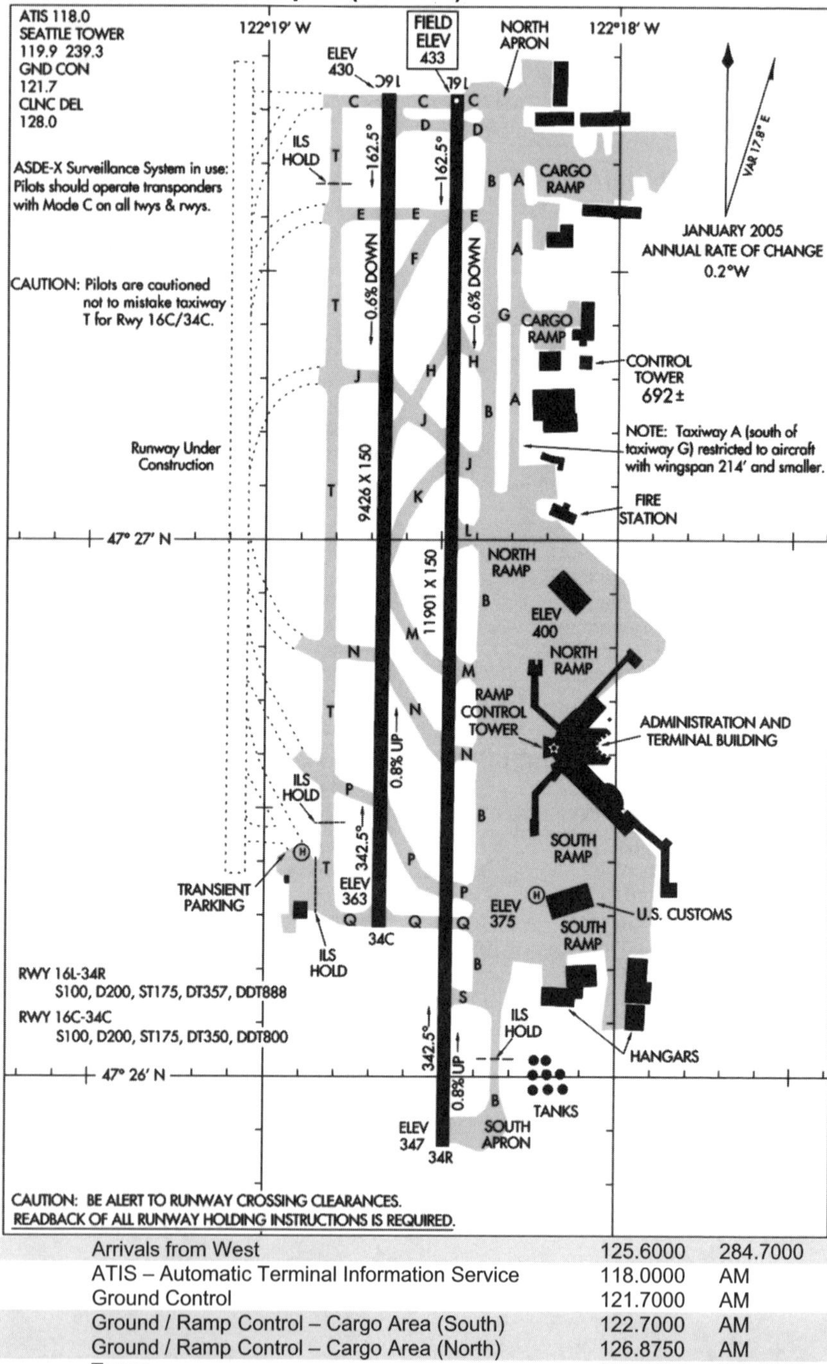

Arrivals from West	125.6000	284.7000
ATIS – Automatic Terminal Information Service	118.0000	AM
Ground Control	121.7000	AM
Ground / Ramp Control – Cargo Area (South)	122.7000	AM
Ground / Ramp Control – Cargo Area (North)	126.8750	AM
Tower	119.9000	239.3000
Approach / Departure – North/East/Southeast (Rwy16)	119.2000	284.7000

Name	Location	Frequency	Frequency/Mode
Approach / Departure – Southwest/West		120.1000	290.9000
Approach / Departure – Northwest (Rwy 34)		120.4000	395.0000
Approach / Departure – SE (Rwy 34) / NW (Rwy 16)		125.9000	290.9000
Approach / Departure – South		126.5000	391.9000
Approach / Departure		123.9000	269.1250
Approach / Departure		338.2000	AM
Clearance Delivery & Pre-Taxi Clearance		128.0000	AM
Gate Control		126.2500	AM
UNICOM		122.9500	AM
Alternate Frequency As Assigned		133.6500	AM
AWOS – Automatic Weather Observation		121.2000	AM
Seattle Seaplanes Seaplane Base	Seattle	122.9000	AM
Skykomish State Airport	Skykomish	122.9000	AM
Skywest Airlines	Seattle	460.6500	465.6500
Southwest Airlines Company	Seattle	460.6500	465.6500
Operations Simplex	Seattle	464.0875	Simplex
Operations Simplex	Seattle	469.0875	Simplex
Air Operations	Seattle	130.8500	AM
Operations Repeater	Seattle	462.4250	467.4250
Operations Repeater	Seattle	464.9250	469.9250
Sun Country Airlines	Seattle	128.8250	AM
Swissport Contract Airline Services	Seattle	452.6250	457.6250
Fueling Operations	Seattle	457.1375	Simplex
451.3875 456.3875 452.0625	Seattle	457.0625	Simplex
United Airlines Inc	Seattle	464.4750	469.4750
Ground Operations [PL 203.5]	Seattle	460.7250	465.7250
Low Power Simplex Frequencies	Seattle		
461.0375 466.0375 461.0625 466.0625 461.0875			
466.0875 461.1375 466.1375 461.2375 466.2375			
461.2625 466.2625 461.2875 466.2875 461.3125			
466.3125 461.6875 466.6875 461.7125 466.7125			
461.8125 466.8125 462.8875 467.8875			
Air Operations	Seattle	129.5000	AM
Air Operations	Seattle	131.0500	AM
	Seattle	451.0875	456.0875
	Seattle	451.9625	456.9625
United Parcel Service (UPS)	Seattle	129.4250	AM
US Air Incorporated	Seattle	464.4500	469.4500
Air Operations	Seattle	131.1500	AM
Vashon Municipal Airport	Vashon	122.9000	AM
Will Rogers Wiley Seaplane Base	Renton	124.7000	AM
UNICOM	Renton	122.8000	AM

Hotels, Resorts, Casinos & Country Clubs

Name	Location	Frequency	Mode / Input
Bellevue Athletic Club	Bellevue	464.4125	Simplex
	Bellevue	464.1125	Simplex
Bellevue Convention Center	Bellevue	453.0375	Simplex
458.0375 453.2125 458.2125 453.6125 458.6125			
Bell Harbor Int'l Conference Center			
Catering Operations [DPL 371]	Seattle	461.2125	Simplex
Kitchen Operations [DPL 627]	Seattle	462.3875	Simplex
Security Personnel [DPL 134]	Seattle	461.1125	Simplex
Best Western Bellevue Inn	Bellevue	461.0625	Simplex
	Bellevue	461.2125	Simplex
Broadmoor Golf Club	Seattle	151.6850	Simplex
Casino Snoqualmie	Snoqualmie	463.6375	Simplex
463.8875 468.6375	Snoqualmie	468.8875	Simplex
Crown Plaza Hotel	Seattle	461.2500	466.2500
461.7500 462.3250 464.4750			
Doubletree Hotels			
Doubletree Plaza Hotel	Kent	461.1750	466.1750
Doubletree Plaza Hotel	Kent	462.0125	Simplex
Doubletree La Posada	Bellevue	451.0875	456.0875
Doubletree La Posada	Bellevue	451.9875	456.9875
461.9375 462.2625 466.9375	Bellevue	467.2625	Simplex
Druid Glenn Golf Course	Covington	151.6550	Simplex
	Covington	151.6850	Simplex
Econo Lodge	Seattle	151.8050	Simplex
Emerald Downs	Auburn	462.1250	Simplex
Foster Golf Links	Seattle	462.7875	Simplex
	Seattle	463.9125	Simplex
Four Seasons Olympic Hotel	Seattle	462.7500	Simplex
Hilton Hotels			
Seattle Airport Hilton	Seattle	151.6850	Simplex
Bellevue Hilton - Paging	Bellevue	465.0000	Simplex
Seattle Hilton	Seattle	462.1875	Simplex
467.1875 462.7875 451.2750 451.5750 452.1250 467.7875			
Holiday Inn Hotels	Seattle	464.3250	
Hotel Monaco	Seattle	452.2250	457.2250
Operations Repeater	Seattle	452.4000	457.4000
Operations Simplex	Seattle	461.6250	Simplex
462.2000	Seattle	463.2250	Simplex
Hotel Vintage Park	Seattle	461.0625	Simplex
	Seattle	463.9125	Simplex
Hyatt Regency Bellevue	Bellevue	464.3750	469.3750
	Bellevue	464.8750	Simplex
Inglewood Country Club	Seattle	151.9250	Simplex
	Seattle	151.9550	Simplex
Leisure Time Resort	Seattle	468.4125	Simplex
	Seattle	469.0875	Simplex
Magic Lanes Casino	Seattle	464.5625	Simplex
	Seattle	469.5625	Simplex
Marriott Hotels	Seattle	464.6750	
Residence Inn Hotel - Seattle	Seattle	154.5700	Simplex

	Residence Inn Hotel - Kent	Kent	461.1625	Simplex
	Residence Inn Hotel - Kent	Kent	464.6125	Simplex
	Residence Inn Hotel – Redmond	Redmond	463.9625	Simplex
	Seattle Marriott Hotel	Seattle	464.7750	
	Courtyard Federal Way	Federal Way	452.1500	Simplex
	Courtyard Federal Way	Federal Way	457.1500	Simplex
Mayflower Park Hotel		Seattle	452.8750	Simplex
	451.3250 451.5000 451.6250 456.6250 463.3250			
Meany Tower Hotel		Seattle	469.4375	
Mercerwood Shore Club		Mercer Island	464.4000	
		Mercer Island	464.9250	
Meridian Valley Country Club		Kent	461.5125	Simplex
	Security	Kent	463.6750	Simplex
	Golf Pro-Shop	Kent	463.5625	Simplex
	466.5125 461.6875 466.6875 461.8625 466.8625			
	461.9125 466.9125 461.9625 466.9625			
Muckleshoot Indian Casino				
Ch1	Housekeeping Operations [PL 67.0]	Auburn	467.7625	Simplex
Ch2	Engineering Operations [PL 79.7]	Auburn	467.8125	Simplex
Ch3	Food & Beverage [PL 123.0]	Auburn	467.7625	Simplex
Ch4	Information Technology [PL 131.8]	Auburn	467.8125	Simplex
Ch5	Security (Old) – Primary [PL 179.9]	Auburn	467.7625	Simplex
Ch6	Security (Old) – Tactical [PL 179.9]	Auburn	467.8125	Simplex
Ch7	Slot Machines [PL 167.9]	Auburn	467.7625	Simplex
Ch8	Shipping & Receiving [PL 173.8]	Auburn	467.7625	Simplex
Ch9	Engineering Working Ch. [DPL 731]	Auburn	464.3625	Simplex
	Paging	Auburn	467.8500	Simplex
Ch1	Security – Primary [PL 151.4]	Auburn	453.1625	458.1625
Ch2	Security – Parking [PL 151.4]	Auburn	453.6125	458.6125
Ch3	Security – Tactical [PL 151.4]	Auburn	453.9375	458.9375
Ch4	Security – 1 Direct [PL 151.4]	Auburn	453.1625	Simplex
Ch5	Security – 2 Direct [PL 151.4]	Auburn	453.6125	Simplex
Ch6	Security – 3 Direct [PL 151.4]	Auburn	453.9375	Simplex
	451.8250 451.8626 451.9375	Auburn	452.3500	457.3500
	452.7000 452.8500 452.7000	Auburn	452.6500	457.6500
	464.5750 464.2500 452.1250	Auburn	452.8000	457.8000
	464.7625 464.8375 452.4625	Auburn	452.3375	457.3375
	Gaming Commission	Auburn	154.9800	Simplex
	Gaming Commission	Auburn	154.0550	Simplex
North Coast Hotel		Seatac	464.3250	
Paramount Hotel		Seattle	468.4500	Simplex
Radisson Hotel Sea Tac		Seattle	464.5250	
	Seattle Airport Radisson Hotel	Seattle	451.7500	456.7500
Red Lion Hotels		Seatac	463.2875	Simplex
	468.2875 463.3125 468.3125 464.7625			
	469.7625 464.7875 467.1125			
Renaissance Madison Hotel		Seattle	463.2500	468.2500
Sea Tac Motor Inn		Seattle	464.5750	
Seattle Executive Inn		Seattle	464.5250	
Seattle Golf Club		Seattle	456.6625	Simplex
	456.6875 456.7125 456.7375	Seattle	457.0375	Simplex
Seattle Inns Inc		Seattle	464.3250	
Seattle Keiro		Seattle	463.3875	Simplex
Seattle Marriott Hotel		Seattle	464.7750	Simplex

Sheraton Hotels

Seattle Sheraton Hotel - Operations	Seattle	464.8250	469.8250
Seattle Sheraton Hotel - Paging	Seattle	465.0000	
Hotel & Towers Operations	Seattle	151.6550	Simplex

154.5400 154.5700 154.6000

Seattle Tennis Club	Seattle	463.3875	Simplex
461.9750 466.9750 464.4125	Seattle	463.6625	Simplex
Seattle Trade Center	Seattle	461.3125	Simplex
	Seattle	464.7125	Simplex
	Seattle	466.0375	Simplex
Silvercloud Hotel at the Stadiums	Seattle	452.1625	Simplex
452.6625 452.7125 457.6625	Seattle	457.7125	Simplex
Sorrento Hotel	Seattle	464.4125	Simplex
469.4125 467.3875 467.4125	Seattle	462.2125	Simplex
Stouffer Madison Hotel	Seattle	464.6750	
	Seattle	463.2500	
Summit at Snoqualmie Ski Resort	Snoqualmie	151.6550	Simplex

151.6850 151.7150 151.7450 151.7750 151.8050
151.8350 151.8650 151.8950 154.5150 154.5400
155.2050 155.2350 155.2650 155.2950

Operations Repeaters	Snoqualmie	152.3150	157.5600

462.0250 462.0500 462.0750

Todays Hotel Seattle Corp	Seattle	464.4750	
Tulalip Casino	Tulalip	451.4625	Simplex

452.1625 452.6375 452.6875

Twin Rivers Golf Club	Fall City	151.7150	Simplex
University Inn	Seattle	461.1000	
Washington Athletic Club	Seattle	462.8000	Simplex
Washington St Conv & Trade Center	Seattle		
Security Primary [PL 100.0]	Seattle	453.2875	Simplex
Operations	Seattle	453.2625	Simplex

463.2125 468.2125 463.7375 468.7375 463.8375
468.8375 468.2625 468.5125 469.3125

West Coast Hotels

Plaza Park Suites	Seattle	464.3250	
West Coast Sea Tac Hotel	Seatac	464.8250	
Grande Hotel	Seattle	461.0500	466.0500

451.8625 452.2375 456.8625 457.2375

Vance Hotel	Seattle	451.2375	Simplex

451.4375 451.7125 452.0875 464.6375

West Seattle Golf Course	Seattle	462.2125	Simplex
	Seattle	467.2125	Simplex
Westin Hotel	Seattle	462.8750	Simplex
Security [PL 218.1]	Seattle	464.6000	Simplex

464.5250 464.7750 464.9250 451.8750 452.0250

Willows Run Golf Course	Redmond	451.3000	456.3000
Wyndham Garden Hotel	Seattle	463.7875	Simplex

468.7875 464.4375 469.4375

Malls & Retail Stores

Bellevue Place		Bellevue	464.7750	469.7750
		Bellevue	464.8750	469.8750
LTR Trunked Radio System				
451.7250 451.8250 451.9625		Bellevue	463.5750	
Bellevue Square Mall		Bellevue	464.3750	469.3750
	[DPL 065]	Bellevue	464.7750	Simplex
Security Operations	[DPL 065]	Bellevue	464.4750	Simplex
464.9250		Bellevue	452.6250	Simplex
LTR Trunked Radio System		Bellevue	452.9750	457.9750
451.9375 452.2125 452.6250		Bellevue	463.6000	468.6000
JC Penney Store		Bellevue	461.6125	Simplex
451.2375 451.3375 451.4375		Bellevue	451.5375	Simplex
JC Penney Store		Bellevue	452.0875	Simplex
GAP Store		Bellevue	467.8500	Simplex
467.8750 467.9000		Bellevue	467.9250	Simplex
Macy's Department Store		Bellevue	462.2750	467.2750
Macy's Department Store		Bellevue	461.6500	Simplex
Macy's Department Store		Bellevue	466.6500	Simplex
Nordstrom Department Store		Bellevue	460.9125	Simplex
460.9125 463.4375 468.4375 463.5625 468.5625				
463.8375 468.8375 464.0875 469.0875 464.2625				
469.2625 464.6125 465.9125 469.6125				
Computer City Stores		Tukwila	461.4125	Simplex
466.0875 467.1375		Tukwila	468.8375	Simplex
Corporate Express		Seattle	461.7875	Simplex
463.3875 463.5875		Seattle	464.9125	Simplex
Costco Wholesale Warehouses		Fife	151.6550	Simplex
151.7150 151.7750		Fife	151.8050	Simplex
151.8350 151.8650		Fife	154.6000	Simplex
Fred Meyer Stores		Seattle	461.3250	
Home Depot Stores		Countywide	461.2125	Simplex
463.2875		Countywide	464.6375	Simplex
Northgate Mall		Seattle	461.2000	466.2000
Mall Management / Security		Seattle	464.7750	469.7750
Macy's Department Store		Seattle	451.7875	456.7875
Macy's Department Store		Seattle	461.5750	466.5750
JC Penney Store		Seattle	461.6125	Simplex
JC Penney Store		Seattle	464.0125	Simplex
JC Penney Store		Seattle	466.1375	Simplex
Target Store (across road) – Ch1		Seattle	461.0375	Simplex
Target Store (across road) – Ch2		Seattle	466.2875	Simplex
Target Store (across road) – Ch3		Seattle	469.4875	Simplex
Target Store (across road) – Ch4		Seattle	462.9125	Simplex
Pacific Place Mall		Seattle	452.4250	457.4250
Ch1 Security		Seattle	451.3500	456.3500
Ch2 Parking Garage Operations		Seattle	453.3625	Simplex
REI Store		Seattle	461.6375	466.6375
SeaTac Mall	[PL 100.0]	Federal Way	464.7750	469.7750
JC Penney Store		Seatac	461.6125	Simplex
Southcenter Mall	[DPL 043	Tukwila	462.1125	467.1125
Nordstrom Store	[PL 146.2]	Tukwila	460.9125	Simplex
South Seattle Mall				

			Location	Frequency	Output
	Macy's Department Store		Federal Way	464.4750	469.4750
Supermall of the Great Northwest			Auburn	466.7625	Simplex
Ch1	Management	[PL 127.3]	Auburn	462.0125	Simplex
Ch4	Housekeeping	[PL 71.9]	Auburn	461.7625	Simplex
Ch5	Security	[PL 127.3]	Auburn	462.8625	Simplex
			Auburn	463.5375	468.5375
			Auburn	463.9625	468.9625
			Auburn	464.5875	469.5875
	463.3875 463.4875 463.5625		Auburn	463.3625	Simplex
	463.8875 464.3625 467.7875		Auburn	464.0875	Simplex
	467.0125 468.3625		Auburn	469.0875	Simplex
	GAP Store		Auburn	467.8500	Simplex
	467.8750 467.9000		Auburn	467.9250	Simplex
Town Center Associates			Renton	461.0125	Simplex
University Village Shopping Center			Seattle	463.7375	468.7375
			Seattle	464.8375	469.8375
			Seattle	463.4875	Simplex
Westlake Center			Seattle	464.8875	Simplex
	Security		Seattle	464.9125	Simplex
	469.8875 469.9125 469.9375		Seattle	464.9375	Simplex
	GAP Store (Westlake area)		Seattle	467.8500	Simplex
	467.8750 467.9000		Seattle	467.9250	Simplex
	Macy's Store (Westlake area)		Seattle	461.5750	466.5750

Media

			Location	Frequency	Output
American Production Services			Seattle	173.2250	Simplex
			Seattle	173.3250	Simplex
Classic Radio Inc			Issaquah	450.1125	Simplex
			Issaquah	450.2250	455.2250
EZ New Orleans / Seattle Inc			Issaquah	450.5625	455.5625
			Issaquah	450.9250	455.9250
	Operations Simplex		Issaquah	450.1625	Simplex
	450.2750 450.3500 450.3750 450.9000				
Hearst Newspapers			Seattle	452.9750	
KBSG 97.3 FM Traffic		[PL 100.0]	Seattle	450.7750	Simplex
KCPQ FOX-13 TV		[PL 141.3]	Seattle	450.1875	455.1875
KCTS Public Television			Seattle	450.2875	
KING Radio FM 98.1		[PL 156.7]	Seattle	161.7300	Simplex
KING-5 NBC Television		[PL 156.7]	Seattle	450.6125	Simplex
KIRO 7 Television			Seattle	151.6550	Simplex
	News Chopper		Seattle	161.6400	Simplex
	Operations Simplex	[PL 110.9]	Seattle	161.7000	Simplex
	Operations Repeater	[PL 110.9]	Seattle	863.0375	818.0375
KJR Radio Incorporated			Issaquah	450.6375	
			Seattle	450.9500	
KOMO 4 ABC Television		[PL 146.2]	Seattle	161.7600	Simplex
	Air Patrol	[PL 146.2]	Seattle	450.4125	455.4125
KSTW UPN-11 Television		[PL 100.0]	Seattle	450.4875	455.4875
Metro Networks Communications					
	Traffic Reporting Ops	[PL 141.3]	Seattle	450.6875	455.6875
			Seattle	455.3125	
Seattle Times			Seattle	173.3250	

Operations Repeater	[PL 151.4]	Issaquah	452.2250	457.2250
Operations Repeater	[PL 88.5]	Seattle	453.0000	458.0000
Operations Repeater		Seattle	464.3250	469.3250
Operations Simplex		Seattle	464.0125	469.0125

468.5125 452.9625 452.9875 457.9625 457.9875

Tribune Television		Seattle	450.9800	455.9800

Medical & Hospitals

	4x4 Rescue Council	[PL 127.3]	Seattle	045.9600	
	Future Repeater Ops	[PL 123.0]	Seattle	047.6600	045.9600
	Operations Simplex	[PL 123.0]	Seattle	155.1750	
	47.6600 Link	[PL 162.2]	Seattle	462.9875	467.9875
	4X4 SIMPLEX	[PL 162.2]	Seattle	467.9375	
	Aircom Systems Inc		Seattle	157.4500	
	American National Red Cross		Countywide	150.7750	Simplex
	Future Repeater	[PL 123.0]	Countywide	047.6200	045.9200
F1	Operations Simplex		Countywide	047.4200	Simplex
F2	Operations Simplex	[PL 123.0]	Countywide	047.6200	Simplex
F3	Operations Simplex		Countywide	047.5000	Simplex
F4	Operations Simplex		Countywide	047.4600	Simplex
F5	Operations Simplex	[PL 127.3]	Countywide	045.9200	Simplex
	Links to Low Band Freqs	[Varies]	Countywide	463.6375	468.6375
	Auburn General Hospital		Auburn	463.2500	468.2500
	Operations		Auburn	465.8875	Simplex
		[PL 107.2]	Auburn	155.3400	
	Hospital Operations		Auburn	TG: 42960	King
	Auburn Regional Medical Center		Auburn	155.3400	Simplex
			Auburn	460.5625	465.5625
	Capitol Hill Family Health Center		Seattle	464.2250	469.2250
			Seattle	464.3750	469.3750
	Childrens Hospital / Medical Center		Seattle	151.9250	Simplex
		[PL 94.8]	Seattle	155.3400	Simplex
	Hospital Operations		Seattle	TG: 60880	King
	Engineering Ops		Seattle	464.9500	
			Seattle	461.2250	
			Seattle	464.5750	
	Operations Repeaters		Seattle	151.4900	Repeater

151.5950 151.8950 151.9550 461.5375
461.7375 462.0125 463.2875

Orthopedic Hospital		Seattle	464.9500	469.9500
Community Memorial Hospital		Enumclaw	155.3400	
Critical Air Medicine Inc		Issaquah	453.1750	
Eastside Hospital		Redmond	464.3750	469.3750
Enumclaw Community Hospital		Enumclaw	TG: 60912	King
Operations Repeater		Enumclaw	464.0000	469.0000
Evergreen Hospital Medical Center		Kirkland	461.1750	
	[PL 110.9]	Kirkland	155.3400	
		Kirkland	TG: 60944	King

462.0375 466.4876 466.8125 467.0375

German Shepard Search Dogs				
Operations Simplex	[PL 110.9]	Kirkland	155.2200	Simplex
Operations Simplex	[PL 118.8]	Kirkland	155.2650	Simplex
Good Samaritain Hospital		Puyallup	TG: 61520	King

Agency / Talkgroup	PL	Location	Frequency / TG	Input / Mode
Group Health Central / Cooperative		Seattle	464.2250	
	[PL 123.0]	Seattle	155.3400	Simplex
	[PL 103.5]	Seattle	464.3750	469.3750
	[PL 107.2]	Tiger Mtn	464.6500	469.6500
	[DPL 134]	Bellevue	464.3750	469.3750
	[PL 186.2]	Countywide	464.3875	469.3875
	[PL 192.8]	Countywide	464.3750	469.3750
		Issaquah	463.6500	
		Seattle	464.3750	
Central Campus Operations		Seattle	TG: 61008	King
Eastside Campus Operations		Issaquah	TG: 60976	King
Harborview Medical Center		Seattle	152.0075	Simplex
		Seattle	155.2800	Simplex
HEAR – Amb. to Hospital	[PL 100.0]	Seattle	155.3400	Simplex
		Seattle	463.5875	
Operations		Seattle	TG: 61040	King
Administration		Seattle	TG: 35696	King
Operations – 2		Seattle	TG: 35856	King
Operations – 3		Seattle	TG: 35888	King
Operations – 4		Seattle	TG: 35920	King
Operations – 5 – Special Projects		Seattle	TG: 35792	King
Security Personnel		Seattle	TG: 35728	King
Tactical		Seattle	TG: 35760	King
		Seattle	935.6375	896.6375
		Seattle	936.7250	897.7250
Highline Community Hospital		Seattle	035.6400	Simplex
		Burien	155.3400	Simplex
		Burien	464.3250	469.3250
		Burien	453.1750	Simplex
		Burien	464.6750	Simplex
		Burien	TG: 61072	King
Riverton Campus		Riverton	TG: 61200	King
Hospitals / E.O.C. Calling		Countywide	TG: 61744	King
King County EMS				
Administration		Countywide	TG: 25168	King
OPS-1 Operations		Countywide	TG: 25200	King
OPS-2 Operations		Countywide	TG: 25232	King
Hutchinson Cancer Research Center		Seattle	461.3500	466.3500
		Seattle	461.5000	466.5000
		Seattle	461.9250	466.9250
King County Exp. Search & Rescue		Skykomish	047.4600	Simplex
		Countywide	047.5000	Simplex
Intercom / 47.4600 LINK	[PL 141.3]	Countywide	463.0625	468.0625
47.5000 LINK	[PL 162.2]	Countywide	463.0625	468.0625
King County SAR Dogs	[PL 123.0]	Seattle	155.2350	Simplex
Madigan Army Hospital		Fort Lewis	TG: 61616	King
Mary Bridge Hospital		Tacoma	TG: 61648	King
HEAR	[DPL 134]	Tacoma	155.3400	Simplex
Northwest Hospital	[PL 141.3]	Seattle	155.3400	Simplex
466.7375 469.5875	469.7375	Seattle	464.4750	469.4750
		Seattle	TG: 61104	King
Overlake Hospital Medical Center		Bellevue	043.6800	Simplex
Security		Bellevue	452.1500	457.1500
HEAR – Amb. to Hospital	[PL 82.5]	Bellevue	155.3400	Simplex
		Bellevue	TG: 61136	King
		Bellevue	462.3000	467.3000

Name	PL	Location	Frequency	Input/Type
		Bellevue	461.8750	466.8750
		Bellevue	464.5250	469.5250
		Bellevue	463.9750	468.9750
Pacific Medical Center		Seattle	151.8650	Simplex
155.2800 155.3400 157.4500				
Pacific Northwest Trackers Assn		Seattle	155.2800	Simplex
Pacific Regent Healthcare		Bellevue	461.1750	
Providence Medical Center		Seattle	TG: 61168	King
Mount Saint Vincent		Seattle	462.7500	Simplex
Hospital Engineering		Seattle	464.2750	Simplex
Hospital Operations		Seattle	464.5250	Simplex
Regional Hospital Command / Coord		Countywide	TG: 61680	King
Riverton General Hospital	[PL 151.4]	Seattle	155.3400	Simplex
Saint Clare Hospital		Issaquah	461.2750	
Saint Francis Community Hospital		Federal Way	TG: 61264	King
HEAR	[PL 162.2]	Federal Way	155.3400	Simplex
466.2625 466.9750 467.1375		Federal Way	462.4500	Simplex
Operations Repeater		Federal Way	464.8000	469.8000
Saint Joseph's Hospital		Tacoma	TG: 61552	King
Seattle Explorer SAR	[PL 123.0]	Countywide	047.4600	Simplex
Seattle Mountain Rescue		Countywide	463.1625	468.1625
	[PL 100.0]	Countywide	155.2050	Simplex
	[PL 100.0]	Countywide	155.1600	Simplex
Shannon Ambulance	[PL 131.8]	Seattle	155.2800	Simplex
Snoqualmie Valley Hospital		Snoqualmie	TG: 61232	King
Stevens Memorial Hospital		Edmonds	TG: 61488	King
Swedish Hospital Medical Centers		Seattle	453.0750	
Security	[PL 103.5]	Seattle	464.8750	469.8750
Swedish Hospital - HEAR	[PL 79.7]	Seattle	155.3400	
Swedish Hospital - Ballard		Ballard	TG: 61296	King
Swedish Hospital - Seattle		Seattle	TG: 51328	King
453.1250 461.0250 463.2625 463.7250 464.1125				
469.1125 464.1500 464.3500 464.6500 464.5250				
Swedish Hospital at Issaquah		Issaquah	462.2125	Simplex
462.4125 451.7125		Issaquah	462.3375	Simplex
University Medical Center - HEAR		Seattle	TG: 61360	King
HEAR	[PL 85.4]	Seattle	155.3400	Simplex
Valley Medical Center		Renton	464.8250	Simplex
HEAR	[PL 100.0]	Renton	155.3400	Simplex
HEAR		Renton	TG: 61424	King
Security	[PL 118.8]	Renton	464.5250	
453.0750 452.2125 452.3875 154.5400				
Veteran's Hospital - HEAR		Seattle	155.3400	Simplex
		Seattle	TG: 61392	King
Virginia Mason Hospital		Seattle	464.9250	469.4250
463.5250 463.5500 464.8000		Seattle	461.7750	466.7750
463.3000 463.2500		Seattle	461.3500	Simplex
HEAR		Seattle	TG: 61456	King
HEAR	[PL 167.9]	Seattle	155.3400	Simplex
		Seattle	453.0250	
Washington Exp. S & R	[PL 123.0]	Issaquah	047.5000	Simplex

Miscellaneous

Amtrak Station			Seattle	160.2750	160.7550
Aviation Methods Inc			Seattle	460.8875	Simplex
	461.4375 466.4375 466.8875				
Bill & Melinda Gates Foundation			Seattle	464.6500	Simplex
Elliott Bay Marina Inc			Ballard	151.8350	Simplex
			Ballard	154.6000	Simplex
Emerald Harbor Marine Inc			Seattle	463.3000	
Harbor Island Terminal					
Ch22	Emergency Incidents		Seattle	463.6000	Simplex
Ch23	Gate Security		Seattle	463.5250	Simplex
Ch32	Transportation Supervisor		Seattle	463.9500	Simplex
Ch33	Rail Supervisors		Seattle	463.9750	Simplex
Ch46	Security – Primary		Seattle	469.5500	Simplex
Ch47	Security – Tactical		Seattle	464.9250	Simplex
Holaday Parks Inc			Seattle	464.1375	
			Seattle	465.8875	
King County Metro Buses					
F-1	Bus Voice		Seattle	452.6500	457.6500
F-2	Bus Voice		Seattle	452.7250	457.7250
F-3	Bus Voice		Seattle	453.3750	458.3750
F-4	Bus Voice		Seattle	453.5250	458.5250
F-5	Data-1 AVL/Misc Data		Seattle	452.3750	457.3750
F-6	Data-2 AVL/Misc Data		Seattle	452.8000	457.8000
F-7	Supervisor-1 Police & Maintenance		Seattle	452.2750	457.2750
F-8	Supervisor-2 Police & Maintenance		Seattle	452.3500	457.3500
	Operations Simplex		Seattle	457.3250	Simplex
Metropolitan Improvement District – Downtown Seattle Management					
	Operations Repeater		Seattle	452.6750	457.6750
	Operations Simplex		Seattle	457.6625	Simplex
	457.6875 457.7125		Seattle	457.7375	Simplex
Microsoft Corporation			Redmond	451.5250	Simplex
1-1	SEC-1	[PL 103.5]	Redmond	451.3125	456.3125
1-2	SEC-2	[PL 114.8]	Redmond	451.3375	456.3375
1-3	FAC-1	[PL 127.3]	Redmond	451.4125	456.4125
1-4	FAC-2	[PL 141.3]	Redmond	451.4625	456.4625
1-5	AV	[PL 156.7]	Redmond	451.5125	456.5125
1-6	ITG-1	[PL 173.8]	Redmond	451.5625	456.5625
1-7	ITG-2	[PL 192.8]	Redmond	451.6125	456.6125
1-8	DSG	[DPL 152]	Redmond	451.6375	456.6375
1-9	TRN-1	[PL 210.7]	Redmond	461.1125	466.1125
1-10	TRN-2	[PL 110.9]	Redmond	463.3375	466.3375
2-1	COMN	[PL 110.9]	Redmond	462.4500	467.4500
2-2	DIR-1	[PL 203.5]	Redmond	452.1375	Simplex
2-3	DIR-2	[PL 203.5]	Redmond	457.1375	Simplex
2-4	DIR-3	[PL 203.5]	Redmond	452.4125	Simplex
2-5	DIR-4	[PL 203.5]	Redmond	457.4125	Simplex
2-6	SEC-3	[PL 94.8]	Redmond	451.6625	456.6625
3-1	CH-01	[PL 100.0]	Redmond	464.5000	Simplex
3-2	CH-02	[PL 100.0]	Redmond	466.2375	Simplex
3-3	CH-03	[PL 114.8]	Redmond	464.5375	469.5375
3-4	CH-04	[PL 100.0]	Redmond	461.4125	466.4125
3-5	CH-05	[PL 203.5]	Redmond	461.7125	466.7125
3-6	CH-06 Security	[PL 173.8]	Redmond	463.7625	468.7625

3-7	CH-07	[PL 210.7]	Redmond	461.1125	466.1125
3-8	CH-08	[PL 162.2]	Redmond	463.5625	468.5625
3-9	CH-09	[PL 173.8]	Redmond	461.9375	466.9375
3-10	CH-10	[PL 110.9]	Redmond	462.4500	467.4500
4-1	CH-11	[PL 173.8]	Redmond	464.5000	Simplex
4-2	CH-12	[PL 173.8]	Redmond	466.2375	Simplex
4-3	CH-13	[PL 110.9]	Redmond	463.3375	468.3375
4-4	WID-1	[DPL 054]	Redmond	451.4000	456.4000
4-5	WID-2	[DPL 054]	Redmond	451.6750	456.6750

462.2000 462.4500 463.3375 468.3375 463.7625
468.7625 464.4875 469.4875 464.5375 469.5375
464.6875 469.6875 464.8375 469.8375 464.9375

Tukwila	Tukwila	452.6875	457.6875
452.0375	Tukwila	452.3125	Simplex

National Ski Patrol System	North Bend	151.7750	Simplex
Northwest Film Group Inc	Seattle	463.6875	Simplex
	Seattle	467.8625	Simplex
	Seattle	467.8875	Simplex
Southern Marine Inc	Seattle	154.6000	Simplex

Washington Motor Escort

F1	Operations Repeater	[PL 107.2]	Seattle	461.0750	466.0750
F2	OPS-1 Operations Rptr.	[PL 107.2]	Seattle	461.0750	461.0750
F3	OPS-2 Operations Rptr.	[PL 107.2]	Seattle	466.5875	466.5875
F4	OPS-3 Operations Rptr.	[PL 107.2]	Seattle	466.8375	466.8375
F5	OPS-4 Operations Rptr.		Seattle	464.8500	464.8500
F6	Operations Repeater	[PL 131.8]	Seattle	461.0500	466.0500

Recreation

5th Avenue Theatre	Seattle	461.1500	Simplex

461.1750 461.2000 461.4000 462.0750 463.2000

A Contemporary Theatre	Seattle	151.6550	Simplex
	Seattle	151.7750	Simplex
Alpental Volunteer Ski Patrol	Snoqualmie P.	160.1100	Simplex
Bellevue Convention Center	Bellevue	453.0375	Simplex

453.2125 453.6125 458.0375 458.2125 458.6125

Burke Museum	Seattle	451.8625	Simplex
	Seattle	451.9375	Simplex
	Seattle	451.9875	Simplex

Experience Music Project Museum

Operations Repeater			Seattle	461.4875	466.4875
Operations Simplex			Seattle	452.7500	
451.5000	451.7750	451.9250	Seattle	452.1250	
452.3000	452.3250	452.4000	Seattle	452.4250	
452.5000	462.2250	462.2500	Seattle	462.2750	
462.3500	462.5000		Seattle	462.5250	

Frye Art Museum	Seattle	461.5625	Simplex
	Seattle	466.7125	Simplex
Full House Sports & Entertainment	Seattle	463.6875	Simplex

464.4875 464.8250 464.8625 466.1875
467.8125 468.2125 468.7375

Fun Forest Amusements	Seattle	461.6625	Simplex
	Seattle	462.1625	Simplex
GES Exposition Services	Seattle	462.0875	Simplex

		Seattle	464.2875	Simplex
		Seattle	464.9125	Simplex
Girl Scouts - Totem Council		Preston	463.8250	
		Carnation	464.8750	469.8750
Henry Art Gallery		Seattle	464.3250	469.3250
		Seattle	464.7750	469.7750
Museum Of Flight		Seattle	463.3500	Simplex
Ch2 Operations	[DPL 031]	Seattle	462.0875	Simplex
Ch3 Education	[DPL 031]	Seattle	462.1500	Simplex
Facilities		Seattle	462.0250	467.0250
Kitchen	[DPL 043]	Seattle	464.6250	Simplex
462.0250 462.0875 462.1500 463.2500				
463.5250 464.6250 464.7250				
Museum of History & Industry		Seattle	463.7375	468.7375
Pacific Science Center Museum				
Operations Primary		Seattle	461.4625	466.4625
Operations Secondary		Seattle	466.4625	Simplex
Qwest Field (Seattle Seahawks Venue)		Seattle		
TRUNKED RADIO SYSTEM (Unknown): TrunkTracker Setting: Unknown				
451.4500 451.8375 452.1250 452.7500 461.0750				
National Football League		Seattle	462.2125	Simplex
462.4125 456.5875 467.2125		Seattle	467.4125	Simplex
467.4875 462.2625 462.4625		Seattle	452.3125	Simplex
467.2625 467.4625 457.3125		Seattle	462.2875	Simplex
462.4875 452.4875 467.2875		Seattle	451.2875	Simplex
457.4875 452.7875 462.3125		Seattle	456.2875	Simplex
452.6375 457.7875 467.3125		Seattle	451.3625	Simplex
457.6375 452.8875 462.3375		Seattle	456.3625	Simplex
452.6875 457.8875 467.3375		Seattle	451.4125	Simplex
457.6875 452.9875 462.3625		Seattle	456.4125	Simplex
457.9875 467.3625 451.5375		Seattle	462.1875	Simplex
467.3875 451.5875				
Safeco Field / Seattle Mariners Baseball				
Operations Repeater		Seattle	452.0750	457.0750
Operations Repeater		Seattle	462.4500	467.4500
Operations Repeater		Seattle	464.2000	469.2000
Operations Simplex		Seattle	452.1500	Simplex
452.2500 452.7000 461.2250		Seattle	461.6750	Simplex
461.8000 463.2000 462.5500		Seattle	462.6750	Simplex
Operations Simplex		Seattle	462.9750	Simplex
Security		Seattle	TG: 24912	Seattle
Seafair Boat Club		Seattle	154.6000	Simplex
Seafair Summer Festival – Hydroplace Race, Air Show and Events				
Repeater		Seattle	464.8125	469.8125
Repeater		Seattle	463.2000	468.2000
Repeater		Seattle	464.7125	469.7125
Simplex		Seattle	464.2625	Simplex
464.3625 464.9375 469.7625		Seattle	464.3125	Simplex
469.1625 469.0375 469.2625		Seattle	469.3125	Simplex
469.3625 469.7375 469.7875		Seattle	469.8125	Simplex
469.8625 469.9125		Seattle	464.0375	Simplex
Seattle Art Museum		Seattle	461.5750	Simplex
Seattle Center		Seattle	453.9375	Simplex
453.9625 453.9875 458.9875				
Seattle Opera		Seattle	461.2625	Simplex
466.7375 469.7375 466.8625 469.8625 469.7375				

Seattle Symphony Orchestra / Benaroya Hall			466.5875	Simplex
Operations Repeater		Seattle	464.3750	469.3750
Operations Simplex		Seattle	463.9500	Simplex
Seattle Woodland Park Zoo		Seattle	TG: 5520	Seattle
Seattle Yacht Club		Seattle	469.0875	Simplex
469.5375 469.6125 469.9125				
Six Flags Enchanted Park		Enumclaw	464.3000	469.3000
461.3375 463.2625 463.3625 463.6625				
466.3375 468.2625 468.3625 468.6625				
Stadium Exhibition Center		Seattle	451.8750	456.8750
		Seattle	452.1750	457.1750
		Seattle	452.9750	457.9750
White River Amphitheater		Auburn	463.5500	Simplex
464.2000 464.2500 464.4000 464.8000				
YMCA Of Greater Seattle		Seattle	461.4375	Simplex
		Seattle	461.9625	Simplex

Schools

Auburn School District 408		Auburn	TG: 43568	King
Buses and Security	[PL 146.2]	Auburn	151.8350	Simplex
466.0625 466.1375 466.2125 466.2875 466.3625				
466.4375 466.5125 466.5875 466.6625 466.7375				
466.8125 466.8875 467.8875 462.8875 151.7450				
Bellevue Christian School		Medina	461.8500	Simplex
		Medina	463.5500	Simplex
		Medina	464.4250	Simplex
Bellevue Community College		Bellevue	453.6750	458.6750
Security/Operations	[PL 85.4]	Bellevue	453.2750	
		Bellevue	464.1375	Simplex
Bellevue School District 405		Bellevue	463.8500	
		Bellevue	463.8750	
Common Talkgroup		Bellevue	TG: 17776	EPSCA
Group Calls		Bellevue	TG: 17584	EPSCA
Maintenance Personnel		Bellevue	TG: 17680	EPSCA
School Net - 1		Bellevue	TG: 17648	EPSCA
School Net - 2		Bellevue	TG: 17616	EPSCA
Security / Warehouse		Bellevue	TG: 17712	EPSCA
Transportation		Bellevue	TG: 17744	EPSCA
Blanchet High School		Bellevue	466.5875	Simplex
		Bellevue	467.1625	Simplex
Bush School, The		Seattle	451.6125	Simplex
Carbonado School District 19		Carbonado	151.8050	Simplex
Cornish College		Seattle	461.8750	466.8750
Enumclaw School District 216				
Buses	[DPL 261]	Enumclaw	155.2650	Simplex
Federal Way School Dist 210		Federal Way	TG: 35120	King
Security	[PL 82.5]	Federal Way	151.9250	Simplex
Buses	[PL 114.8]	Federal Way	154.5150	Simplex
Operations Repeater	[PL 141.3]	Federal Way	463.9125	468.9125
Forest Ridge School of the Sacred Heart			451.7375	Simplex
Security & Maintenance		Bellevue	462.3375	Simplex

Agency	PL	Location	Frequency	Mode
Security & Maintenance		Bellevue	462.3625	Simplex
Green River Community College		Auburn	151.6850	Simplex
		Auburn	151.7150	Simplex
Highline Community College		Seattle	155.2200	
Security		Des Moines	464.8750	469.8750
		Des Moines	464.3500	
		Des Moines	451.2250	456.2250
Highline Public Schools	[PL 136.5]	Seattle	154.5150	
		Seattle	852.5875	807.5875
Maintenance		Seattle	TG: 35376	King
Security		Seattle	TG: 35152	King
Transportation		Seattle	TG: 35344	King
Issaquah School District		Issaquah	TG: 35184	King
School Busses	[PL 167.9]	Issaquah	463.3500	468.3500
468.3375 469.2125 469.4375		469.6375	467.9250	Simplex
Schools Security		Issaquah	464.1250	469.1250
John F. Kennedy High School			153.3950	Simplex
			154.4900	Simplex
Kent School District 415	[PL 114.8]	Kent	154.5400	Simplex
Operations		Kent	TG: 43600	King
159.6000 159.7800 159.9600		160.1400		
Kent-Meridian High School		Kent	453.3375	Simplex
453.7875 458.3375 458.7875				
Security		Kent	TG: 43728	King
School Operations	PL 114.8]	Kent	154.6000	Simplex
Transportation 1 – Primary		Kent	TG: 43792	King
Transportation 2 – Alternate		Kent	TG: 43824	King
King Schools		Seattle	155.2350	
Lake Washington School District				
Buses	[DPL 125]	Kirkland	151.6550	Simplex
Mercer Island School District		Mercer Island	463.3750	
North Seattle Community College		Seattle	463.8750	468.8750
461.0500 461.1250 462.4250		463.6500		
Northshore School District 417		Bothell	151.6550	
151.7150 151.7450 151.8350 151.8950				
Buses	[PL 151.4]	Bothell	155.2650	Simplex
Renton School District 403				
Buses	[DPL 346]	Renton	151.9550	Simplex
Buses	[DPL 346]	Renton	151.9850	Simplex
		Renton	TG: 43632	King
Simplex		Renton	462.2125	Simplex
467.2125 462.2875 467.2875		Renton	462.3625	Simplex
467.3625 462.4125 462.4625		Renton	467.4625	Simplex
Renton Technical College		Renton	451.6250	456.6250
Riverview School District 407		Carnation	151.8950	Simplex
		Carnation	TG: 35216	King
451.5000 451.7875 451.8875		451.9375	451.9625	
Seattle Academy		Seattle	452.3250	Simplex
		Seattle	457.3250	Simplex
Seattle Central Community College		Seattle	151.6250	Simplex
		Seattle	151.6400	Simplex
Security		Seattle	461.6000	466.6000
Engineering		Seattle	464.9000	469.9000
		Seattle	151.5350	159.9750

Agency / Use		Location	Frequency	Input / Mode
Seattle Pacific University		Seattle	151.6850	Simplex
		Seattle	151.8350	Simplex
Operations Repeater		Seattle	451.7500	456.7500
462.0375 463.8625 464.0375 467.0375 469.0375				
Seattle School District 1		Seattle	151.7750	Simplex
Administration		Seattle	TG: 8784	Seattle
Common		Seattle	TG: 8816	Seattle
Operations		Seattle	TG: 8720	Seattle
Buses		Seattle	TG: 8848	Seattle
Security		Seattle	TG: 8752	Seattle
Contract Bus Service		Seattle	461.3000	466.3000
Seattle University		Seattle	463.6875	Simplex
Security		Seattle	461.8250	466.8250
Security F-1		Seattle	462.6500	
Shoreline Community College		Seattle	464.1500	Simplex
	[PL 110.9]	Seattle	151.8050	Simplex
Security F-1	[PL 110.9]	Seattle	151.9250	Simplex
		Seattle	464.9750	
Operations Repeater		Seattle	451.3125	456.3125
Operations Repeater		Seattle	451.4125	456.4125
Operations Repeater		Seattle	451.5125	456.5125
451.9750 464.4250 464.7500 469.4250 469.7500				
Shoreline School District		Seattle	151.7750	Simplex
452.7125 457.7125				
Shorewood High School		Shoreline	466.1875	Simplex
468.5875 469.0875 469.5375				
Showalter Middle School		Seattle	464.3250	Simplex
464.3750 464.4250 464.8250				
Skykomish School District		Skykomish	TG: 35248	King
Snoqualmie Valley School District				
Buses	[DPL 043]	Snoqualmie	151.8050	Simplex
		Snoqualmie	TG: 35280	King
South Central School District		Seattle	TG: 43664	King
South Seattle Community College		Seattle	151.5800	Simplex
159.8250 159.9450 160.0950		Seattle	160.1700	Simplex
		Seattle	160.1850	Simplex
Tahoma School District		Maple Valley	151.6850	Simplex
Buses	[DPL 261]	Maple Valley	151.7750	Simplex
		Maple Valley	TG: 43696	King
151.7450 151.8650		Maple Valley	153.0200	160.1100
Tukwila School District	[DPL 246]	Tukwila	159.4950	Simplex
University Of Washington				
Plant Operations		Seattle	154.5150	Simplex
Plant Operations		Seattle	154.6000	Simplex
Plant Operations		Seattle	155.0400	Simplex
Computer Science Lab		Seattle	151.6250	Simplex
Applied Physics Lab		Seattle	151.6250	Simplex
Operations Simplex		Seattle	469.6875	Simplex
Family Housing		Seattle	151.6850	Simplex
Operations		Seattle	453.3375	458.3375
Health Sciences		Seattle	467.7625	Simplex
467.8125 467.8500 467.8750 467.9000			467.9250	Simplex
Maintenance	[PL 162.2]	Seattle	453.6500	458.6500
Qwest Field Radios		Seattle	453.0125	Simplex

Qwest Field Radios | Seattle | 458.0125 | Simplex

TRUNKED RADIO SYSTEM (Unknown): TrunkTracker Setting: Unknown
453.0500 453.9500 460.4500

Unknown Use Simplex | Seattle | 462.4375 | Simplex
462.4625 467.5125 451.3375 | Seattle | 469.5125 | Simplex
451.4625 451.7375 452.0875 | Seattle | 452.1375 | Simplex
452.5125 452.6875 464.7125 | Seattle | 463.8250 | Simplex
Seattle | 462.8500 | Simplex

Vashon Island School District 322 | Vashon Isl. | 461.5875 | Simplex
463.8875 464.0125 466.8375
Vashon Isl. | TG: 35312 | King

Vashon Island School District 402 | Vashon Isl. | 151.7450 | Simplex
Vashon Isl. | 151.8950 | Simplex
Vashon Isl. | 160.4250 | Simplex

Zion Preparatory Academy | Seattle | 452.6625 | Simplex
456.4625 451.4625 452.0625 | Seattle | 457.6625 | Simplex

Security Services

Aronson Security Group | Seattle | 451.4125 | Simplex
451.6875 451.9625 | Seattle | 452.1625 | Simplex

Agency	Tone	Location	Frequency	Input/Mode
Eagle Guard Service Inc		Issaquah	462.1000	467.1000
Guardian Security Systems		Seattle	462.0375	Simplex
Hi-Way Prevention Patrol	[PL 131.8]	Enumclaw	464.3000	469.3000
North County Alarm/Patrol	[PL 167.9]	Seattle	154.5400	Simplex
Northwest Protective Svc.	[PL 156.7]	Issaquah	463.9250	468.9250
Pacific Rim Security	[PL 107.2]	Federal Way	464.1000	469.1000
Pierce County Security		Enumclaw	452.0000	457.0000
Pinkerton Security & Investigation		Humphrey	464.7000	469.7000
Puget Sound Security Patrol		Issaquah	461.7000	466.7000
Tactical Response Invest. & Patrol		Seattle	464.2375	Simplex
466.5625 466.8875 469.7625				
WA Merchant Patrol	[PL 210.7]	Federal Way	461.5500	466.5500
Western Security Patrol	[DPL 546]	Kent	462.1250	466.1250
Valley Security Patrol	[PL 103.5]	Issaquah	461.9000	466.9000

Public Utilities

Cedar River Water District	Cedar River	TG: 41840	King
Coal Creek Utility District	New Castle	045.5600	
Highline Water District	Highline	TG: 33648	King
King County Water District #11	Countywide	TG: 41904	King
Mutual Aid - North	Countywide	TG: 63344	King
Mutual Aid - South	Countywide	TG: 63376	King
King County Water District 42	Seattle	047.9000	
King County Water District No 90	Renton	048.2200	
Lakehaven Utility District	Federal Way	158.1900	
158.8200 151.5950 152.9670			
Northshore Utility District	Kirkland	158.1900	
	Bothell	158.8950	
Puget Sound Energy	Issaquah	451.0500	456.0500
Conventional Frequencies	Countywide	153.4100	Simplex

153.4400 153.6050 153.6200 153.6500 153.6800
158.1300 158.1450 173.3500 451.0500 451.2500

TRUNKED RADIO SYSTEM (EDACS): TrunkTracker Setting: ED

Sites at Auburn and High Point

935.4500 935.4625 935.4750 935.4875 935.5000
935.9250 935.9500 935.9625 935.9750 936.0000
936.3875 936.4000 936.4125 936.4250 936.4375
936.4500 936.4625 936.4750 936.4875 936.5000
937.3875 937.4000

Seattle Municipal Utilities	Seattle	451.1750	
Seattle City Light	North Bend	852.0375	
	North Bend	852.7875	
	Seattle	851.2875	
	Seattle	853.5375	
Shoreline Wastewater Management	Seattle	155.7450	
Southwest Suburban Sewer District	Seattle	045.0800	Simplex
	Seattle	045.5200	Simplex
Tanner Electric Cooperative	North Bend	047.9600	Simplex
Val Vue Sewer District	Riverton	045.0800	Simplex
	Riverton	045.5200	Simplex
Washington Natural Gas Company			
(part of Puget Sound Energy)	Seattle	048.1400	Simplex
	Seattle	048.1800	Simplex
Woodinville Water District	Woodinville	154.0850	Simplex

Weather Information

National Weather Service – NOAA Weather Radio

Conditions & Forecasts – Seattle	162.5500	Simplex
Conditions & Forecasts – Puget Sound Marine	162.4250	Simplex

King County SAME Alert Code: 053033

KITSAP COUNTY

Government & Public Safety

Bainbridge Island City Government		Bainbridge	158.8425	153.7925
Public Works Department	[PL 203.5]	Bainbridge	153.8750	158.8425
Bainbridge Island Fire	[DPL 023]	Bainbridge	033.7000	033.0600
Tac-3 Fireground	[DPL 047]	Bainbridge	033.6200	Simplex

Fire Station Locations & Apparatus

21	Bainbridge Island	8895 Madison Ave N	E21, M21, A21, WT21, WT21A, R21
22	Bainbridge Island	7934 NE Bucklin Rd	E22, A22, WT22
23	Bainbridge Island	12985 Phelps Rd	E23, A23, WT23

Bainbridge Island Police: See Sheriff North Dispatch

Bremerton City Government				
Public Works	[PL 110.9]	Bremerton	155.8950	154.8900
Water Resources Division		Bremerton	453.8875	458.8875
Housing Authority		Bremerton	156.0750	Simplex
Bremerton Fire Department	[PL 110.9]	Bremerton	154.1450	Simplex

Fire Station Locations & Apparatus

1	Bremerton	817 Pacific Ave & 8th	E1, A1, L1, DC, Boat
1	Bremerton	5005 Kitsap Way/Shorewood	E2, M2, E2A, M2A
1	Bremerton	3031 Olympus Dr & 31st	E3, M3, E3A

Bremerton Police Department

Ch1	Operations Repeater	[PL 110.9]	Bremerton	155.7750	153.8150
Ch2	Car to Car + Public Works	[PL 110.9]	Bremerton	155.8950	154.8900

Unit Identifications

Charles	Central	King	K-9 Unit	Tom	Traffic
David	Downtown	Nora	North	William	West
Edward	East	Robert	Roving		
George	Gang Force	Sam	South		

A 'four' preceeds the letter designator. Four is the Unit Designator for all Bremerton Police Units. Numbers after the letters are employee numbers.

Disposition Codes

A Arrest Made	J F.I.R. Written	Q Unable to Locate
B Criminal Citation Issued	K Other Report Made	R No Action Needed
C No Arrest Made	M No Report / Unfounded	S Location Not Found
D Evidence Submitted	N Non-crim. Referral	T Transportation Provided
H Followup Report Made	O Oral Warning	U Assistance Rendered
I Non-criminal Citation	P Public Order Restored	V Cancelled by Radio

German Shepherd Search Dogs		Countywide	155.2650	Simplex
Kitsap County Fire	[PL Varies]	Countywide	154.3700	
TAC-2 County Fire	[PL Varies]	Countywide	153.7700	
TAC-3 County Fire	[PL 110.9]	Countywide	159.0150	
Red Net		Countywide	153.8300	Simplex
Vehicle Repeaters	[PL 192.8]	Countywide	453.3125	458.3125

Kitsap Fire District 4 – Suquamish

41	Meadowvale	Old Military Rd	E41, WT41, R41, M41, A41
42	Island Lake	Central Valley Rd NE	E42, R42
43	Brownsville	Ogle Rd NE	E43, A43
44	Tracyton	NW Tracy Ave	E44, A44
45	North Perry	Trenton Ave NE	E45

Kitsap Fire District 1 – Silverdale

51	Silverdale	Silverdale Way NW	E51, L51, R51, M51, A56, M51A
52	Olympic View	Olympic View Rd NW	E52
53	Seabeck	Seabeck Hwy NW	E53
54	Hintzville	Hintzville Rd NW	E54
55	Lake Tahuya	Kingsway NW	E55
56	Seabeck Nicholas	Seabeck Hwy NW	E56, M56, A56, WT56, E56A

North Kitsap Fire & Rescue

81	Kingston	26642 Miller Bay Rd SE	E81, M81, BR81, PIO81, Marine
82	North Kitsap	Bond Rd NE	E82
83	North Kitsap	Indianola Rd NE	E83, WT89
84	Suquamish	18533 Augusta Ave NE	E84, A84, WT84
85	S Kingston	12651 NE Marine View Dr	E85, WT85
86	(Port Gamble FD)	31941 Little Boston Rd NE	
88	North Kitsap	Hwy 104	E88
89	Hansville	4900 NE Twin Spats Rd	E89, A89, WT89

Poulsbo Fire Department

71	Poulsbo (HQ)	911 NE Liberty Rd	E71, M71, WT71, BR71, U71
72	Hood Canal	28882 Falkner Rd NE	E72, A/M 72
73	Keyport	1863 NE Pacific Ave	E73, A/M 73, BR73
77	Pioneer Hill		E77, A77, WT77

South Kitsap Fire & Rescue

6	Sunnyslope	6000 SW Rhododendron	E6, WT6
7	Wye Lake	7750 Sylvan St	E7, WT7
8	Orchard Heights	1974 Fircrest Drive SE	E8, A8, M8
9	Yukon Harbor	7433 SE Mile Hill Dr	E9, Air 9
10	Banner	5629 Banner Rd SE	E10, A10, WT10
11	Port Orchard	6258 Bethel Rd SE	E11, WT11, Brush 11, Duty Chief
12	Olalla	8696 Olalla Valley Rd SE	E12, WT12, BR12
13	Nelson	5770 SE Nelson Rd	E13, WT13
14	Burley	12685 Bethel Burley Rd SE	E14, M14
15	Minterbrook	13591 Glenwood Rd SE	E15, WT15

16	Gorst	4045 W State Route 3	E16, A16, M16
17	Glenwood	6560 Glenwood Rd SE	E17, A17, WT17
18	Bremerton Airport	8850 State Route 3	E18, Crash 18
19	Navy Yard City	400 S. Charlotte	E19, Mass Casualty Unit
20	Rocky Point	1550 Rocky Point Rd	E20

Kitsap County Government Countywide 453.2375 458.2375
453.4625 453.6125
TRUNKED RADIO SYSTEM (Motorola): TrunkTracker Setting: Type II 800
867.7375 868.2750 868.5500 868.8000
Public Works [PL 71.9] Countywide 867.6625
453.7750 868.3000 868.8500 867.9125
Storm Water / Flood Control Countywide 155.9550 453.7750
Wastewater Services Countywide 153.9650 155.7450
Unknown Use Countywide
158.9700 156.1650 460.0750 460.1750
460.4000 460.4250 460.4500 460.5250
155.9250 155.5350 155.6850 154.9950
155.0250

Kitsap County Search & Rescue Countywide 155.2050 Simplex
Explorer Search & Rescue Countywide 155.2350 Simplex

Kitsap Public Utilities [PL 100.0] Countywide 155.9550 153.9650

Kitsap County Alternate Communications System
Amateur Radio Emrg. Svc. [PL 110.9] Countywide 145.4300 144.8300

Kitsap Transit Authority Bremerton 453.9875 458.9875
TRUNKED RADIO SYSTEM TrunkTracker Setting: Type 1 Preset 12 (Block 7: Size 4)
852.5375 853.2875 854.0375

[A]	Routed (Regular Routed Bus Service)	Countywide	TG: 700-01
[B]	Route (Private Communications)	Countywide	TG: 700-02
[C]	Worker/Driver Buses	Countywide	TG: 700-03
[D]	Access/School Buses	Countywide	TG: 700-04
	Access Channel A – Dispatch	Countywide	TG: 8464
	Access Channel B – Private Call	Countywide	TG: 8496

Kitsap County Sheriff
Sheriff South [PL 110.9] Countywide 158.8800 153.9200
South also dispatches Port Orchard Police
Sheriff North [PL 110.9] Countywide 155.4300 158.9700
North also dispatches Poulsbo, Bainbridge Island, Suquamish, Little Boston
Bainbridge Island Police Tactical Bainbridge 158.8200 Simplex
Bainbridge Island Police Tactical Bainbridge 155.6700 Simplex
Sheriff Tactical [PL 110.9] Countywide 154.9650 153.9950

Kitsap County Sheriff Unit Identifications: Unit Designator / Patrol Area

The numbers following the Unit Designator is the officers employee number.
All Tri/Cities units begin with a numeric designator for area: Poulsbo (6), Port Orchard (7), Bainbridge Island (8), Suquamish Reservation (9), Little Boston Reservation (11).

Charles	Westpark (BPD 0100-1700 / KCSO from 1700-0100)	John	Southeast	Robert	Roving Units – usually designated areas at start of shift
Edward	East	King	K-9 Unit	Sam	Sergeant
Frank	Silverdale	Lincoln	North, NE, NW	Tom	Traffic
George	Southwest	Nora	North		
Henry	West	Ocean	Extra Security Assignments, School Security		

Patrol Areas

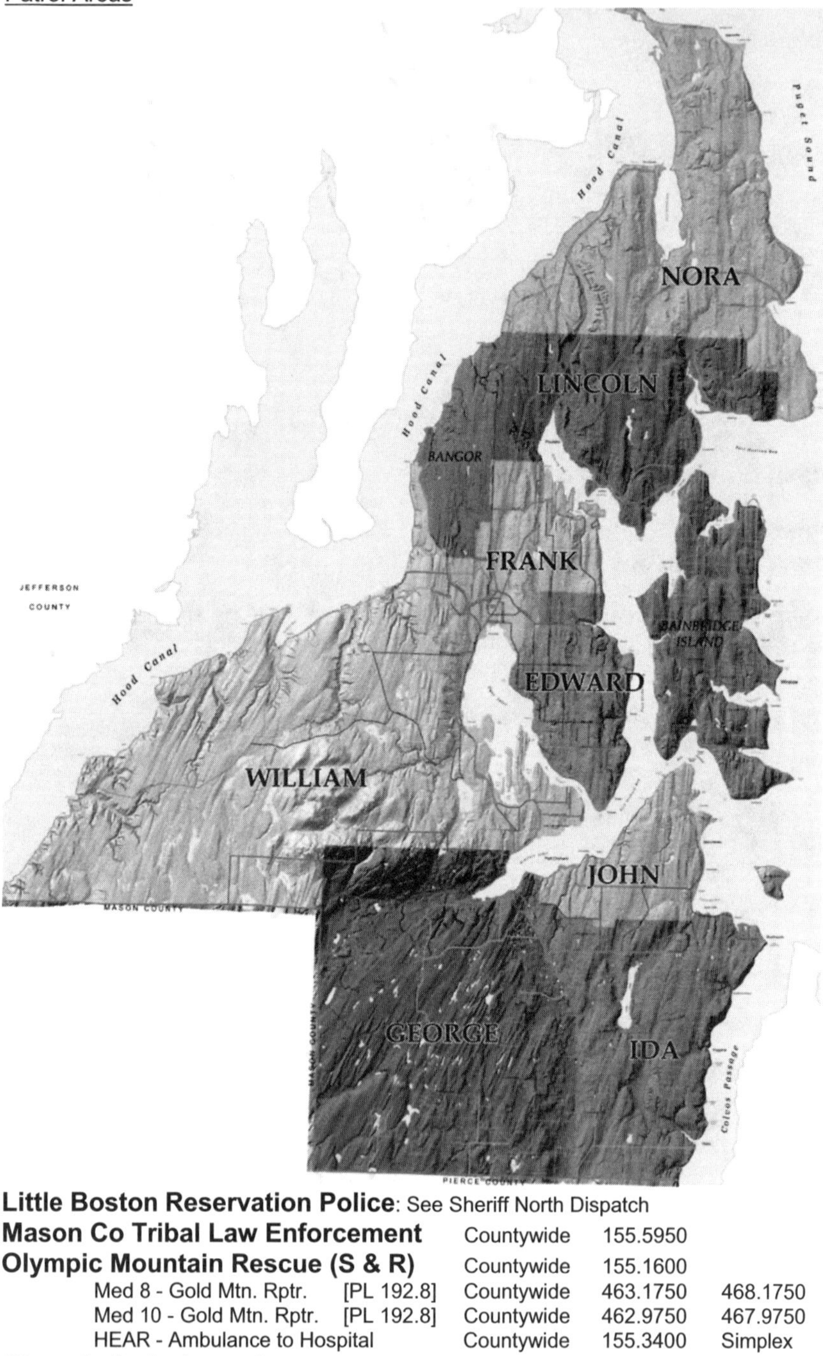

Little Boston Reservation Police: See Sheriff North Dispatch				
Mason Co Tribal Law Enforcement		Countywide	155.5950	
Olympic Mountain Rescue (S & R)		Countywide	155.1600	
Med 8 - Gold Mtn. Rptr.	[PL 192.8]	Countywide	463.1750	468.1750
Med 10 - Gold Mtn. Rptr.	[PL 192.8]	Countywide	462.9750	467.9750
HEAR - Ambulance to Hospital		Countywide	155.3400	Simplex
Olympic Ambulance Inc	[PL 192.8]	Bremerton	462.9500	467.9500
OSCCR		Countywide	156.1350	

Agency		Location	Freq	Freq 2
Port of Bremerton		Port Orchard	154.0250	
Port of Kingston		Kitsap	453.3375	458.3375
Port Orchard Public Works	[PL 110.9]	Port Orchard	154.9400	Simplex
Port Orchard City Police		Port Orchard	156.0300	159.1500
Also see Sheriff South Dispatch				
Tactical		Port Orchard	156.2400	Simplex
Tactical (+ Public Works)	[PL 110.9]	Port Orchard	154.9400	Simplex
Poulsbo City Government		Poulsbo	156.0150	
		Poulsbo	155.9550	
Poulsbo City Police		Poulsbo	156.0150	
Also see Sheriff North Dispatch				
Red Net Mutual Aid		Countywide	153.8300	
Suquamish Reservation Police: Also see Sheriff North Dispatch				
Tribal Law Enforcement Agencies		Countywide	158.7300	
Washington Dept. of Nat. Resources		Countywide	151.4150	
Washington State Patrol				
Note: Not all WSP units or dispatchers transmit PL tone at all times				
Ch15 Kitsap / Mason Counties	[PL 110.9]	Countywide	154.6650	158.8950
Ch1 Statewide / Aircraft Ops	[PL 100.0]	Statewide	155.9700	159.0750
Winslow Police Department		Bainbridge Isl.	158.8200	Simplex

Military

Bangor Naval Submarine Base

Home of: Submarine Group 9, Trident Training & Refit Facilities, Seabee Unit 418

Channel	Location	Freq	Freq 2
Base Operations – Repeater	Silverdale	140.6250	142.0500
Base Operations – Simplex	Silverdale	140.5750	Simplex
Crane Operations	Silverdale	138.5750	Simplex
Fire Department	Silverdale	140.8250	Simplex
Hospital Communications	Silverdale	143.6000	Simplex
Marine Operations – Repeater	Silverdale	141.9750	142.6250
Marine Operations – Simplex	Silverdale	140.1500	Simplex
Marine Operations – Simplex	Silverdale	140.2000	Simplex
Marine Operations – Simplex	Silverdale	141.9500	Simplex
Naval Investigative Service (NIS)	Silverdale	140.0750	Simplex
Naval Investigative Service (NIS)	Silverdale	140.7750	Simplex
Paging System	Silverdale	143.6750	Simplex
Public Works Communications	Silverdale	140.9750	Simplex
Radiation Control	Silverdale	140.9500	Simplex
Security – Primary (Repeater)	Silverdale	140.0000	140.9250
Security – Secondary	Silverdale	138.7500	Simplex
Security – Harbor Area	Silverdale	140.7500	Simplex
Security – Fleet	Silverdale	140.2250	Simplex
Security – Docks	Silverdale	143.5000	Simplex
Security – Subbase	Silverdale	140.9250	Simplex
Subbase – Service	Silverdale	142.0750	Simplex
Subbase – Administration	Silverdale	138.9750	Simplex
Subbase – Administration	Silverdale	140.5000	Simplex
Transportation Services	Silverdale	143.5250	Simplex
Other Frequencies In Use	Silverdale	138.8250	Simplex

138.9500	139.5250	140.0250	140.1750
140.4750	140.5250	140.6500	140.8750
140.9000	141.5750	142.0250	142.2625

Puget Sound Naval Shipyard

Home of: Naval Station Bremerton, Cruiser Homeport, Bremerton Naval Hospital,

Pacific Fleet Submarine Logistics Support, Fleet Supply Center

Emergency Services	Bremerton	140.4500	Simplex
Hospital – Helipad Operations	Bremerton	350.0000	AM
Hospital – Operations	Bremerton	142.6000	Simplex
Hospital – Security	Bremerton	140.1500	Simplex
Paging Systems	Bremerton	142.5750	Simplex
Paging Systems	Bremerton	142.7250	Simplex
Railroad Operations	Bremerton	148.3250	Simplex
Public Works & Electrical Maintenance	Bremerton	142.6000	Simplex
Security & Law Enforcement	Bremerton	138.7250	Simplex
Security & Law Enforcement	Bremerton	139.5750	Simplex
Security & Law Enforcement	Bremerton	140.2500	Simplex

Puget Sound Federal Fire Department

29	Manchester	Beach Dr, Naval Fuel Depot	E21, HM21, Foam 21, E21A
27	Bremerton	Naval Station Bremerton	E27, A27, HM27, A28, CMD 27
28	Bremerton	Naval Station Bremerton	E28, L28, HM28
61	Bangor	Naval Submarine Base	E61, R61, A61, CMD 61, E61A
62	Bangor	Naval Submarine Base	L62, R62
63	Jackson Park	Naval Support Activity	E63 (Dispatched by Bremerton FD)
79	Keyport	Naval Weapons Center	E79, R79

Miscellaneous

Bainbridge Island Schools [PL 110.9]	Bainbridge Isl	155.2050	Simplex
Bremerton National Airport	Bremerton	123.0500	AM
Seattle Center Approach/Departure	Bremerton	127.1000	AM
Bremerton School District	Bremerton	151.9250	Simplex
	Bremerton	154.5700	Simplex
Bremerton Sun Newspaper	Bremerton	461.5500	
CBS Incorporated	Bremerton	450.2500	
Central Kitsap School District	Kitsap	154.6000	Simplex
School Busses [DPL 065]	Kitsap	151.7750	Simplex
463.4375 464.3250 464.4250			
464.5750 464.8750 464.9250			
Harrison Memorial Hospital	Bremerton	461.2125	Simplex
	Bremerton	466.2125	Simplex
Paging	Bremerton	462.7500	Simplex
Hilder Pearson Elementary	Bremerton	464.9750	
	Bremerton	151.7450	Simplex
Islandwood Environmental Education Center			
Operations Repeater	Bainbridge	461.4625	466.4625
Operations Simplex	Bainbridge	468.4125	Simplex
Operations Simplex	Bainbridge	468.6875	Simplex
J C Penney Stores	Silverdale	468.2125	Simplex
	Silverdale	468.8125	Simplex
KCPQ Inc	Bremerton	450.1875	Simplex
Kingston Jr High School	Kingston	151.8650	Simplex
Kitsap Mall			
Maintenance	Silverdale	462.0125	Simplex
Security	Silverdale	461.9375	Simplex
KITZ Television	Silverdale	161.6400	Simplex
	Silverdale	161.7000	Simplex
	Silverdale	161.7300	Simplex
Martha & Mary Health Service	Poulsbo	154.6000	Simplex

Agency	Location	Frequency	Notes
	Poulsbo	158.3400	Simplex
National Hot Rod Association	Bremerton	464.8250	Simplex
468.5250 468.6875 461.0750			
North Kitsap School District	Kingston	461.5500	
	Kingston	463.6250	
	Poulsbo	151.7450	Simplex
Northgate Inns Inc	Bremerton	461.2750	
Northwest Protective Service	Bremerton	463.9250	468.9250
Olympic College	Bremerton	461.5500	
463.9000 151.7450 151.8650 159.8850 160.0950			
Port of Poulsbo Seaplane Base	Poulsbo	122.9000	AM
Race Comm Electronics	Bremerton	151.8950	Simplex
	Bremerton	461.4500	
	Bremerton	461.6250	
Silver Spur Saddle Club	Bremerton	154.6000	Simplex
South Kitsap School District	Port Orchard	464.8500	
462.8250 464.8500 154.5700	Port Orchard	151.9250	Simplex
School Bus Operations	Port Orchard	151.6550	Simplex
Suquamish Tribal Gaming Agency	Bremerton	463.6250	
	Bremerton	463.7250	
	Bremerton	464.0500	
Suquamish Village Square Inc	Suquamish	461.6875	Simplex
461.7625 466.6875 466.7625			
Target Stores Inc	Silverdale	467.7625	Simplex
	Silverdale	469.0625	Simplex
Totem Girl Scout Council	Port Orchard	151.9250	Simplex
Trophy Lake Club	Countywide	463.5500	468.5500
Wal-Mart Stores	Bremerton	154.6000	Simplex
Wing Point Golf And Country Club	Bainbridge Isl	151.8050	Simplex
	Bainbridge Isl	151.8650	Simplex

Utilities

Agency	Location	Frequency	Notes
Annapolis Water District	Port Orchard	045.1600	Simplex
Cascade Natural Gas Corp	Bremerton	048.5000	Simplex
North Perry Avenue Water District	Bremerton	153.5150	Simplex
Puget Sound Energy Inc	Bremerton	158.1300	Simplex
153.4100 153.6350 451.0500			
Puget Sound Energy Inc	Bremerton	936.3875	
TRUNKED RADIO SYSTEM (EDACS): TrunkTracker Setting: ED			
936.4000 936.4125 936.4250 936.4375 936.4500			
936.4625 936.4750 936.4875 936.5000 937.4250			
Silverdale Water District 16	Silverdale	453.5375	458.5375
United Telephone Co of the NW	Poulsbo	451.5000	

Weather Information

National Weather Service – NOAA Weather Radio

Conditions & Forecasts – Seattle 162.5500 Simplex

Kitsap County SAME Alert Code: 053035

KLICKITAT COUNTY

Government & Public Safety

Agency	Description	PL	Location	Frequency	Output
Columbia River Intertribal Fish Com			Roosevelt	460.1625	465.1625
			Roosevelt	156.2250	151.1300
			Dallesport	155.5425	153.7925
Dallesport Fire District: See Klickitat County Fire Districts					
Goldendale City Fire: See Klickitat County Fire Districts					
Goldendale City Government			Goldendale	153.9800	Simplex
	155.0700 155.4150		Goldendale	154.0850	Simplex
Goldendale Police: See Klickitat County Sheriff EAST					
Klickitat County Fire Districts			Countywide	153.7700	
	Dispatch Repeater	[PL 162.2]	Countywide	154.1300	153.7700
			Countywide	154.4150	153.8900
			Countywide	154.3550	
			Countywide	154.4000	
Tac 1	Goldendale Tactical		Countywide	154.0700	Simplex
	Centerville Fire District 11		Centerville	158.8125	153.8675
Klickitat County Sheriff			Goldendale	452.5000	457.5000
	WEST Operations	[PL 114.8]	Countywide	153.8450	158.8350
	EAST/CENTRAL	[PL 192.8]	Countywide	154.0250	158.9100
	Alternate / Tactical		Countywide	154.9500	Repeater
Klickitat County Government			Goldendale	155.1150	
			White Salmon	155.9850	
			Countywide	156.0600	Simplex
			Bingen	153.8450	Repeater
Klickitat Valley Hospital Ambulance			Goldendale	154.5400	Simplex
Tri County Communications			Goldendale	460.3250	465.3250
			Goldendale	155.7300	153.9350
			Goldendale	155.5500	153.8000

Washington State Patrol	Roosevelt	154.6650	155.5200
Note: Not all WSP units or dispatchers transmit PL tone at all times			
Ch7 Klickitat / Okanogan Counties	Countywide	154.6800	Simplex
Ch1 Statewide / Aircraft Ops [PL 100.0]	Statewide	155.9700	159.0750
White Salmon City Government	White Salmon	155.0850	
White Salmon Fire Department: See Klickitat County Fire Districts			
White Salmon City Police	White Salmon	155.5650	

Klickitat County Law Enforcement Radio Codes

10-4	Message received	10-42	Ending shift
10-17	En route	10-63	Prepare to copy
10-23	Arriving	10-82	Prisoner in custody
10-24	Clear	10-86	Deceased Person
10-41	Beginning shift		

Miscellaneous

Big Horn Wind Project Power Generator			
Operations Repeater	Bickleton	452.3875	457.3875
Chenowith School Dist 9	Dallesport	461.2000	
Columbia Basin Electric Cooperative	Roosevelt	047.7000	048.3800
Educational Service District 112	White Salmon	151.8050	Simplex
Goldendale Airport	Goldendale	122.9000	AM
Goldendale School Dist 404	Goldendale	151.6850	Simplex
	Goldendale	155.2950	Simplex
Henkle Middle School	Goldendale	151.6550	Simplex
	Goldendale	151.7450	Simplex
Hood River County School District	White Salmon	464.6250	
Klickitat Public Utility District	Goldendale	451.3875	456.3875
Operations Repeater	Countywide	153.4550	158.2500
452.2000 457.2000			
Klickitat Valley Broadcasting Svcs	Goldendale	161.6400	Simplex
Klickitat Valley Hospital	Goldendale	155.3400	Simplex
155.3850 461.0750 453.3500			
Portland General Electric	Wishram	048.4600	Simplex
Puget Sound Energy	Goldendale	451.0500	456.0500
	Goldendale	451.2500	456.2500
Skyline Hospital	White Salmon	152.0075	Simplex
Operations Repeater	White Salmon	155.8650	150.7750
USA Media Group	Goldendale	152.3000	Simplex
Wasco County School District 12	Dallesport	461.2000	
Wasco Electric Cooperative Inc	Centerville	153.6500	Simplex
White Salmon Schools 405-17	White Salmon	464.6250	469.6250

Weather Information

National Weather Service – NOAA Weather Radio

Conditions & Forecasts – Goldendale	162.5250	Simplex
Conditions & Forecasts – Richland	162.4500	Simplex
Conditions & Forecasts – The Dalles	162.4000	Simplex

Klickitat County SAME Alert Code: 053039

LEWIS COUNTY

Government & Public Safety

Centralia Fire Department		Centralia	154.1450	Simplex
Centralia City Government				
Utilities / Light Department		Centralia	158.1900	Simplex
Utilities / Light Department		Centralia	153.4250	Simplex
Utilities / Light Department		Centralia	153.5300	Simplex
Utilities / Light Department		Centralia	153.6950	Simplex
Centralia Police Department		Centralia	154.0400	Simplex
Dispatch		Centralia	156.1800	159.0000
Tactical		Centralia	155.0100	Simplex
Chehalis City Government		Chehalis	154.1000	Simplex
453.0875 453.8875		Chehalis	458.0875	Simplex
458.8875 458.9125		Chehalis	453.9125	Simplex
Chehalis Fire Department		Chehalis	154.1900	Simplex
		Chehalis	158.7600	Simplex
Fire Apparatus: E48-1, E48-2, E48-3, E48-4, Aid 48, Tower 48				
Chehalis Police Department		Chehalis	155.5950	156.0300
		Chehalis	155.0100	Simplex
		Chehalis	155.3700	Simplex
		Chehalis	158.8350	Simplex
Lewis County Sheriff's Office				
Dispatch	[PL 88.5]	Countywide	155.6250	Repeater
Lewis County Fire District 2		Toledo	154.0850	Simplex
Lewis County Fire District 6		Chehalis	153.9650	Simplex
		Chehalis	159.0600	Simplex

Lewis County Fire Districts (All)		Countywide	154.1900	Simplex

Fire Districts

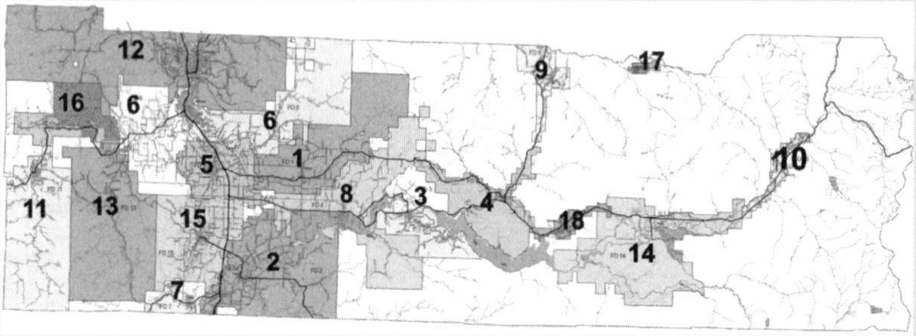

1	Onalaska Area	8	Salkum Area	15	Winlock Area
2	Toledo Area	9	Mineral Area	16	Doty-Dryad Area
3	Mossyrock Area	10	Packwood Area	17	Paradise Area
4	Morton Area	11	PeEll Area	18	Glenoma Area
5	Napavine Area	12	Centralia Rural Area	36	City of Centralia
6	Greater Lewis County	13	Boistfort-Curtis Area	48	City of Cehahlis
7	Vader Area	14	Randle Area		

Lewis County Fire District 12		Centralia	154.3250	Simplex
Dispatch	[PL 123.0]	Centralia	154.9725	159.0975
		Centralia	154.4300	Simplex

Fire Stations & Apparatus

12-1	Fords Prairie	E12-1, M12-1, E12-1A, R12-1, BR12-1, Tech R12-1, Air12, E12-1A, Aid 12-1
12-2	Big Hanaford Road	E12-2
12-4	Seminary Hill Road	E12-4, WT12-4
12-5	Cooks Hill Road	WT12-5
12-6	Lincoln Creek Road	E12-6, WT12-6, Aid 12-6, BR 12-6
12-7	Independence Road	E12-7, WT12-7
12-8	Garrard Creek Road	

Lewis County Fire District 14	Randle	156.1050	155.8050
	Randle	155.9550	Simplex

Lewis County Fire District 15 (Winlock and surrounding area)

15-1	Main Station	609 NW Kerron Ave, Winlock
15-2	Grand Prairie	241 N Military Road
15-3	West District	541 Nelson Road

Lewis County Government				Chehalis	154.9800	Simplex
				Centralia	154.9950	Simplex
				Centralia	155.0100	Simplex
	453.2875	458.2875	453.5875	Countywide	155.1000	155.7450
	458.2375	458.5875	158.9250	Chehalis	155.4150	156.0000
	159.0300	156.2250		Packwood	155.7150	156.0300
	453.9125	453.9625	453.2375	Chehalis	155.8050	Simplex
				Mineral	159.3150	Simplex
Morton City Government				Morton	154.0550	Simplex
Mossyrock City Water & Sewer				Mossyrock	153.8150	Simplex
Napavine City Government				Napavine	155.0400	158.9550
Port of Centralia Public Safety				Centralia	154.0625	Simplex

Washington State Patrol

District 5: Lewis, Klickitat, Cowlitz, Clark, Skamania, Wahkiakum Counties

Ch14 Lewis County Dispatch	Countywide	155.8500	Simplex
Winlock Fire Department	Winlock	154.1900	Simplex

Airports

Chehalis-Centralia Airport	Chehalis	122.8000	AM
Seattle Approach / Departure	Chehalis	121.1000	AM
Seattle Center Approach / Departure	Chehalis	124.2000	317.6000
Ed Carlson Memorial Field / South Lewis County Airport			
Common Traffic Advisory (CTAF)	Toledo	122.9000	AM
Packwood Airport	Packwood	122.9000	AM
Strom Field Airport	Morton	122.9000	AM

Miscellaneous

Adna School District 26	Adna	159.4950	Simplex
Administration & Maintenance	Adna	159.6600	Simplex
Administration & Maintenance	Adna	160.1250	Simplex
Centralia School District 401	Centralia	452.9000	457.9000
155.2350	Centralia	151.7450	Simplex
Chehalis School District	Chehalis	152.8700	Simplex
152.9300 152.9600	Chehalis	159.7200	Simplex
159.7500 160.0650	Chehalis	153.0350	Simplex
Circuit City Distribution Center	Chehalis	464.4250	Simplex
	Chehalis	464.5250	Simplex
Cispus Learning Center	Randle	155.0850	Simplex
	Randle	155.7450	Simplex
Energy Northwest	Packwood	153.4250	158.1300
Fred Meyer Distribution Center	Chehalis	461.9500	Simplex
463.5000 463.6250 463.6500	Chehalis	463.8500	Simplex
Gibson House Casino	Centralia	452.0375	Simplex
452.1375 452.3125	Centralia	457.6375	Simplex
Laidlaw School Bus Transportation	Boistfort	461.2000	466.2000
Morton School District	Morton	152.9150	159.6000
Mount Rainier Scenic Railroad	Mineral	160.6350	Simplex
	Mineral	161.3850	Simplex
Onalaska School District	Onalaska	157.6200	Simplex
Providence Hospital	Centralia	462.1750	467.1750
154.5150	Centralia	151.7150	Simplex
Public Utility District 1	Randle	047.9800	048.3800
Railcar Washington Manufacturing	Chehalis	461.1500	Simplex
	Chehalis	464.0250	Simplex
Tenino School District 402	Tenino	159.5100	Simplex
159.6000 159.6675	Tenino	159.8400	Simplex
159.9300 159.8025	Tenino	160.1400	Simplex
Toledo School District 237	Toledo	155.2050	Simplex
Vader School District 308	Vader	159.7950	Simplex

Weather Information

National Weather Service – NOAA Weather Radio		
Conditions & Forecasts – Capitol Peak	162.4750	Simplex
Lewis County SAME Alert Code: 053041		

MASON COUNTY

Government & Public Safety

Agency / Channel		Tone	Location	Frequency	Frequency / Mode
American Medical Response Ambulance				852.2125	807.2125
Mason County Fire Departments			Belfair	156.1050	Simplex
Ch3	Belfair Fire Dispatch	[PL 110.9]	Shelton	153.7400	Simplex
Ch4	Red Net		Shelton	153.8300	Simplex
	Shelton Dispatch	[PL 110.9]	Shelton	154.1900	Simplex
	North Mason County		Shelton	154.3700	
			Shelton	155.0850	
			Shelton	155.1450	
Mason County Government					
	Facilities		Shelton	460.5625	Simplex
	Health Service		Shelton	453.7375	Simplex
	Road Maintenance		Shelton	151.0550	159.1200
Mason County Medic One			Shelton	461.1500	
			Shelton	155.4000	Simplex
			Shelton	463.6500	
Mason County Sheriff / Shelton PD			Shelton	453.3250	
			Shelton	453.6750	
TAC 1 Primary Operations		[PL 114.8]	Countywide	460.2250	465.2250
TAC 2 Alternate			Countywide	460.1000	Simplex
			Countywide	155.5950	Simplex
Shelton City Public Works			Shelton	153.8450	Simplex
Skokomish Department of Public Safety					
	Law Enforcement & EMS		Hoodsport	155.7300	155.3550
Squaxin Island Indian Tribe			Countywide	155.5950	
			Shelton	158.7300	

Washington State Patrol

Note: Not all WSP units or dispatchers transmit PL tone at all times

Ch15	Kitsap / Mason Counties	[PL 110.9]	Countywide	154.6650	158.8950
Ch1	Statewide / Aircraft Ops	[PL 100.0]	Statewide	155.9700	159.0750

Miscellaneous

American Protective Services	Tacoma	461.7750	
	Tacoma	461.9250	
Lucky Dog Casino	Skokomish	451.3375	Simplex
451.7375 452.0625	Skokomish	452.2875	Simplex
Mary M Knight School District 311	Elma	152.3750	
Mason County Public Utilities District	Shelton	048.3600	047.8600
Mason General Hospital	Allyn	152.0075	
	Allyn	155.3400	
	Allyn	155.4000	
Mission Creek Youth Camp [PL 156.7]	Belfair	158.8350	Simplex
North Mason School District	Belfair	451.8375	456.8375
	Belfair	451.8625	Simplex
	Belfair	452.2375	Simplex
Sand Hill Elementary	Sand Hill	464.2000	Simplex
	Sand Hill	464.4500	Simplex
Sanderson Field Airport	Shelton	122.8000	AM
Seattle Approach / Departure	Shelton	121.1000	AM
Shelton School District	Shelton	464.8000	
School Bus Operations	Shelton	461.9750	466.9750
School Bus Operations	Shelton	464.6000	469.6000
Squaxin Island Gaming Commission	Shelton	151.6850	
	Shelton	461.1125	Simplex
	Shelton	466.1125	Simplex
WA Explorer Search & Rescue	Belfair	047.5000	

Public Utilities

Mason County Public Utility District	Countywide	043.4000	047.8600
Operations Repeater	Countywide	452.7500	457.7500

153.5150 048.3600 047.8600 451.2000 460.5125

Weather Information

National Weather Service – NOAA Weather Radio

Conditions & Forecasts – Seattle	162.5500	Simplex
Conditions & Forecasts – Capitol Peak	162.4750	Simplex

Mason County SAME Alert Code: 053045

PACIFIC COUNTY

Government & Public Safety

Bay Center Fire District 6: See Pacific County Fire Districts
Ilwaco City Fire: See Pacific County Fire Districts

Long Beach City Police Department		Long Beach	159.2100	
		Long Beach	460.3000	
		Long Beach	460.2000	
Long Beach City Street Department		Long Beach	156.1950	
Long Beach Fire Department		Long Beach	154.4450	
Maritime Fire & Safety Association		Countywide	150.9800	154.5850
Operations Repeater		Countywide	159.4800	158.4450
Naselle Fire Protection District 4		Naselle	154.2200	
Pacific County Government		Chinook	453.3375	Simplex
		Raymond	453.6750	458.6750
County Transit		Countywide	453.8250	458.8250
Highways Department	[PL 136.5]	Raymond	453.2000	
Pacific County Fire Districts		Countywide	154.4150	
Countywide Fire Dispatch		Countywide	154.1900	
Countywide Fire Incidents		Countywide	154.2050	
		Countywide	154.8225	156.2175
Pacific County Sheriff		Countywide	460.4500	465.4500
North Dispatch	[PL 136.5]	South Bend	460.0750	465.0750
South Dispatch		Long Beach	460.2250	465.2250
Back-up Tac	[PL 136.5]	Countywide	460.3500	465.3500
Raymond City Government		Raymond	153.8000	

Agency		Location	Frequency	Input/Mode
Raymond City Fire Department		Raymond	155.2800	Simplex
		Raymond	155.2950	Simplex
		Raymond	155.3400	Simplex
Raymond City Police	[PL 136.5]	Raymond	460.3500	465.3500
Shoalwater Bay Indian Tribe		Raymond	155.7750	
Emergency Management		Raymond	155.0850	Simplex

Washington State Patrol

Note: Not all WSP units or dispatchers transmit PL tone at all times

District 8: Clallam, Jefferson, Grays Harbor, Pacific, Mason, Kitsap Counties

Ch	Description	PL	Location	Frequency	Input/Mode
Ch16	Grays Harbor / Pacific		Countywide	154.6950	Simplex
Ch1	Statewide / Aircraft Ops	[PL 100.0]	Statewide	155.9700	159.0750

Miscellaneous

Agency	Location	Frequency	Mode
Ocean Beach Hospital	Ilwaco	152.0075	Simplex
	Ilwaco	155.3400	Simplex
	Ilwaco	155.2200	Simplex
Ocean Beach School District 101	Ilwaco	151.7450	
Port Of Ilwaco	Ilwaco	461.8500	
	Ilwaco	464.5500	Simplex
Airport	Ilwaco	122.9000	AM
Public Utility District No 2	Ilwaco	047.8400	Simplex
047.9400 048.4400			
Raymond City	Raymond	155.2950	Simplex
Raymond School District 116	Raymond	155.1600	Simplex
Shoalwater Bay Bingo & Casino	Raymond	153.1100	
South Bend School District 118	South Bend	155.2200	Simplex
	South Bend	155.2650	Simplex
Surfside Homeowner's Association	Ocean Park	151.7150	Simplex
	Ocean Park	151.7750	Simplex
Willapa Harbor Airport	South Bend	122.9000	AM
Willapa Harbor Hospital	South Bend	155.3400	Simplex
	South Bend	460.6625	Simplex
Willapa Valley School District 160	Menlo	152.4200	
Willapa Valley Water District	Raymond	453.6625	Simplex

Weather Information

National Weather Service – NOAA Weather Radio

Description	Frequency	Mode
Conditions & Forecasts – Astoria	162.4000	Simplex
Conditions & Forecasts – Capitol Peak	162.4750	Simplex
Conditions & Forecasts – Olympic Nat'l Park	162.4250	Simplex

Pacific County SAME Alert Code: 053049

PIERCE COUNTY

Government & Public Safety

Pierce County Law Enforcement Radio Callsigns

Callsign	Assignment	Callsign	Assignment
Adam #	Various Administration Units	Paul #	Tacoma PD – Swing, 1 man
Baker #	Tacoma PD – Swing, 2 man	Robert #	Various Reserve Officers
David #	Tacoma PD – Detectives	Robert #	Tacoma PD – Day, 2 man
Dupont #	Dupont PD	Roy #	Roy PD
Eatonville #	Eatonville PD	Ruston #	Ruston PD
Fircrest #	Fircrest PD	Sam #	Various Sergeants
Frank #	Tacoma PD – Day, 1 man	Steilacoom #	Steilacoom PD
Harbor #	Gig Harbor PD	Tom #	Various Traffic Units
Henry #	Tacoma PD – Midnight 2 man	Union #	Tacoma PD – Midnight 1 man
Lakewood #	Lakewood PD	UP #	University Place PD
Lincoln #	Tacoma PD – Lieutenant	William #	Wilkeson PD
Mary #	Tacoma PD – Motorcycle	William #	Tacoma PD – Schools
Nora #	Tacoma PD – Midnight 1 man	X-Ray #	Tacoma PD – Undercover
Paul #	Buckley PD		

American Medical Response (AMR) Tacoma TG: 54096 Tacoma

Bonney Lake Fire Department: See Pierce County Fire Districts

Bonney Lake City Government

			Location	Frequency	
			Bonney Lake	158.9400	
			Bonney Lake	154.0925	
			Bonney Lake	155.9475	
	Road Maintenance		Bonney Lake	150.9950	

Bonney Lake Police Department

			Location	Frequency	
F-1	Operations	[PL 136.5]	Bonney Lake	155.7300	159.1500
F-2	Alternate	[PL 136.5]	Bonney Lake	155.9100	

Buckley Fire Department: See Pierce County Fire Districts

Buckley City Government

	Public Works/Police Tac		Buckley	153.8600	

Buckley Police Department

			Buckley	159.0900	
F-1	Operations	[PL 173.8]	Orting	155.0100	
F-5	Alternate		Buckley	153.8600	
			Buckley	156.0900	

Carbonado Fire Department: See Pierce County Fire Districts

Carbonado Law Enforcement: Contract to Buckley Police

Central Pierce Fire District: See Pierce County Fire Districts

Dupont Police Department

Police Dispatched on Pierce Sheriff WEST

Dupont Fire Department: See Pierce County Fire Districts

Eatonville Fire Department: See Pierce County Fire Districts

Eatonville City Government — Eatonville 154.0850

Eatonville Police Department: Dispatched on Pierce Sheriff EAST

	Car to Car + Public Works		Eatonville	154.0850	Simplex

Edgewood Law Enforcement: Contract to Pierce Sheriff EAST

Fife City Government — Fife 045.6400

Fife Fire Department: See Pierce County Fire Districts

Fife Police Department

		[PL 167.9]	Fife	159.1125	153.7925
	Dispatch & Operations	[PL 167.9]	Fife	155.4900	
	Records (Countywide)		Fife	156.0900	
	LERN		Fife	155.3700	
	Unknown Use		Fife	153.9875	

Fircrest Fire Department: See Pierce County Fire Districts

Fircrest Public Works — Fircrest 153.9800

Fircrest Police Department: Dispatched on Pierce Sheriff WEST

Gig Harbor Police Department: Dispatched on Sheriff WEST

	Car to Car + City Public Works		Gig Harbor	156.0150	

Lakewood Police Department

Ch2	Dispatched by PCSO	[PL 173.8]	Tacoma	155.6400	154.7100

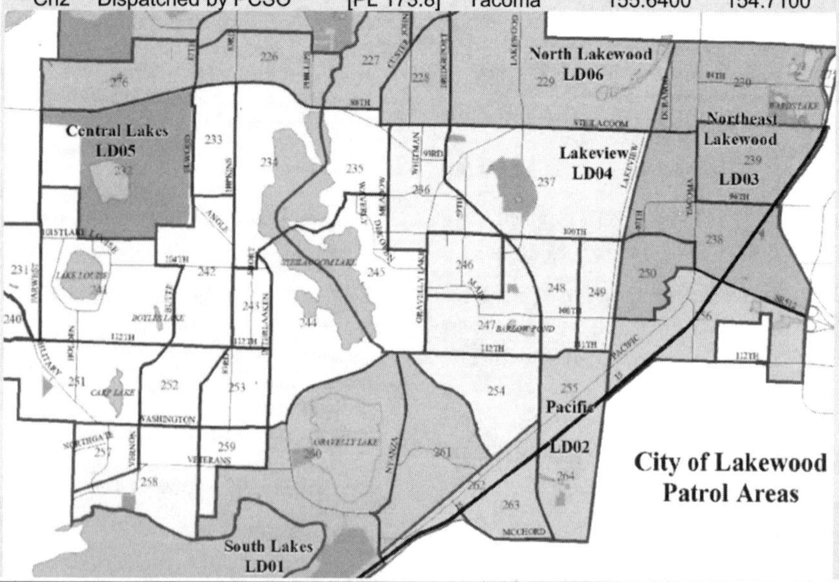

Milton City Government	Milton	153.8600	
Milton Police Department: Dispatched by Fife Police			
Orting Fire Department	Orting	156.2025	153.9275
Orting Police Department			
Dispatched by Buckley	Orting	155.0100	

Pierce County Fire Districts

Fire-1	County Fire Dispatch	[PL 131.8]	Countywide	153.8900	Simplex
Fire-2	Fireground West	[PL 131.8]	Countywide	154.3550	Simplex
Fire-3	Fireground Central + Roy	[PL 131.8]	Countywide	154.1600	Simplex
Fire-4	Fireground Buckley + Puy	[PL 131.8]	Countywide	154.2650	Simplex
Fire-5	Fireground East / Valley	[PL 136.5]	Countywide	154.3850	Simplex
Fire-6	Fireground South	[PL 131.8]	Countywide	154.3250	Simplex
	Fire District 2		Orting	460.6000	465.6000
	Fire District 25		Crystal Mtn	153.8300	Simplex

Pierce County Fire Districts – Fire Station Locations & Apparatus

11	Sumner	Harrison & Narrow	E111, L121, M131
12	Sumner	136th St	E112, E113
20	Lakewood (2)	10928 Pacific Hwy SW	E210, L220, BC2
21	Lakewood (2)	5000 Steilacoom Blvd SW	E211, L221, R221, BC2, M231
22	Lakewood (2)	8517 Washington Blvd SW	E212, M232, Marine 282
23	Lakewood (2)	14505 Grant Ave SW	E213, M233, M239
24	Lakewood (2)	8310 87th Ave SW/Oakbrook	E214, E219
31	University Place (2)	40th St W	E311, E312, E313, L321, R391, BC3, A/M331, A/M332
41	East Pierce (2)	Bonney Lake	E41, M41, BR41
42	East Pierce (2)	Bonney Lake	E42, M42, BR42, E429
43	East Pierce (2)	Bonney Lake	E43
44	East Pierce (2)	Lake Tapps	E44, M44
45	East Pierce (2)	Lake Tapps	E45, A45, WT45, E459
46	East Pierce (2)	Lake Tapps	Marine 46
51	Gig Harbor (2)	Kimball DAL	E511 R591 M531 U541 U543 BC5
52	Gig Harbor (2)	Point Fosdick Dr, Wollochet	E512, WT552, Trench 1
53	Gig Harbor (2)	Kamus Dr, Fox Island	E513, WT553
54	Gig Harbor (2)	Ray Nash Dr, Arletta	E514
55	Gig Harbor (2)	Rosedale St NW, Rosedale	E515, WT555, A515
56	Gig Harbor (2)	144th St NW, Purdy	E516, WT556
57	Gig Harbor (2)	Crescent Valley	E517, WT557
58	Gig Harbor (2)	Bujacich Rd NW, Swede Hill	E518, M518, M538
59	Gig Harbor (2)	62nd Ave NW, Artondale	E519, M519, A539
60	Central Pierce (3)	22nd Ave E, Spanaway	E610, A610, Marine 610, U679
61	Central Pierce (3)	S 114th St, Parkland	E611, L611, M611, E621
63	Central Pierce (3)	17 Ave E, Midland	E613, M613, M623
64	Central Pierce (3)	224 St E, Elk Plain	E614, M614
65	Central Pierce (3)	Park Ave S, Spanaway	E615, M615
66	Central Pierce (3)	128th St E, South Hill	E616, L616, M616, HM616, U616
67	Central Pierce (3)	Canyon Rd E (Summit North)	E617, WT617
68	Central Pierce (3)	Canyon Rd E (Summit South)	E618, M618
69	Central Pierce (3)	110 Ave E (Airport)	E619, E629, M629
81	Edgewood (5)	24th St E & Meridian	E811, A831, U842, WT851, BC8
111	North Puyallup (4)	Goettsch Ave NE, Puyallup	E1111, R1131, WT1151, BC11
121	Buckley (4)	S Cedar Weeron Rd	E1211, A1231
122	Buckley (4)	Wilkeson	E1212, A1232
123	Buckley (4)	Buckley City	E1213
131	Browns Dash Pt (2)	Wa Tau Ga NE, Browns Pt.	E1311, A1331, U1341
132	Browns Dash Pt (2)	Dogwood NE, Dash Pt.	E1312, A1332, BC13
141	Riverside	56th Ave E	E1411, E1412, A1431, U1443, WT1451, Marine 1481, BR141

152	South Pierce (6)	340th & Hwy 7	E1512, A1532, U1542, WT1552
153	South Pierce (6)	Trek Dr & Hwy 6	E1513, A1533, WT1553
154	South Pierce (6)	King Flandor Rd & Hwy 7	E1514, A1534, WT1554
161	Key Peninsula (2)	Waunaminter	E1611, WT1651, E1619
162	Key Peninsula (2)	Key Center	E1612, A1632, U1642, WT1652
163	Key Peninsula (2)	Long Branch	E1613, M1633, U1643, WT1653
164	Key Peninsula (2)	Rocky Bay	E1614, WT1654
165	Key Peninsula (2)	Key Peninsula	E1615, M1635, WT1655, R1695
166	Key Peninsula (2)	Key Peninsula	E1616, A1636, WT1656
171	Roy (3)	3rd & Hwy 507	E1711, A1731/9, BR1741, WT1751
172	Roy (3)	Harts Lake	E1712, WT1752
173	Roy (3)	Lacamas	E1713, A1733, WT1753
174	Roy (3)	McKenna	E1714, A1734/A1730
181	Orting (5)	SR 162, McMillian	E1811, U1840, U1841
182	Orting (5)	Patterson Rd	E1812, WT1852
183	Orting (5)	OK Highway	E1813, WT1853
184	Orting (5)	Orville Rd	E1814
191	Carbonado	5th St & 8th Ave	E1911
201	South Prairie (5)	Washington St	E2011, E2012, E2014, A2031
211	Graham (6)	251st & 50th Ave E	E2111, L2121, M2131, WT2151
212	Graham (6)	78th Ave E	E2112, WT2152
213	Graham (6)	304th St E	E2113, M2133, M2137
214	Graham (6)	149th & Kapowain Hwy	E2114
215	Graham (6)	Support Services	E2115
219	Graham (6)	Boeing	E2149
221	Lake Tapps	W Tapps Dr E	E2211, E2221
222	Lake Tapps	N Tapps Hwy E	E2212, E2222
231	Ashford-Elbe (6)	Ashford	E2311, A2331, R2333, WT2351
232	Ashford-Elbe (6)	Elbe	E2312, BR2314, A2332, WT2352
241	Bonney Lake	214th Ave E	E2411 E2413
242	Bonney Lake	Bonney Lake Blvd	E2412
243	Bonney Lake	N Island Dr & N Vista, Inlet Is.	E2413
251	Crystal Mountain		E2511
261	Greenwater	Greenwater Chevron	E2611
262	Greenwater	Crystal River Ranch	E2612
271	Anderson Island (2)	Anderson Island	E711, A2731, WT2751, Marine2782
421	Dupont (2)		E4211, E4212, A4231
431	Eatonville (6)	Eatonville	E4311, A4331, U4344, U4391
451	Milton		E4511, E4512, E4513, R4531
461	Ruston (1)		E4611, E4612, A4631, WT4651
481	Steilacoom (2)		E4811, A4831, R4831, E4812
521	Buckley (4)		E5211, E5212, A5231, R5241
1	Puyallup	902 7th St NW	E1, E3, BC1, Special Ops
2	Puyallup	3509 27th St E	E2, HM2
3	Puyallup	311 W Pioneer	L3, M3
4	Puyallup	Western WA Fairgrounds	E4, E4A, M4, M4A

Pierce County Fire District 22	Countywide	158.9400	
Sumner Fire Dispatch	Sumner	155.8800	
Pierce County Fire District 23	Elbe	151.1000	
Pierce County Government	Tacoma	453.2250	458.2250
WA State Emergency Mgmnt	Tacoma	045.2000	
WA State Emergency Mgmnt	Puyallup	045.3600	
WA State Emergency Mgmnt	Tacoma	045.4800	
Animal Control/Emerg Mgmnt	Graham	156.2400	
Public Works Department	Tacoma	453.6250	458.6250
Public Works Department	Tacoma	453.8750	458.8750
453.7875 458.3125			

	Transit – Local		Tacoma	935.9000	
	Transit – Express		Tacoma	935.9250	
	Transit – Shuttle		Tacoma	938.4625	
	Transit – Shuttle		Tacoma	939.4500	
	Transit – Maintenance		Tacoma	939.9375	
	Transit – Maintenance		Tacoma	939.9625	
Pierce County Sheriff's Department			Tacoma	151.3550	156.0600
	East Dispatch	[PL 173.8]	Tacoma	154.9500	159.0300
	East Admin	[PL 173.8]	Tacoma	155.6100	158.9100
Ch2	West Dispatch	[PL 173.8]	Tacoma	155.6400	154.7100
	West Admin	[PL 173.8]	Tacoma	155.3100	154.8900
	Countywide Records	[PL 173.8]	Tacoma	156.0900	156.0900
Port of Tacoma			Tacoma	153.7850	
			Tacoma	158.8950	
	Container Ops	[PL 100.0]	Tacoma	453.2875	Simplex
	Container Ops	[PL 123.0]	Tacoma	453.4125	Simplex
	Operations		Tacoma	453.4875	Simplex
	Operations		Tacoma	453.5375	Simplex
	Security Dispatch		Tacoma	453.6625	458.6625
	Freight Maintenance		Tacoma	453.9125	Simplex
	Operations	[PL 103.5]	Tacoma	458.2875	Simplex
	Operations		Tacoma	458.4125	Simplex
	Operations		Tacoma	458.9625	Simplex
	Operations	[PL 179.9]	Tacoma	458.9875	Simplex

Puyallup Fire Department: See Pierce County Fire Districts

Puyallup City Government					
	Puyallup Fair Fair-Com		Puyallup	155.0550	
	Public Works		Puyallup	158.7450	
	Police Tac		Puyallup	158.9550	
Puyallup Police Department					
Ch1	Dispatch	[PL 162.2]	Puyallup	155.1300	
Ch3	Tactical		Puyallup	158.9550	
	LERN		Puyallup	155.3700	
	Countywide Records		Puyallup	156.0900	

TRUNKED RADIO SYSTEM (Unknown): TrunkTracker Setting: Unk

Not yet in use: 868.3500 868.3750 868.4000 868.5000 868.5750

Puyallup Tribal Police		[PL 127.3]	Puyallup	166.6375	168.3750

Roy Police Department: Dispatched on Pierce Sheriff EAST

	Car to Car + City Public Works		Roy	155.7900	
Rural/Metro Ambulance			Tacoma	TG: 54320	Tacoma

Ruston Police Department: Dispatched by Tacoma Police

South Prairie Law Enforcement: Contract to Wilkeson Police

Steilacoom Police Department: Dispatched on Pierce Sheriff WEST

Steilacoom Town Government			Steilacoom	153.8450	
Sumner City Government			Sumner	453.5375	Simplex
	Fire Dispatch	[PL 100.0]	Sumner	155.8800	
	Public Works Department		Sumner	156.1650	Simplex
	Public Works Department		Sumner	151.0550	Simplex
Sumner Police Department		[PL 100.0]	Sumner	154.7850	158.9250
Tacoma City Government					
Ch3	Public Works	[PL 131.8]	Tacoma	453.4250	458.4250
	Water/Sewer Departments		Tacoma	453.5500	458.5500
	Street Maintenance		Tacoma	453.7500	458.7500
	Sanitation		Tacoma	TG: 53168	Tacoma
	Sanitation		Tacoma	TG: 53200	Tacoma

	Sanitation – Recycling	Tacoma	TG: 53360	Tacoma
	Street Maintenance	Tacoma	TG: 53680	Tacoma
	Water/Sewer Departments	Tacoma	TG: 53296	Tacoma
	160.1025 160.0650 153.5300	Tacoma	151.5350	Simplex
	Water/Sewer Department	Tacoma	153.6425	Simplex
City 1	City Government Common Channel	Tacoma	TG: 53008	Tacoma
City 2	City Government Common Channel	Tacoma	TG: 53040	Tacoma
P&F	Police & Fire Common Channel	Tacoma	TG: 54512	Tacoma
	Police & Fire Talk-Around [PL 156.7]	Tacoma	866.1000	Simplex
	City Common Talk-Around [PL 156.7]	Tacoma	867.0500	Simplex

TRUNKED RADIO SYSTEM (Motorola): TrunkTracker Setting: Type II 800

Simulcast Cell:

867.9875* **868.2500*** **868.3250*** 866.7875 866.8125
866.8625 867.1625 867.1875 867.2125 867.2375
867.2625 867.4625 867.9500 866.8375

McNeil Island Intellirepeater

868.6250* 868.6000 866.7625 867.3750 867.4000 867.9250

IF THESE FREQUENCIES CHANGE, FREE UPDATES ARE AT SCANSEATTLE.COM

Tacoma Fire Department

800 MHz Trunked System

Ch1	Dispatch	Tacoma	TG: 54704	Tacoma
Ch2	EMS Tactical	Tacoma	TG: 54672	Tacoma
Ch3	Fire Tactical	Tacoma	TG: 54640	Tacoma
Ch4	Fire Tactical	Tacoma	TG: 54608	Tacoma
Ch5	Fire Tactical	Tacoma	TG: 54576	Tacoma
Ch6	Alternate	Tacoma	TG: 54544	Tacoma
Ch7	Alternate	Tacoma	TG: 54512	Tacoma
Ch8	Alternate	Tacoma	TG: 54480	Tacoma
Ch9	Alternate	Tacoma	TG: 54448	Tacoma
Ch10	Training	Tacoma	TG: 54416	Tacoma
	Events	Tacoma	TG: 54736	Tacoma
	Link with Pierce County FD's	Tacoma	TG: 55120	Tacoma
	Talk-Around 1 Simplex [PL 156.7]	Tacoma	866.0750	Simplex
	Talk-Around 2 Simplex [PL 156.7]	Tacoma	866.1500	Simplex
	Talk-Around 3 Simplex [PL 156.7]	Tacoma	866.5500	Simplex

OLD VHF Conventional Radio System

	Countywide Fire Dispatch	Tacoma	153.8900
F-1	Tacoma F-1 Fire [PL 131.8]	Tacoma	154.1300
F-2	Tacoma F-2 EMS [PL 131.8]	Tacoma	154.2950
F-3	Tacoma F-3 Tactical	Tacoma	159.3900
F-4	Tacoma F-4 Tactical	Tacoma	156.2250

Tacoma Fire Station Locations & Apparatus

1	Tacoma	901 Fawcett Ave	E1, Trk1
2	Tacoma	2701 Tacoma Ave S	E2, Trk2, BC182
3	Tacoma	206 Browns Pt Blvd	E3
4	Tacoma	1453 S 12th St & M St	E4, R4
6	Tacoma	1015 E F St	E6, U49 (Boston Whaler)
7	Tacoma	5448 S Warner & 56th St	E7
8	Tacoma	4911 S Alaska	E8, R2, U48 (Tech Rescue)
9	Tacoma	3502 6th Ave	E9, BC181
10	Tacoma	7247 S Park Ave	E10
11	Tacoma	3802 McKinley Av & 38th St	E11, R5
12	Fife	2015 54th Ave E	E12, Trk4, R3, U44 (HazMat)
13	Tacoma	3825 N 25th St & Proctor	E13, Trk3
14	Tacoma	4701 N 41st St & Gove	E14
15	Tacoma	3510 E 11th St & Alexander	E15, U45 (Water Tender)
16	Tacoma	7217 6th Av & MacArthur	E16, R1
17	Fircrest	302 Regents Blvd	E17, U42 (Mobile Air)

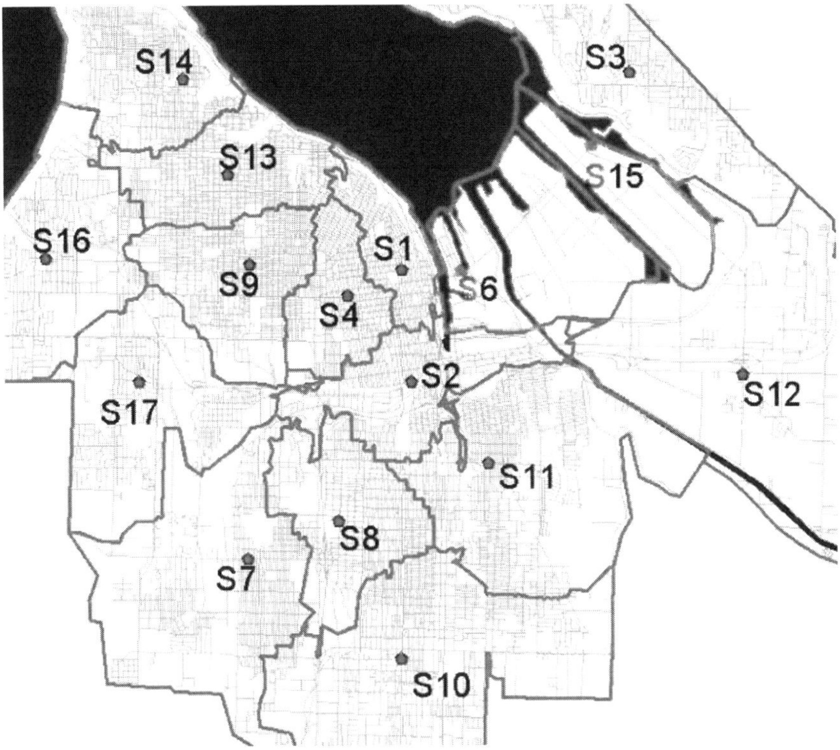

Tacoma Police Department			Tacoma	155.3700	
800 MHz Trunked System					
N PRIM	North Primary (Sectors 1 & 2)		Tacoma	TG: 52176	Tacoma
N TAC 1	North Tactical 1		Tacoma	TG: 52240	Tacoma
N TAC 2	North Tactical 2		Tacoma	TG: 52272	Tacoma
EVNT N	North Events		Tacoma	TG: 52080	Tacoma
S PRIM	South Primary (Sectors 3 & 4)		Tacoma	TG: 52304	Tacoma
S TAC 1	South Tactical 1		Tacoma	TG: 52400	Tacoma
S TAC 2	South Tactical 2		Tacoma	TG: 52432	Tacoma
EVNT S	South Events		Tacoma	TG: 52112	Tacoma
RECORDS	Records		Tacoma	TG: 52208	Tacoma
	Talk-Around 1	[PL 156.7]	Tacoma	866.5750	Simplex
	Talk-Around 2	[PL 156.7]	Tacoma	866.6000	Simplex
	Talk-Around 3	[PL 156.7]	Tacoma	867.0750	Simplex
Tacoma PD OLD UHF Conventional Radio System					
F-1	North Dispatch	[PL 131.8]	Tacoma	460.1500	465.1500
F-2	South Dispatch	[PL 131.8]	Tacoma	460.0500	465.0500
F-3	Records	[PL 131.8]	Tacoma	460.3500	465.3500
F-4	Tac 1 – North TAC	[PL 131.8]	Tacoma	460.3000	465.3000
F-5	Tac 2 – South TAC	[PL 131.8]	Tacoma	453.2250	458.2250
F-6	Tac 3 + Public Works	[PL 131.8]	Tacoma	453.4250	458.4250
F-7	Tac 7 + Water/Sewer	[PL 131.8]	Tacoma	453.5500	458.5500
	Police low-power	[PL 131.8]	Tacoma	453.2625	Simplex
	Police low-power		Tacoma	453.3375	Simplex

Tacoma Police – Sector Maps

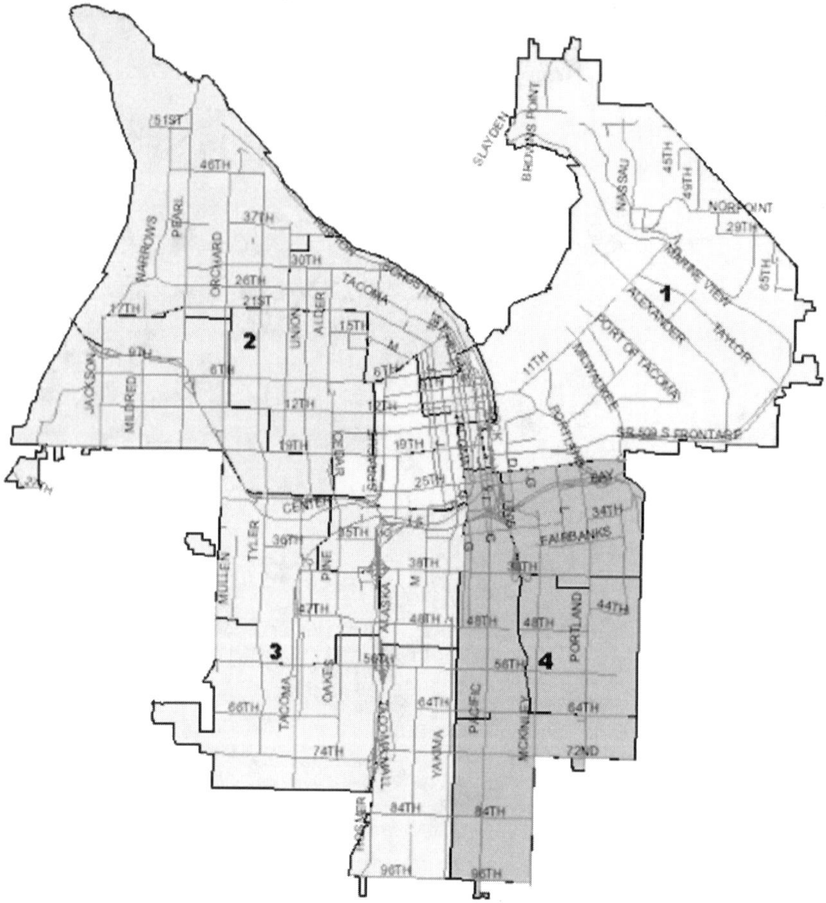

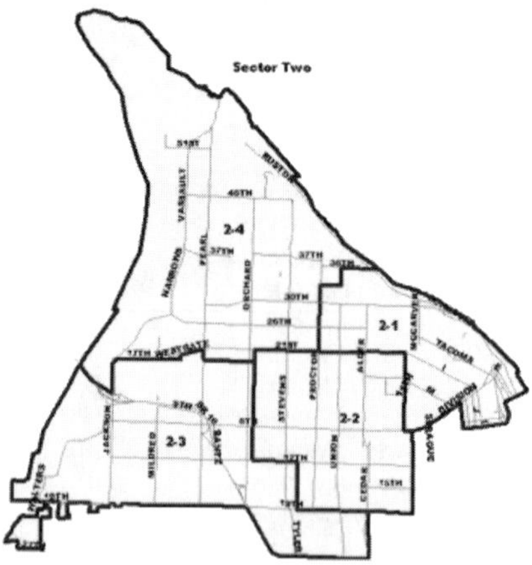

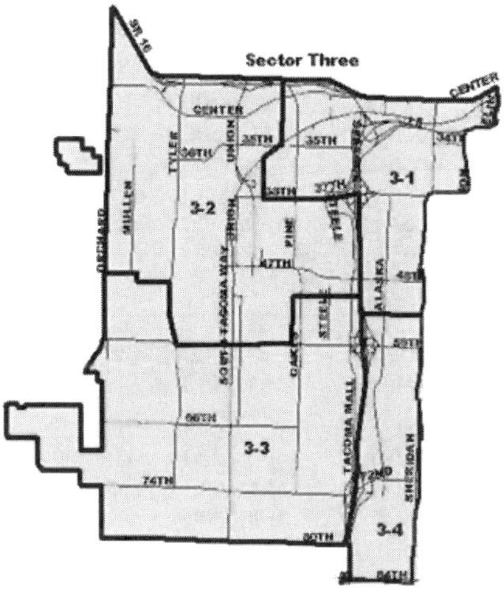

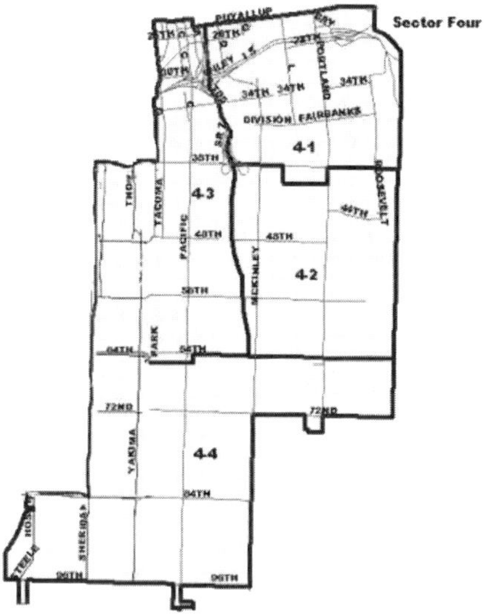

University Place City Government Unviersity Plc 453.0500 458.0500

University Place Police Department: Dispatched on Pierce Sheriff WEST

WA Corrections Center For Women Gig Harbor 155.7000
Gig Harbor 453.8250

Washington State Government Gig Harbor 866.5625
867.0625 867.5125
Steilacoom 867.5125

McNeil Island State Prison [DPL 023] Gig Harbor 868.1375
Steilacoom 155.8350
Buckley 158.8200

Highway Maintenance Tacoma 047.1000
047.0200 047.2000 047.2400

TRUNKED RADIO SYSTEM (LTR): TrunkTracker Setting: LTR
Department of Transportation Statewide LTR system
852.6625 852.6625 854.1625 854.1625

Washington State Patrol

Note: Not all WSP units or dispatchers transmit PL tone at all times
District 1: Pierce & Thurston Counties

Ch4	West Pierce County Disp	[PL 151.4]	Tacoma	155.5200	Simplex
Ch11	Districtwide Car to Car	[PL 151.4]	Countywide	154.9050	Simplex
Ch19	Thurston County Dispatch	[PL 151.4]	Olympia	154.9200	Simplex
Ch31	East Pierce County Disp	[PL 151.4]	East Pierce	154.7400	Simplex
	Thurston County Car-Car	[PL 151.4]	Olympia	155.4600	Simplex

Wilkeson Police Department: Dispatched by Buckley Police

Airports

American Lake Seaplane Base	Tacoma	122.9000	AM
Bandera State Airport	Bandera	122.9000	AM
Pierce County Thun Field Airport	Puyallup	122.7000	AM
Seattle Approach / Departure	Puyallup	126.5000	AM
Clearance Delivery	Puyallup	121.8500	AM
Ranger Creek Airport	Greenwater	122.9000	AM

Shady Acres Airport	Spanaway	122.9000	AM
Spanaway Airport	Spanaway	122.9000	AM
Seattle Approach / Departure	Spanaway	126.5000	AM
Swanson Airport	Eatonville	122.9000	AM
Tacoma Narrows Airport	Tacoma	122.9500	AM
ATIS – Airport Information	Tacoma	124.0500	AM
Tower (Day) / CTAF (Night)	Tacoma	118.5000	253.5000
Ground Control	Tacoma	121.8000	AM
Seattle Approach / Departure	Tacoma	120.1000	AM

Medical & Hospitals

Allenmore Hospital	Tacoma	155.3400	Simplex
	Tacoma	151.7450	Simplex
American Red Cross	Tacoma	047.4200	Simplex
Good Samaritan Hospital	Puyallup	464.8250	469.8250
152.0075 155.2950 155.3400	Puyallup	461.4000	Simplex
461.5000 463.3500 159.7650	Puyallup	464.0000	Simplex
159.9900 462.3000 462.3500	Puyallup	451.6250	456.6250
462.4000 462.4500 462.5250	Puyallup	451.4000	456.4000
467.3000 467.3500 467.4000	Puyallup	467.4500	Simplex
	Puyallup	153.6650	153.6050
Mary Bridge Children's Hospital	Tacoma	155.2950	Simplex
	Tacoma	155.3400	Simplex
MultiCare Medical Center	Tacoma	155.3400	Simplex
	Tacoma	453.1750	
	Tacoma	456.2375	Simplex
Operations Repeaters	Tacoma	151.5500	Repeater
152.9000 153.1100 153.2150 153.5750 464.3750			
Puget Sound General Hospital	Tacoma	155.3400	Simplex
Saint Joseph Hospital	Tacoma	152.0075	Simplex
Security Dispatch	Tacoma	155.2800	Simplex
	Tacoma	155.3400	Simplex
Operations Repeater	Tacoma	151.5950	158.4000
Western State Hospital	Ft Steilacoom	462.9000	Simplex
Security Dispatch [PL 67.0]	Ft Steilacoom	155.4000	Simplex

Military

Fort Lewis / Gray Army Airfield

Home of: 1st Special Forces Group, 9th Infantry Division, 15th Support Brigade

Gray Army Airfield

Approach / Departure Control	Tacoma	120.1000	AM
ATIS – Automatic Terminal Info Service	Tacoma	124.6500	306.2000
Airfield Operations	Tacoma	032.3000	Simplex
Ground Control	Tacoma	121.9000	290.2000
Metro / Dispatcher	Tacoma	134.1000	AM
Tower	Tacoma	119.3250	256.8000
Tower	Tacoma	276.4000	AM
Flight Following (Bullseye)	Tacoma	379.1000	AM
034.6000 141.5000 393.3000			

McChord Air Force Base

Home of: 62nd Airlift Wing, NORAD Western Air Defense Sector, 446th Airlift Wing, 22nd Special Tactics Squadron

ATIS – Automatic Terminal Info Service	Tacoma	109.6000	270.1000

Dispatcher	Tacoma	372.2000	AM
Ground Control	Tacoma	121.6500	275.8000
Approach / Departure Control	Tacoma	126.5000	391.9000
Tower	Tacoma	124.8000	236.6000
Tower	Tacoma	259.3000	AM
NORAD Sector Operations Control Center		228.6000	AM

228.9000 234.6000 235.9000 238.4000 239.7000
252.0000 254.2000 256.6000 260.8000 265.4000
267.0000 270.2000 271.0000 274.4000 275.0000
277.6000 277.8000 281.6000 282.6000 288.4000
292.7000 293.6000 295.8000 300.1250 308.1000
309.4000 316.3000 320.6000 324.0000 327.9000
328.0000 336.6000 341.8000 346.2000 348.2000
351.5000 355.2000 359.8000 362.3000 364.2000
364.3000 364.8000 374.0000 377.0000 387.0000
387.8000 390.2000 394.2000 397.8000 398.0000

TRUNKED RADIO SYSTEM (Motorola): TrunkTracker Setting: Mot Type II UHF

Base: 407.0000 Offset: 25.0 (Digital with some Encryption)

407.3750* 407.4750* 409.2000* 409.3000*
409.4250 409.5000 409.7000 409.9000

Alpha Net	[Dig-P25]	Tacoma	TG: 43216	McChord
Security Police	[Dig-P25]	Tacoma	TG: 41552	McChord
Airfield Operations	[Dig-P25]	Tacoma	TG: 41744	McChord
Aircraft Maintenance (Blue)	[Dig-P25]	Tacoma	TG: 43248	McChord

Miscellaneous

Airlift Northwest	Puyallup	155.2950	Simplex
Alpine Ascents Mountain Guides	Ashford	153.2600	159.9750
BNSF Railway Company	Tacoma	160.9950	160.2150
Canterwood Golf & Country Club		151.6850	Simplex
		151.7450	Simplex
Operations Repeater	Gig Harbor	452.7000	457.7000
Computer City Supercenters	Countywide	461.0875	Simplex
461.4625 466.0875 466.4875			
Crystal Mountain Resort	Crystal Mtn	154.5150	Simplex
154.5700 154.6000 151.6550 151.9250			
Dieringer School District 343	Sumner	155.2350	
Emerald Queen Casino	Tacoma	463.2000	
Slot Attendants [DPL 754]	Tacoma	464.3500	
Security Channel-1 [DPL 754]	Tacoma	464.6250	
Facilities Channel-8 [DPL 754]	Tacoma	464.7000	
461.2250 461.3500 461.7750			
463.5750 464.1500 464.2250			
Foss Waterway Marina	Tacoma	463.5500	Simplex
	Tacoma	464.2500	Simplex
Friends of Explorer Search/Rescue	Tacoma	042.9600	
Golden Pacific Railroad	Tacoma	160.8750	Simplex
Home Depot Stores			
461.2125 463.2875 464.6375			
International Mountain Guides	Paradise	152.4650	157.7250
Crystal Mountain Repeater	Crystal Mtn	151.9250	159.5500
J C Penney Stores			
461.6125 466.2625 466.9625			
Lakewold Gardens	Lakewood	461.5500	Simplex

	Location	Frequency	Mode
	Lakewood	462.0750	Simplex
Metropolitan Park District Tacoma	Countywide	151.5200	Simplex
Northwest Trek Wildlife Park	Eatonville	151.9550	Simplex
Point Defiance Zoo Ops	Tacoma	154.5700	Simplex
151.5800 151.5950 152.4650	Countywide	152.9750	Simplex
Mount Rainier Scenic Railroad	Elbe	160.6350	Simplex
	Elbe	161.3850	Simplex
Pierce Cnty 4x4 Search & Rescue	Puyallup	155.1750	Simplex
Pierce Cnty Public Transportation	Tacoma	452.3250	
	Graham	452.8750	
Pierce County Search & Rescue Council			
047.4600 047.5000 047.5400 047.5800 047.6600			
155.1600 155.1750 155.2050 155.2200 155.2350			
155.2650 155.2800 155.2950 462.9375 463.1875			
Pierce County Security Inc	Tacoma	151.9250	Simplex
Port of Tacoma	Tacoma	153.7850	Simplex
Operations Repeater	Tacoma	453.4875	458.4875
Operations Repeater	Tacoma	453.8125	458.8125
Operations Simplex	Tacoma	453.9875	Simplex
453.0625 453.1625 453.1875 453.2875 453.3625			
453.3875 453.4125 453.5125 453.5375 453.6125			
453.6375 453.6625 453.9125 453.9375 453.9625			
Seattle Times	Orting	453.0000	458.0000
452.9625 452.9875 457.9625 457.9875			
Silver Dollar Casino	Tacoma	464.8250	Simplex
462.2250 462.4500	Tacoma	464.3750	Simplex
South Hill Mall	Puyallup	464.9750	
Security	Puyallup	464.5500	
Tacoma Art Museum	Tacoma	464.2125	Simplex
Tacoma Convention Center	Tacoma	460.2000	465.2000
460.4625 465.1000 465.3750	Tacoma	465.4375	Simplex
465.4625 460.1000 460.3750	Tacoma	460.4375	Simplex
Tacoma Mall	Tacoma	462.3250	467.3250
	Tacoma	451.6500	456.6500
Tacoma News Tribune	Tacoma	453.0000	
461.0375 461.1125 461.2375 461.3125 463.2875			
463.8125 464.0125 464.2125 464.2375 464.2625			
464.4125 464.5125 466.3125 466.3125			
Target Stores Inc	Tacoma	467.7625	Simplex
469.0625 461.6625 461.8625 461.6625 461.8625			
Toys R Us Stores			
457.5375 457.5875 468.7625 469.6625			
WA Explorer Search & Rescue	Tacoma	047.5000	Simplex
	Tacoma	047.5400	Simplex
Western Washington Fair Assoc.	Puyallup	151.9250	Simplex

Schools

	Location	Frequency	Mode
Annie Wright School	Tacoma	461.8375	Simplex
461.9875 463.3375 463.5875	Tacoma	464.9375	Simplex
Bellarmine Preparatory School			
464.3125 467.8500 464.1625			
Bethel School District #403		151.8950	Simplex

Agency	Location	Frequency	Frequency/Mode
151.7450 152.9000		151.9250	Simplex
Operations Repeater	Spanaway	152.3450	157.6050
Buses [PL 123.0]		151.6850	Simplex
Charles Wright Academy	University Plc	461.5875	Simplex
461.6875 452.3875 452.5125	University Plc	461.5375	Simplex
Chief Leschi Indian School	Puyallup	461.2750	466.2750
462.8375 466.2125 466.5875	Puyallup	467.1625	Simplex
Clover Park High School	Countywide	467.8375	Simplex
	Countywide	467.8875	Simplex
Clover Park School District No 400	Tacoma	151.6550	Simplex
School Buses	Tacoma	047.4600	Simplex
Clover Park Technical College	Tacoma	151.7750	Simplex
151.8650 151.9250 154.5150 154.5400			
Operations Repeater	Tacoma	152.4200	157.6800
Dieringer School District 343	Sumner	155.2350	Simplex
Eatonville School District 404	Eatonville	151.7750	Simplex
Fife School District Number 417	Milton	464.0250	
Operations Repeater	Fife	452.2250	457.2250
451.5625 461.0375 461.8000 462.8875			
462.9125 464.4125 464.5375			
Franklin Pierce School District 402	Tacoma	151.7150	Simplex
Orting School District #344	Orting	466.1875	Simplex
460.8875 465.8875 466.0875	Orting	160.0350	Simplex
466.0625 466.4625	Orting	469.5625	Simplex
Pacific Lutheran University	Tacoma	151.8650	Simplex
Security	Tacoma	151.7450	Simplex
	Tacoma	461.5000	466.5000
	Tacoma	464.2250	469.2250
	Tacoma	151.8050	159.8550
Peninsula School District 401	Gig Harbor	155.2650	Simplex
Pierce College	Tacoma	464.5250	469.5250
Puyallup School District	Puyallup	468.4375	Simplex
Buses	Puyallup	154.5400	Simplex
School Operations	Puyallup	154.5150	Simplex
School Operations	Puyallup	154.5700	Simplex
461.9875 467.0375 466.9875			
Edgemont Junior High School	Edgewood	152.9300	159.6000
Stahl Junior High School	Puyallup	153.0350	159.7800
Emerald Ridge High School	Puyallup	153.3200	158.1600
Operations Repeater	Sumner	152.4650	157.7250
	Puyallup	160.0200	Simplex
Steilacoom Historical Schools	Steilacoom	155.2050	Simplex
Sumner School District	Sumner	469.4125	Simplex
Buses [DPL 156]	Sumner	151.8650	Simplex
464.4125 468.2875 469.2125			
	Sumner	153.0200	159.6450
Tacoma Community College	Tacoma	462.8625	Simplex
466.0625 463.4500 462.0625 461.9125		452.6625	Simplex
461.9000 461.7125 461.6375 461.5625		452.7625	Simplex
460.8875 465.8875 461.7625 461.0625			
Tacoma School District #10	Tacoma	154.6000	Simplex
151.8350 463.3875 464.6875 464.6125		462.8125	Simplex
464.5375 464.4875 464.4375 464.0375			
461.7875 461.4125 461.1375 463.9875			

	Tacoma	155.1000	158.8650
University of Washington Tacoma	Tacoma	463.4125	Simplex
	Tacoma	463.8125	
University of Puget Sound	Tacoma	151.8050	Simplex
	Tacoma	461.1500	Simplex
Janitorial	Tacoma	461.7375	Simplex
Paging System	Tacoma	157.7400	Simplex
Security Personnel	Tacoma	462.7250	
Facilities Maintenance	Tacoma	452.4250	
	Tacoma	464.4250	Simplex
University Place School District	University Plc	451.4500	456.4500
Operations Repeaters	University Plc	451.7250	456.7250
463.9000 464.1250 464.3250 464.7750			
White River School District	Buckley	151.5800	159.9150
Buses [DPL 464]	Buckley	151.8050	Simplex
	Buckley	152.3600	Simplex

Utilities

Elmhurst Mutual Power & Light	Tacoma	048.2800	Simplex
Firgrove Mutual Inc	Puyallup	153.5000	Simplex
Harbor Water Company Inc	Gig Harbor	037.5200	Simplex
Lakeview Light & Power Co Inc	Tacoma	153.4700	Simplex
Lakewood Water District	Tacoma	048.9000	048.3200
Ohop Mutual Light Company	Eatonville	047.7600	Simplex
	Eatonville	048.5400	Simplex
Parkland Light & Water Co	Tacoma	451.1000	
Peninsula Light Co Inc	Key Center	047.7800	Simplex
	Key Center	047.9200	Simplex
Puget Sound Energy Inc	Tacoma	153.4100	
153.4400 153.5450 153.6800 158.1450			
TRUNKED RADIO SYSTEM (EDACS): TrunkTracker Setting: ED			
Tacoma Tower Site:	Tacoma		
936.3875 936.4000 936.4125 936.4250 936.4375			
936.4500 936.4625 936.4750 936.4875 936.5000			
Sumner Tower Site:	Sumner		
936.9000 936.9125 936.9250 936.9500			
Southeast Tacoma Mutual Water	Tacoma	153.5000	
Spanaway Water District	Spanaway	451.0750	
Summit Water & Supply Company	Tacoma	153.5000	
Tacoma City Utilities	Tacoma	048.0200	
048.4800 153.7100 158.2500 451.0250 451.1250			
153.2450 153.3350 159.5850	Tacoma	159.9000	Simplex
Tanner Electric Cooperative	Anderson Isl	047.9600	
Washington Natural Gas Company	Tacoma	048.1600	

Weather Information

National Weather Service – NOAA Weather Radio

Conditions & Forecasts – Seattle	162.5500	Simplex
Conditions & Forecasts – Capitol Peak	162.4750	Simplex
Pierce County SAME Alert Code: 053053		

SAN JUAN COUNTY

Government & Public Safety

Cascade Ambulance Service				
Shared Operations Repeater		Eastsound	464.3000	469.3000
San Juan County Fire Department		Eastsound	154.3850	Simplex
Dispatch: Orcas Island	[PL 192.8]	Friday Harbor	154.2050	153.8300
Dispatch: Friday Harbor	[No PL]	Friday Harbor	155.1000	
Dispatch: San Juan Island	[PL 203.5]	San Juan Is	155.2800	
Dispatch: Lopez Island	[PL 203.5]	Lopez Is	155.1000	
Fire District 3 (Future TRS)		Friday Harbor	155.1450	153.8075
Fire District 3 (Future TRS)		Friday Harbor	155.7975	156.1575
San Juan County Government		Friday Harbor	155.1000	
Public Works Department		Friday Harbor	453.4625	Simplex
		Friday Harbor	458.4000	
		Friday Harbor	458.6750	
Public Works Department		Friday Harbor	851.0125	806.0125
San Juan County Sheriff	[PL 118.8]	Friday Harbor	453.2500	458.2500
1## San Juan Island; 2## Orcas Island; 3## Lopez Island; Z#: Detectives				
San Juan Island Emergency Medical		Friday Harbor	155.2800	Simplex

Washington State Patrol

Note: Not all WSP units or dispatchers transmit PL tone at all times

District 7: Snohomish, Skagit, Whatcom, San Juan, Island Counties

Ch13	West Snohomish County	[PL 141.3]	Everett	155.6550	154.8450
Ch17	Districtwide Car to Car	[PL 141.3]	Countywide	154.9350	Simplex
Ch18	Skagit / Island Counties	[PL 141.3]	Countywide	155.5050	159.0450
Ch20	Whatcom County	[PL 141.3]	Bellingham	154.7550	155.2500
	East Snohomish County	[PL 156.7]	Skykomish	154.7550	Simplex

Miscellaneous

Name	Location	Frequency	Notes
Bellingham Herald	Eastsound	462.5500	
Custer Water Association	Rosario	463.3500	
Doe Bay Water Users Association	Lopez Island	048.3600	
	Eastsound	464.2375	469.2375
Friday Harbor Airport/Seaplane Base	Friday Harbor	128.2500	AM
Whidbey Approach / Departure	Friday Harbor	118.2000	AM
Aeronautical Radio Inc at Friday Habor	Friday Harbor	131.5250	AM
Inns at Friday Harbor	Friday Harbor	151.8350	Simplex
KGMI Inc	Eastsound	450.0200	
	Eastsound	450.3500	
KIRO TV	Eastsound	863.0375	
Lopez Island Airport	Lopez	128.2500	AM
Mt Baker School District	Eastsound	463.4250	
Orcas Island Airport	Eastsound	128.2500	AM
Orcas Island School District	Eastsound	463.7250	468.7250
	Eastsound	467.8500	
Orcas Power & Light	Friday Harbor	047.7400	
	Friday Harbor	048.3600	
	Friday Harbor	160.0950	
Port Of Friday Harbor	Friday Harbor	154.5700	Simplex
	Friday Harbor	154.6000	Simplex
Roche Harbor Seaplane Base	Roche Harbor	128.2500	AM
Rosario Resort	Eastsound	463.3500	468.3500
462.2000 462.2250 462.2500	Eastsound	462.8000	Simplex
451.1875 452.7625 451.2375	Eastsound	451.3125	Simplex
	Eastsound	451.3375	Simplex
Seaplane Base	Eastsound	128.2500	AM
San Juan Island School District	Friday Harbor	154.5400	Simplex
Security Specialists Plus Inc	Eastsound	461.3250	
Semiahmoo Hotel Company Inc	Eastsound	461.2500	
Valley Broadcasters Incorporated	Orcas Island	450.8500	
Whatcom Security Agency Inc	Eastsound	463.7250	
Wights Marine Services Inc	Eastsound	463.4500	
YMCA Of Greater Seattle	Eastsound	464.3000	

Weather Information

National Weather Service – NOAA Weather Radio

	Frequency	Notes
Conditions & Forecasts – Puget Sound Marine	162.4250	Simplex

San Juan County SAME Alert Code: 053055

SKAGIT COUNTY

Government & Public Safety

Skagit County Law Enforcement Callsigns – First Letter Indicates Agency

D	Anacortes PD	K	La Conner PD	U	Skagit County Sheriff
G	Sedro Woolley PD	H	Burlington PD	X	Mount Vernon PD
I	Swinomish Tribal PD	N	Concrete PD	Z	Upper Skagit Tribal PD

Anacortes City Government		Anacortes	154.2650	
154.4300 155.9400 158.8650		Anacortes	453.5625	Simplex
		Anacortes	453.7000	458.7000
		Anacortes	453.1625	458.1625
Anacortes Police Dept.	[DPL 723]	Anacortes	851.2375	
Repeater Frequency		Anacortes	453.9500	458.9500
Burlington City Government		Burlington	154.0850	
Burlington Fire Department: See Skagit County Fire				
Burlington Police Dept.	[PL 118.8]	Burlington	453.9000	458.9000
Dispatched on Mt. Vernon PD Frequency when no dispatcher available				
Concrete City Government		Concrete	154.2650	154.4300
Public Works Department		Concrete	154.9800	
Concrete Fire Department: See Skagit County Fire				
Concrete Police: By Sheriff	[PL 110.9]	Concrete	453.4000	458.4000
Hamilton Town Government / Fire		Hamilton	154.2650	
		Hamilton	154.4300	
La Conner Police Dept.	[PL 118.8]	La Conner	453.6500	
Now Dispatched by Sheriff's Office on 453.400				
Mount Vernon City Government / Fire		Mount Vernon	154.1000	
Public Works	[PL 136.5]	Mt Vernon	155.0850	
		Mt Vernon	154.8225	

Mount Vernon Fire Department: See Skagit County Fire					
Mount Vernon Police Dept.		[PL 118.8]	Mt Vernon	453.6250	458.6250
NW Air Pollution Authority			Mt Vernon	155.0850	
Port Of Skagit			Burlington	158.8950	
Sedro Woolley City			Sedro Woolley	158.9550	
Sedro Wooley Police Department			Sedro Woolley	460.3500	465.3500
	Dispatched on Burlington PD when open; Sheriff frequency at other times				
F-4	Tactical Simplex	[PL 118.8]	Sedro Woolley	453.5250	Simplex
Skagit County Fire					
F-1	County Fire Dispatch	[PL 136.5]	Countywide	154.4300	Simplex
F-2	Tac 2 Operations	[PL 136.5]	Burlington	154.2650	Simplex
F-3	City Fire – Mount Vernon	[PL 136.5]	Mt Vernon	154.2350	Simplex
Tac 3	Tactical		Mt Vernon	155.6850	154.8225
HEAR	Medic Dispatch	[PL 136.5]	Countywide	155.3400	Simplex

Skagit County Fire Station Locations & Apparatus

Station / Location	Radio	Apparatus
Alger #1	Tac-2	E1411, A1419
Alger #2	Tac-2	E1421, R1427, A1429
Allen	County Fire	E521, WT526, R527
Anacortes #1	County Fire	E2911, E2912, L2915, R2917, M13, M15
Anacortes #2	County Fire	E2921, M14
Bayview	Tac-2	E1211, E1212, WT1216, R1217, A1219
Big Lake #1	Tac-2	E911, E912, U918, A919
Big Lake #2	Tac-2	E921
Big Lake #3	Tac-2	E931
Birdsview	County Fire	E1011, WT1016
Burlington	Tac-2	E611, E612, E613, L615, WT616, R617, U618, A619
Cedardale	Tac-2	E321, E322, R329
Clear Lake 1	Tac-2	E411, E412, R417
Clear Lake 2	Tac-2	E421
Concrete	County Fire	E1, E2, E5
Conway	Tac-2	E311, E312, WT316, R317
Day Creek	County Fire	E1611, WT1616, U1618
Edison	County Fire	E511, R517
Grassmere	County Fire	E1021, WT1026
Guemes Island	County Fire	E1711, A1719
Hamilton	County Fire	E2511, A2519
Hickson	County Fire	E821, E822, WT826
Hope Island	Tac-2	E1331, R1337
La Conner	Tac-2	E1321, E1322, WT1326, R1327
Lake Cavanaugh	County Fire	E711, E712, WT716, R717, Marine 718
Lake McMurray	County Fire	E1511, WT1516, R1517, A1519
Lyman	County Fire	E841, R847
Marblemount	County Fire	E1911, WT1916, R1927
McLean Road	County Fire	E211, E212, E213, WT216, R217, A219
Mount Erie #1	County Fire	E1111, E1112, WT1116, R1117, A1119
Mount Erie #2	County Fire	E1121, E1122, R1127
Mount Vernon 1	City Fire	E111, E112, L115, R117, U118
Mount Vernon 2	City Fire	E122, L125, U128
Mount Vernon 3	City Fire	E131, L132, U138, A139
Prarie	County Fire	E831, R837, A839
Punkin Center	County Fire	E851, E852, WT856, R857
Rockport	County Fire	E1031, WT1037
Samish Island	County Fire	E531, R537
Sedro Woolley	County Fire	E5511, U5510, E5512, WT5516, R5517, U5518, A5519
Summit Park	Tac-2	E1311, E1312, WT1316, R131

Skagit County Medic One Ambulance Locations

Med 1	ALS	UGH Sedro Woolley	Med 10	ALS	Concrete Back-Up
Med 2	ALS	SVH Mt. Vernon	Med 11	BLS	Puget Power Diablo
Med 3	BLS	Marblemount	Med 12	BLS	Upriver Back-Up
Med 4	ALS	SVH Mt. Vernon	Med 13	ALS	Anacortes Sta 1
Med 5	BLS	UGH Sedro Woolley	Med 14	ALS	Anacortes Sta 2
Med 6	BLS	SVH Mt. Vernon	Med 15	ALS	Anacortes Back-Up
Med 7	ALS	Concrete			

Skagit County Government		Mt Vernon	867.7500	
155.7300		Countywide	460.2000	465.2000
453.2500 453.3250 453.3625		Countywide	453.6750	
453.8125 456.3000 853.3125		Countywide	460.1375	465.1375
		Countywide	155.6325	159.0150
		Lyman	155.8050	154.9575
		Lyman	155.7675	158.9175
		Lyman	159.1500	155.9925
		Lyman	153.7850	159.0150
Public Works		Countywide	867.7500	822.7500
Public Works		Countywide	868.6375	823.6375
Public Works	[PL 136.5]	Mt Vernon	151.0550	156.0600
Flood Control Council		Mt Vernon	154.1000	
Flood Control / Dike District		Countywide	158.8350	Simplex
Transit System		Lyman	868.8875	823.8875
Skagit County Sheriff	[PL 100.0]	Countywide	453.4000	458.4000
Cony – Fidalgo Island	[PL 118.8]	Fidalgo Island	453.4000	458.4000
Administration	[PL 100.0]	Countywide	155.3100	
Jail Operations		Mt Vernon	453.9750	Simplex
Major Incidents + Search & Rescue		Countywide	155.8425	151.4375
Swinomish Tribal Police		La Conner	453.5000	458.5000
Dispatch	[PL 114.8]	Swinomish	166.6375	
Upper Skagit Tribal Police	[PL 114.8]	Upper Skagit	166.6375	

Washington State Patrol

Note: Not all WSP units or dispatchers transmit PL tone at all times

District 7: Snohomish, Skagit, Whatcom, San Juan, Island Counties

Ch17	Districtwide Car to Car	[PL 141.3]	Countywide	154.9350	Simplex
Ch18	Skagit / Island Counties	[PL 141.3]	Countywide	155.5050	159.0450

Miscellaneous

Anacortes Airport	Anacortes	122.8000	AM
CTAF	Anacortes	128.2500	AM
Avalon Golf Club		154.6000	Simplex
Concrete Municipal Airport	Concrete	122.9000	AM
Glenhaven Lakes Club Inc		461.2375	Simplex
Salish Fireworks	Sedro Wooley	463.2125	Simplex
Seattle Times	Mount Vernon	453.0000	
Skagit Regional Airport	Burlington	123.0750	AM
Whidbey Approach/Departure	Burlington	120.7000	AM
Skyline Seaplane Base	Anacortes	128.2500	AM
Skagit Speedway	Mount Vernon	151.9250	Simplex
Operations Simplex	Countywide	151.9550	Simplex
461.9875 462.1125 463.2250 466.9875 467.1125			
Skagit Valley Casino (Upper Skagit Indian Tribe)			
Casino Operations	Bow	462.8625	Simplex
464.0375 465.8875 466.0375 466.1625 466.2125			
466.2375 466.2875 466.3375 466.3875 466.4125			
466.4375 466.5125 466.5375 466.5625 466.5875			
Swinomish Casino	Countywide	463.6125	Simplex
Target Stores Inc	Mount Vernon	461.6625	Simplex
461.8625 467.7625 469.0625 469.8875			
Thousand Trails	Countywide	154.6000	Simplex
451.5375 463.3125			

Medical & Hospitals

American Red Cross	Anacortes	047.4200	
Island Hospital	Anacortes	155.2800	Simplex
463.8625 468.8625	Anacortes	155.3400	Simplex
	Anacortes	154.5400	Simplex
Skagit Valley Hospital	Mount Vernon	154.5400	Simplex
Countywide EMS Primary Dispatch	Mount Vernon	155.2800	Simplex
	Mount Vernon	155.3400	Simplex
United General Hospital	Sedro Woolley	451.3375	Simplex
	Sedro Woolley	451.5000	Simplex

Schools

Anacortes School District 103		Anacortes	151.7750	Simplex
Buses	[PL 85.4]	Anacortes	151.7150	Simplex
464.4875 466.1875 466.2625 467.8125 469.5125				
Burlington Edison Schools	[PL 107.2]	Burlington	155.2650	Simplex
Operations Repeater		Burlington	452.1250	457.1250
Operations Repeater		Burlington	462.2000	467.2000
451.1875 451.2875 451.4125 451.4875				
456.1875 456.2875 456.4125 456.4875				
Concrete School District	[PL 127.3]	Concrete	155.2050	Simplex
La Conner School District	[PL 100.0]	La Conner	155.2350	Simplex
Mount Vernon School Dist No 320		Mount Vernon	464.3125	Simplex
469.3125		Mount Vernon	452.6625	457.6625
Oak Harbor High School		OakHarbor	154.5150	Simplex
Sedro Woolley School District				
School Buses	[PL 103.5]	Sedro Woolley	155.2050	Simplex
Sedro Woolley High School		Sedro Woolley	457.3125	Simplex
Sedro Woolley High School		Sedro Woolley	457.3625	Simplex
Sedro Woolley High School		Sedro Woolley	457.8125	Simplex
Sedro Woolley High School		Sedro Woolley	457.8625	Simplex
Skagit Valley College		Mount Vernon	154.5150	Simplex
Security		Mount Vernon	452.6500	Simplex
		Mount Vernon	452.6250	457.6250
Stanwood School District #401		Mt Vernon	864.6375	Simplex
Western Washington University		Bellingham	462.1875	Simplex
		Bellingham	467.1875	Simplex

Public Utilities

Cascade Natural Gas Corp	Mount Vernon	048.5000	
Public Utility District	Burlington	048.0600	048.5400
Puget Sound Energy Inc	Anacortes	153.4100	Simplex
153.4250 153.5450 153.6200 153.6350 158.1300			
158.1450 451.0500 451.2500 451.6125 456.6125			
Skagit Co Flood Ctrl Council	Burlington	154.1000	
Skagit County Public Utilities	Concrete	158.2500	
	Countywide	453.4000	

Weather Information

National Weather Service – NOAA Weather Radio

Conditions & Forecasts – Puget Sound Marine	162.4250	Simplex
Skagit County SAME Alert Code: 053057		

SKAMANIA COUNTY

Government & Public Safety

Columbia River Inter-Tribal Fisheries Enforcement				
		Milla	155.5425	153.7925
North Bonneville Town	[PL 103.5]	N Bonneville	453.9000	
Skamania County EMS				
Ch1		Stevenson	453.1000	458.1000
Ch2		Stevenson	453.6000	458.6000
Skamania County Fire		Stevenson	154.1600	
Ops-1 “Lookout” West County	[PL 88.5]	Stevenson	460.6250	465.6250
Ops-2 “Defiance” Central County	[PL 94.8]	Stevenson	460.6250	465.6250
Ops-3 “Strunk” Washougal Area	[PL 100.0]	Stevenson	460.6250	465.6250
Simplex Operations	[PL 103.5]	Stevenson	460.6250	Simplex
Car-to-Car	[PL 107.2]	Stevenson	460.6250	Simplex
Simplex Countywide	[PL 100.0]	Stevenson	453.0625	Simplex
District 5 Simplex Ops	[PL 103.5]	Stevenson	465.5625	Simplex
Tactical		Countywide	453.4250	Simplex
Tactical		Countywide	458.1250	Simplex
Skamania County Government		Stevenson	039.8200	
		Stevenson	155.8050	
Roads Flagging Simplex Channel		Stevenson	453.7125	Simplex
F-1 Roads “Lookout” West	[PL 88.5]	Stevenson	453.8250	458.8250
F-2 Roads “Defiance” Central	[PL 94.8]	Stevenson	453.8250	458.8250
F-3 Roads “Simplex”	[PL 100.0]	Stevenson	453.8250	Simplex
HEARS		Stevenson	155.3400	Simplex
Skamania County Sheriff		Stevenson	155.3700	
North County		Stevenson	453.6750	458.6750
South County	[PL 103.5]	Stevenson	453.2500	458.2500

	Stevenson	453.5000	
Stevenson Town Government	Stevenson	045.1200	
	Stevenson	453.7750	458.7750
	Stevenson	460.0750	Simplex
Yacolt Town Government	Cougar	153.9500	155.6550
Yacolt Fire Department	Cougar	453.3000	458.3000
	Swift	155.1600	
	Swift	155.2650	

Washington State Patrol

Note: Not all WSP units or dispatchers transmit PL tone at all times

District 5: Lewis, Klickitat, Cowlitz, Clark, Skamania, Wahkiakum Counties

Ch7	Clark / Skamania County [PL 136.5]	Countywide	154.6800	155.5800

Miscellaneous

Public Utility District 1	Skamania	048.0200	048.5400
Skamania Lodge	Stevenson	451.2375	456.2375
	Stevenson	452.1625	457.1625
Union Pacific Railroad	Skamania	160.3650	160.2900

Weather Information

National Weather Service – NOAA Weather Radio

Conditions & Forecasts – Portland	162.5500	Simplex
Conditions & Forecasts – The Dalles	162.4000	Simplex

Skamania County SAME Alert Code: 053059

SNOHOMISH COUNTY

Government & Public Safety

Snohomish County Regional 800 MHz Trunked Radio System

Scanner Setting. Motorola Type II 800 Standard (Mixed Analog & APCO P25 Digital)

IF THESE FREQUENCIES CHANGE, FREE UPDATES ARE AT SCANSEATTLE.COM

Subregional Simulcast Cells

Note. Each talkgroup in the lists in this section operates at all times on one or both of the subregional simulcast cells (East and West) listed below, based generally on geographic focus of the talkgroup. To avoid missed conversations, use the Simulcast Cell for the cell covering the part of the county to which the talkgroup you are interested in is focused. Note, however, that most talkgroups can "roam" between cells and onto the intellirepeaters listed below when needed to provide clear communications with radios that are not receiving that talkgroup's primary cell well. System control channels are noted below by a * and bold. If your scanner is capable of Control Channel Only Mode (see your manual), only control channels need to be entered for each cell. The other (voice) frequencies will be tuned automatically by your scanner using data provided on the control channel.

West: **West Snohomish County Simulcast Cell**
Covers: Everett, Mill Creek, Mukilteo, Lynnwood, Edmonds, Mountlake Terrace.
868.3375* 867.9375* 867.1750* 866.2625 866.7750
866.8000 866.8250 867.2000 867.4125

East: **East Snohomish County Simulcast Cell**
Covers: General all parts of Snohomish County not covered by the West cell.
868.9250* 868.7500* 867.6750 868.3625 868.3875
868.5875 868.6125

Subregional Intellirepeaters

Deer Creek Intellirepeater
866.3000 866.7000 866.8500 867.7750 868.6625

North Mountain Intellirepeater
866.3000 866.7000 866.8500 867.7750 868.6625

Agency	Location	Frequency	Output
Airlift Northwest	Arlington	155.2950	Simplex
Arlington Fire Department	Arlington	154.1600	Simplex
See SnoPac North County			
Arlington Town Government	Arlington	155.0550	Simplex

Arlington Police Department: Dispatched on Snohomish County Sheriff NORTH

Bothell City Fire: See King County listing

Bothell City Government: See King County listing

Bothell Police Department: See King County listing

Agency	Location	Frequency	Output
Brier City Government	Brier	155.1450	156.2250

Brier Police Department

See Snohomish County Law Enforcement Communications (Sno-Comm)

Darrington Fire Department: See SnoPac North County

Agency	Location	Frequency	Output
Darrington Town Government	Darrington	155.0550	Simplex
Events	Darrington	TG: 61808	Snohomish

Darrington Police Department: Dispatched on Snohomish County Sheriff NORTH

Agency	Location	Frequency	Output
Edmonds City Government	Edmonds	453.4375	458.4375

Edmonds Fire Department: See SnoComm South County

Edmonds Police Department

Snohomish County Law Enforcement Communications (Sno-Comm)

Everett City Government

Use	Location	Frequency	Output
Public Works	Everett	153.9800	Simplex
Public Works	Everett	153.9050	156.2400
Housing Authority	Everett	461.9250	466.9250
Housing Authority	Everett	TG: 62224	Snohomish
Housing Authority – Conventional	Everett	852.5125	807.5125

Everett Fire Department (SnoPac)

Channel	Use	Location	Frequency	Output
Ch1	Dispatch & Operations	Everett	154.3850	Simplex
Ch2	Tactical / Fireground	Everett	154.2800	Simplex
Tac-1	Fire Tactical	Everett	TG: 592	Snohomish

Everett Police Department

Channel	Use	Location	Frequency	Output
EV POL 1	Patrol Dispatch	Everett	TG: 48	Snohomish
EV POL 2	Patrol Records Access	Everett	TG: 80	Snohomish
EV POL 3	Alternate / Tactical	Everett	TG: 112	Snohomish
EV POL 4	Alternate / Tactical	Everett	TG: 144	Snohomish
EV TAC 1	Car to Car	Everett	TG: 176	Snohomish
EV TRAIN	Training	Everett	TG: 208	Snohomish
EV CID	Criminal Investigations	Everett	TG: 240	Snohomish
EV SIU	Special Investigations Unit	Everett	TG: 45904	Snohomish
EV ERT	Emergency Response Team	Everett	TG: 45968	Snohomish

OLD VHF System

Channel	Use	Location	Frequency	Output
Ch1	Dispatch & Operations	Everett	154.9950	156.2100
Ch2	Records Access – South	Everett	153.8000	154.7250
Ch3	Dispatch Direct	Everett	154.9950	Simplex
Ch4	Alternate + City Public Works	Everett	153.9800	Simplex
Ch5	LERN	Everett	155.3700	Simplex
Ch6	Records Access – North	Everett	153.8000	154.7250
Ch7	Records Access Direct	Everett	153.8000	Simplex
Ch8	Car to Car + City Public Works	Everett	153.9050	Simplex

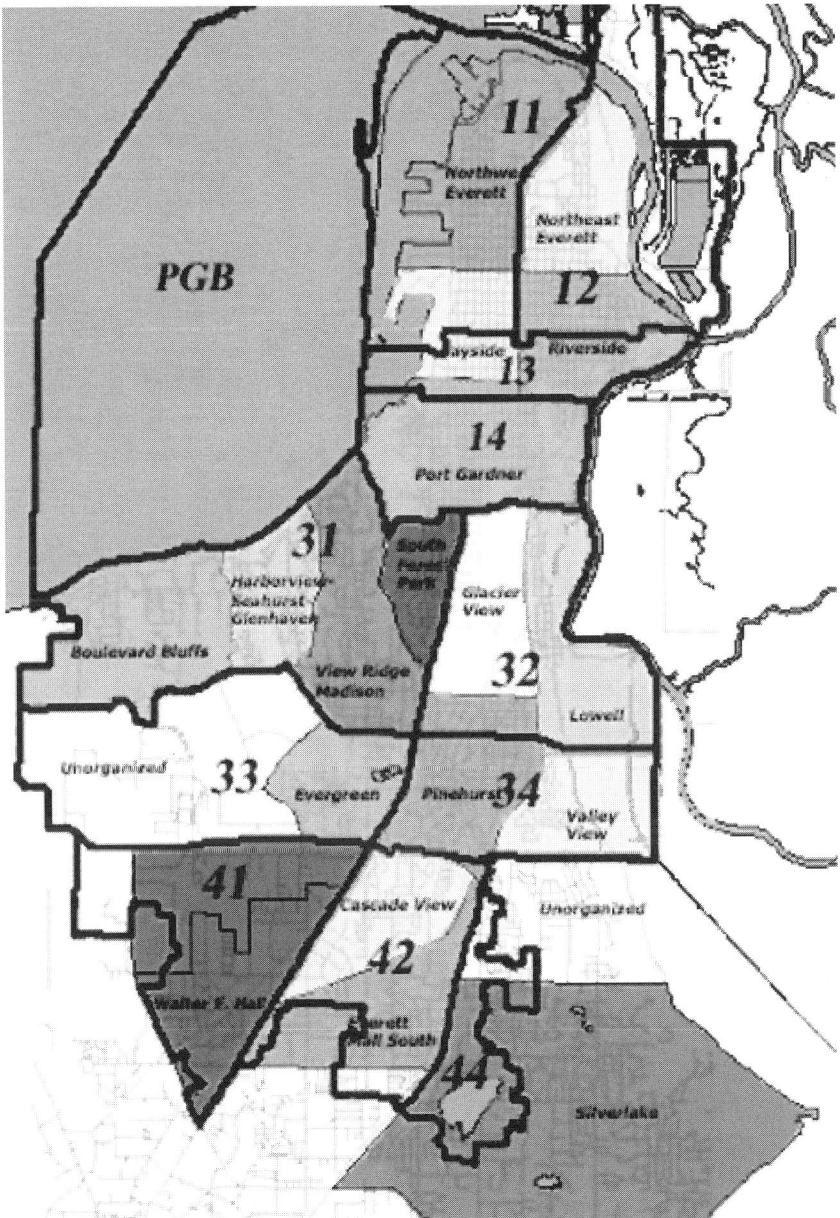

Gold Bar Fire Protection: See SnoPac North, East County

Gold Bar Town Government Gold Bar 153.8150 Simplex

Gold Bar Police Department: Dispatched on Snohomish County Sheriff EAST

Granite Falls Fire Protection: See SnoPac North County

Granite Falls Police: Dispatched on Snohomish County Sheriff NORTH

Index Fire Protection: See SnoPac East

Index Law Enforcement: Contract to Snohomish County Sheriff EAST

Lake Stevens Fire Protection: See SnoPac North County

Lake Stevens Police Department: Dispatch on Snohomish County Sheriff EAST

Lynnwood City Government	Lynnwood	155.8650	Simplex
Parks Department	Lynnwood	453.3125	Simplex
Parks Department	Lynnwood	458.3125	Simplex

Lynnwood Fire Department: See SnoCom South County

Lynnwood Police Department

Snohomish County Law Enforcement Communications (Sno-Comm)

Marysville Fire Department: See SnoPac North County

Marysville City Government	Marysville	153.8750	Simplex
Public Works	Marysville	152.9450	Repeater
153.1250 151.5350 151.5800	Marysville	151.5200	Repeater
	Marysville	151.6850	Repeater
Marysville Police Department	Marysville	158.7750	Simplex
153.8750 158.7750			
MPD POL 1 Dispatch	Marysville	TG: 9616	Snohomish
MPD POL 2 Car to Car	Marysville	TG: 9648	Snohomish
MPD POL 3 Administration / TAC	Marysville	TG: 9680	Snohomish

Mill Creek Police Department: Dispatched on SnoCom 1

Car to Car Simplex	Mill Creek	158.7300	Simplex
Mill Creek Fire Protection	Mill Creek	154.3250	Simplex

Monroe Fire Department: See SnoPac East County

Monroe Police Department: Dispatched on Snohomish County Sheriff EAST

Mountlake Terrace Fire Department: See SnoComm South County

Mountlake Terrace Police Department

Snohomish County Law Enforcement Communications (Sno-Comm)

Mukilteo Fire Department: See SnoPac North County

Mukilteo City Government	Mukilteo	156.0000	Simplex

Mukilteo Police Department: Dispatched on Snohomish County Sheriff SOUTH

Port of Edmonds	Edmonds	155.9850	Simplex
Port of Everett	Everett	467.7875	Simplex
	Everett	154.5700	Simplex
	Everett	154.6000	Simplex
Rural / Metro Ambulance	Countywide	155.2800	Simplex

TRUNKED RADIO SYSTEM (Motorola): TrunkTracker Setting: Type II 900 (E2-900)

935.3375 935.3500 935.3625 935.3750

Dispatch & Operations	Countywide	TG: 4816	RMA
Shannon Ambulance	Everett	155.2800	Simplex
Snohomish City Government	Snohomish	155.1000	Simplex

Snohomish City Police: Dispatched on Snohomish County Sheriff EAST

Snohomish County Government

453.8250 851.8375 852.5875 853.3375 866.1875

Emergency Management	Arlington	154.0550	Simplex
Road Maintenance	Everett	151.1000	156.1050
Denney Juvenile Justice Center	Everett	159.1275	155.9325
Senior Services	Mukilteo	862.4125	816.4125
Senior Services	Mukilteo	864.7125	818.7125
Executives	Countywide	TG: 59344	Snohomish
Medical Examiner	Countywide	TG: 60016	Snohomish
ALERT Team Responses	Countywide	TG: 60496	Snohomish
Portable Repeater	Countywide	453.3500	458.3500

Shared Trunked Talkgroups

R MUTUAL AID	Public Safety Mutual Aid & Command	TG: 45488	Snohomish
RNTF	Regional Narcotics Task Force	TG: 45520	Snohomish
LERN MON	Patch to WSP King Lake Base	TG: 60624	Snohomish
ISL SO MA	Patch to Island County Sheriff	TG: 60656	Snohomish
SK SO MA	Patch to Skagit County Sheriff	TG: 60688	Snohomish
ALL GOV	All Governments Link	TG: 61072	Snohomish
SCHOOL	All Schools – Shared	TG: 61168	Snohomish
SNO PSAP NET	All 911 Centers (For Announcements)	TG: 61584	Snohomish
S EVENT	South Area Events	TG: 61680	Snohomish
W EVENT	West Area Events	TG: 61744	Snohomish
ALL EVENT 1	Countywide Events	TG: 61936	Snohomish
ALL EVENT 2	Countywide Events	TG: 62000	Snohomish
EMER	SnoPac Emergency	TG: 1072	Snohomish
EOC HAIL	Emergency Operation Centers Calling	TG: 60912	Snohomish

Radio Maintenance

S COMM 1	System Support – Primary	TG: 61328	Snohomish
S COMM 2	System Support – Alternate	TG: 61360	Snohomish
S COMM 3	System Support – Alternate	TG: 61392	Snohomish
S COMM 4	Digital Tests – Clear	TG: 45584	Snohomish
S COMM 5	Digital Tests – Encrypted	TG: 45616	Snohomish
DEMO 1	Tests & Demonstrations	TG: 61424	Snohomish
DEMO 2	Tests & Demonstrations	TG: 61456	Snohomish
DEMO 3	Tests & Demonstrations	TG: 61488	Snohomish
DEMO 4	Tests & Demonstrations	TG: 61520	Snohomish

Snohomish County Fire Protection Districts

1 W DISP	Dispatch – West	TG: 496	Snohomish
2 E DISP	Dispatch – East	TG: 528	Snohomish
3 N DISP	Dispatch – North	TG: 560	Snohomish
4 M DISP	Dispatch – Marysville FD	TG: 9872	Snohomish
5 DISP 5	Dispatch – SnoCom FD's	TG: 5360	Snohomish
6 DISP 6	Dispatch – SnoCom FD's	TG: 5392	Snohomish
TAC 1 WP	West Response – Primary	TG: 592	Snohomish
TAC 2 WS	West Response – Secondary	TG: 624	Snohomish
TAC 3 EP	East Response – Primary	TG: 656	Snohomish
TAC 4 ES	East Response – Secondary	TG: 688	Snohomish
TAC 5 NP	North Response – Primary	TG: 720	Snohomish
TAC 6 NS	North Response – Secondary	TG: 752	Snohomish
TAC 7 MP	Marysville Response – Primary	TG: 9904	Snohomish
TAC 8 MS	Marysville Response – Secondary	TG: 9936	Snohomish
TAC 21	SnoCom Response	TG: 5424	Snohomish
TAC 22	SnoCom Response	TG: 5456	Snohomish
TAC 23	SnoCom Response	TG: 5488	Snohomish
TAC 24	SnoCom Response	TG: 5520	Snohomish
EMER-SCF	SnoCom Emergency	TG: 5712	Snohomish
ADMIN	Fire Administration Countywide	TG: 59664	Snohomish
TRAIN	Fire Training Countywide	TG: 59696	Snohomish

PREV	Fire Prevention Countywide		TG: 59728	Snohomish
MARSHAL	Fire Marshal Countywide		TG: 59952	Snohomish
MA SNO 1	Countywide Fire Mobilization		TG: 59472	Snohomish
MA SNO 2	Countywide Fire Mobilization		TG: 59504	Snohomish
PSOPS N3	SnoPac Mutual Aid		TG: 60432	Snohomish
PSOPS N4	SnoCom Mutual Aid		TG: 60464	Snohomish
	Snohomish County FD #5	Sultan	460.5000	465.5000

SnoCom South (West) County Areas

Ch1	South Dispatch	[PL 127.3]	Countywide	154.0100	Simplex
Ch2	South Fireground	[PL 127.3]	Countywide	154.3250	Simplex

SnoPac North County Areas

Ch1	North Dispatch	[PL 127.3]	Countywide	154.4300	Simplex
Ch2	North Fireground	[PL 127.3]	Countywide	153.7700	Simplex
Ch3	Tactical	[PL 127.3]	Countywide	155.7600	Simplex
Ch4	Tactical	[PL 127.3]	Countywide	153.8150	Simplex
Ch5	Fire Mutual Aid		Statewide	153.8300	Simplex

SnoPac East County Areas

Ch1	East Dispatch	[PL 127.3]	Countywide	154.2350	Simplex
Ch2	East Tactical	[PL 127.3]	Countywide	153.7700	Simplex
Ch3	Tactical	[PL 127.3]	Countywide	155.7600	Simplex
Ch4	Tactical	[PL 127.3]	Countywide	153.8150	Simplex

Hospital Patches

H-COMMON	All Hospitals Common	Countywide	TG: 60080	Snohomish
STEVENS	Stevens Hospital	Countywide	TG: 60112	Snohomish
PROVIDENCE	Providence Hospital	Countywide	TG: 60144	Snohomish
VALLEY	Valley General Hospital	Countywide	TG: 60176	Snohomish
CASCADE	Cascade Valley Hospital	Countywide	TG: 60208	Snohomish
DARRINGTON	Darrington Clinic	Countywide	TG: 60240	Snohomish

Snohomish County cities Law Enforcement Comm (Sno-Comm)

SC POLICE 1	SnoCom PD Dispatch	Countywide	TG: 4816	Snohomish
SC POLICE 2	Lynnwood Police Dispatch	Countywide	TG: 4880	Snohomish
SC POLICE 3	Future Police Dispatch	Countywide	TG: 4944	Snohomish
SC TAC 1	SnoCom PD Car to Car	Countywide	TG: 4848	Snohomish
SC TAC 2	SnoCom PD Car to Car	Countywide	TG: 4912	Snohomish
SC ERT	Emergency Response Teams	Countywide	TG: 46192	Snohomish
SC CID	Criminal Investigations	Countywide	TG: 46256	Snohomish
SC NARC	Narcotics Enforcement	Countywide	TG: 46320	Snohomish
MAIN	Cities Police Dispatch	Countywide	TG: 2416	Snohomish
TAC	Cities Police TAC	Countywide	TG: 2448	Snohomish
ADMIN	Cities Police Administration	Countywide	TG: 2480	Snohomish
E ADMIN	Edmonds/Woodway Admin	Countywide	TG: 4976	Snohomish
L ADMIN	Lynwood Administration	Countywide	TG: 5008	Snohomish
M ADMIN	Mountlake Ter/Brier Admin	Countywide	TG: 5040	Snohomish

Tac-1	Dispatch & Operations	[PL 127.3]	Countywide	155.0250	Simplex
Tac-2	Car to Car / Alternate	[PL 127.3]	Countywide	155.8200	Simplex
Tac-3	Alternate (Night: Lynwood)	[PL 127.3]	Countywide	155.1450	Simplex
Tac-4	Law Enforcement Radio Net		Countywide	144.3700	Simplex
Tac-5	Edmonds Public Works / PD Alternate		Edmonds	158.9250	Simplex

Sno-Comm Radio Codes

10-3 Use Tac 2
10-6 Busy
10-8 Clear Traffic
10-9 Repeat Message
10-11 Out of Vehicle
10-19 Enroute
10-15 Person in Custody
10-21 Phone Call
10-22 Disregard
10-28 Registration Check
10-29 Wanted Check
10-47 On Scene
10-49 Tow Truck
10-51 Medical Examiner
10-60 Bomb Scare
10-84 Stolen Vehicle
F Felony Warrant
M Misdemeanor Warrant
Code-1 Back-up Routine
Code-2 Back-up Immediate
Code-3 Back-up Emergency
Code-4 Situation OK

Snohomish County Sheriff's Office

SAR	Search & Rescue Operations	Countywide	TG: 1616	Snohomish
SOUTH	South – Dispatch	Countywide	TG: 1648	Snohomish
SOUTH TAC	South – Tactical	Countywide	TG: 1936	Snohomish
EAST	East – Dispatch	Countywide	TG: 1680	Snohomish
EAST TAC	East – Tactical	Countywide	TG: 1968	Snohomish
NORTH	North – Dispatch	Countywide	TG: 1712	Snohomish
NORTH TAC	North – Tactical	Countywide	TG: 2000	Snohomish
DATA	Warrant Inquiries	Countywide	TG: 1776	Snohomish
TAC 3	Sheriff Training	Countywide	TG: 1840	Snohomish
NARC	Narcotics Enforcement	Countywide	TG: 46032	Snohomish
ERT	Emergency Response Team	Countywide	TG: 46064	Snohomish
CID	Criminal Investigations	Countywide	TG: 46096	Snohomish
	NORTH Operations	Countywide	158.8500	159.4650
	SOUTH Operations	Countywide	159.2100	154.8000
	EAST Operations	Countywide	159.1650	Simplex
Ch1	Tactical / Car to Car	Countywide	155.4150	Simplex
Ch5	Records Access	Everett	154.0550	Simplex

Snohomish County Sheriff Radio Codes

10-4 Message Received	10-15 Person in Custody	10-33 Warrant
10-6 Confidential Message	10-19 Return to Station	F Felony
10-8 In Service / Clear	10-21 Phone Call	M Misdemeanor
10-9 Repeater Message	10-22 Disregard	10-47 On Scene
10-10 Out of Service	10-28 Registration Check	10-98 Mental Problem

Snohomish County Government

Juvenile Justice Center	Everett	159.1275	Repeater
County Center	Everett	155.7600	Simplex

Stanwood Fire Department: See SnoPac North County

Stanwood Town Government	Stanwood	155.7450	Simplex

Stanwood Police Department: Dispatched on Snohomish County Sheriff NORTH

Tactical / Car to Car	Stanwood	154.8750	Simplex
Stilliguamish Tribal Police [PL 151.4]	Arlington	153.8150	Simplex
Sultan Town Government	Sultan	155.1000	Simplex

Sultan Police Department: See Snohomish County Sheriff EAST

Car to Car + City Public Works	Sultan	155.1000	Simplex
Tulalip Tribal Government	Marysville	453.2625	Simplex
Tulalip Tribal Police	Marysville	154.9725	155.5575

Washington State Patrol

Note: Not all WSP units or dispatchers transmit PL tone at all times

District 7: Snohomish, Skagit, Whatcom, San Juan, Island Counties

Ch13	West Snohomish County	[PL 141.3]	Everett	155.6550	154.8450
Ch17	Districtwide Car to Car	[PL 141.3]	Countywide	154.9350	Simplex
	East Snohomish County	[PL 156.7]	Skykomish	154.7550	Simplex

Woodway Police Department

See Snohomish County Law Enforcement Communications (Sno-Comm)

Snohomish County Fire Station Locations & Apparatus

(Abreviations – Ev Everett; SCS SnoCom South; SPN SnoPac North; SPE SnoPac East)

#	Department (Radio)	Station Location	Apparatus
1	Everett (Ev)	Rucker Ave	E1, M1, L1, BC1
2	Everett (Ev)	16th St	E2, A2, M2
3	Everett (Ev)	13th St	E3

4	Everett (Ev)	Glenwood Ave	E4
5	Everett (Ev)	Beverly Blvd	E5, M5
6	Everett (Ev)	Evergreen Way	E6, M6, L6, A6
7	Everett (Ev)	Silverlake Rd	E7
8	Everett (Ev)	Oakes Ave	E8
11	SCFD #1 (SCS)	Meridian Ave, Silver Lake	L11, M11, M11A
12	SCFD #1 (SCS)	108th & 35th Ave SE	E12, M12
13	SCFD #1 (SCS)	Puget Park Dr, Silver Firs	E13, M13
14	SCFD #1 (SCS)	68th Ave W, Blue Ridge	E14, A14
15	SCFD #1 (SCS)	44th Ave W, Civic Center	E15, L15, M15, A15
16	Edmonds (SCS)	196th St SW	E16, A16, U16, A16A
17	Edmonds (SCS)	5th Ave N or 275 6th Ave N	E17, A17, E17A
	Stevens Hospital (SCS)		M17
18	SCFD #1 (SCS)	228th St SW, Brier	E18, M18
19	SCFD #1 (SCS)	58th Ave W	E19, M19
20	Edmonds (SCS)	88th Ave W	E20, L20, A20, U20
21	Alderwood M (SCS)	13th Ave W, Martha Lake	E21, M21, Decon 21
22	Alderwood M (SCS)	Darnson (Danvers) Rd	E22, M22, SQ22
23	Alderwood M (SCS)	Serene Way, Lake Serene	E23, E23A, M23
24	Mukilteo (SPN)	5th St	E24, A24
25	Mukilteo (SPN)	47th Pl W	E25, A25
26	Paine Field (Ev)	109th St SW, County Airport	E26, M26, X1, X2
27	Hat Island (SPN)	Saratoga Dr	E27, WT27, A27, BR27, U27
28	NAS Everett	W Marine View Dr W	E28
29	Everett Navy Base	W Marine Dr	E29, E29a
30	Boeing Field (Ev)		E30, X30, R30, HM30, A30, E31
31	Monroe (SPE)	Village Ct	E31 WT31 L31 M31 S31 A31 BR31
32	Monroe (SPE)	132nd St SE	E32, M32
34	Arlington		E34
35	Arlington		E35
36	Arlington Heights	228th St NE	E36
37	Oso (SPN)	SR 530 NE	E37
38	Darrington (SPN)	Swede Heaven Rd	E38
39	Darrington (SPN)	Seemann St	E39, R39, A39, MSO 39
40	Snohomish (SPE)	Foster Slough Rd, Forbes Hill	E40, S40
41	Snohomish (SPE)	Maple Ave, Snohomish City	E41, A41
42	Snohomish (SPE)	171st Ave SE, Three Lakes	E42, R42, A42
43	Snohomish (SPE)	Avenue D, North of city	E43, A43, WT43, R43, S43, HM43
44	Bothell	Bothell	E44
45	Bothell	Bothell	E45
46	Arlington (SPN)	N MacLeod Ave	E46, WT46, R46, A46, Air46, E46A
47	Arlington (SPN)	188th St NE	E47, L47, A48, M48
49	Arlington Hts (SPN)	Stilaguamish Ave	E49, WT49, M49, A49
51	Sultan FD (SPE)	Alder St	E51, WT51, A51, E51A
53	Gold Bar (SPE)	Lewis Ave, downtown	E53, R53, A53, BR53
54	Gold Bar (SPE)	SR 2 SE, east	E54, A54, U54
55	Index (SPE)	Avenue A	E55
59	Boeing FD	Tulalip	E59
60	Tulalip FD (SPN)	Waterworks Rd	E60, WT60, A60, RM 22
61	Marysville (SPN)	Grove St	E61, M61, A61, A612, E61A
62	Marysville (SPN)	Marysville	E62, A62
63	Marysville (SPN)	Marysville	E63, M63
64	Lake Goodwin (SPN)	Lake Loma	E64
65	Lake Goodwin (SPN)	Lake Goodwin	E65, WT65, A65
68	Getchell (SPN)	Getchell	E68, A68, WT68, E69, WT69
71	Clearview (SPE)	Hwy 9 & 180th, Clearview	E71, M71, A71, WT71, BR71, L71
72	Clearview (SPE)	35th & 180th	E72, L72, M72, A72, SQ72
73	Clearview (SPE)	45th Ave SE & 224th St SE	E73, A73
74	Clearview (SPE)	Maltby area	E74, M74

75	Clearview (SPE)	Lost Lake	E75, A75
76	Clearview (SPE)	Mill Creek area	E76, M76
77	Clearview (SPE)	Clearview	E77, A77
81	Lake Stevens (SPN)	Lake Stevens	E81, E82, A81, WT81, A82, M82
85	Lake Roesiger (SPN)		E85, A85, E85A
87	Granite Falls (SPN)		E87, A87, WT87, U87, E87A
89	Robe Valley (SPN)	Robe Valley	E89, A89
90	Bryant (SPN)	Bryant	E90, WT90, A90, BR90
91	Bryant (SPN)	Bryant	E91, SQ91
92	Bryant (SPN)	Bryant	E92, E92A, U92
93	Silvana (SPN)	Silvana	R93
94	Silvana (SPN)	Silvana	E94, A94, WT94, U94
95	Silvana (SPN)	Silvana	E95, A95, WT95
96	Freeborn/W (SPN)	Freeborn	E96, A96, Air 96, WT96
97	Freeborn/W (SPN)	Warm Beach	E97, A97, WT97
99	Stanwood (SPN)	Stanwood	E99, M99, SQ99

Snohomish County Law Enforcement Radio Designators

X 1 #	Sheriff – North unit	Boy	Brier Police
X 2 #	Sheriff – South unit	Edward	Edmonds Police
X 3 #	Sheriff – East unit	J	Jail Wagon & Judicial Services
9 ##	Darrington Police	K	Mill Creek Police
11 ##	Granite Falls Police	Lincoln	Lynnwood Police
12 ##	Monroe Police	Mary	Motor unit
13 ##	Mukilteo Police	Nora	Narcotics Enforcement
14 ##	Stanwood Police	Nora	Animal Control (Sheriff)
15 ##	Sultan Police	Tom	Mountlake Terrace Police
10 ##	Gold Bar Police	Tom	Traffic (Sheriff)
16 ##	Arlington Police	William	Woodway Police
17 ##	Snohomish (city) Police	Z	Sheriff SWAT
19 ##	Lake Stevens Police		

Airports

Arlington Municipal Airport (AWO - Runway Titles: 11/29; 16/34))			
Common Traffic Advisory Frequency	Arlington	122.9000	AM
Seattle Center Approach / Departure	Arlington	128.5000	306.9000
Camano Island Airfield (Runways: 15/34)	Stanwood	122.9000	AM
Skagit Weather / AWOS-3	Burlington	121.1250	AM
Boeing Company	Everett	451.9375	456.9375
	Everett	461.4875	466.4875
	Everett	461.6375	466.6375
	Everett	461.9875	466.9875
Test Frequencies (Seattle & Everett)	Seattle	123.1500	AM

123.2250 123.2500 123.2750 123.3250 123.3500
123.4250 123.4750 124.5500 227.8000 252.2000
266.4000 287.2000 291.8000 292.1000 299.9500
315.0000 341.6000 349.6000 361.5000 363.3500
384.5000 397.1000

TRUNKED RADIO SYSTEM (Motorola – "Boeing"). Setting: Type II 800

These frequencies used at many sites through Northwest Washington.

853.7625* 851.5125 852.2625 853.0125 853.4625

Boeing Security	Countywide	TG: 8976	Boeing
Boeing Fire	Countywide	TG: 9104	Boeing
Boeing Fire	Countywide	TG: 28592	Boeing
Darrington Muni. Airport (Runway: 10/28)	Darrington	122.9000	AM
Firstair Field Airport (Runway Titles: 7/25)	Monroe	122.9000	AM
Flightline Services at Snohomish Airport	Everett	122.9500	AM
Operations Simplex	Everett	160.1100	Simplex
Harvey Field Airport (Runway Titles: 14/32)	Snohomish	123.0000	AM
Sky Harbor Airport (Runway Titles: 7/25)	Sultan	122.8000	AM
Snohomish Flying Service	Snohomish	464.4750	Simplex
Snohomish County Paine Field (PAE - Runway Titles: 11/29; 16L/34R; 16R/34L)			
CTAF + West Arrivals	Everett	132.9500	AM
Airport Terminal Info Service (ATIS)	Everett	128.6500	AM
Unicom	Everett	122.9500	AM
Control Tower + East Arrivals	Everett	120.2000	256.7000
Ground Control (0700-2100)	Everett	121.8000	339.8000
Clearance Delivery	Everett	126.7500	AM
Seattle Approach / Departure	Everett	128.5000	306.9000
OPS 1 Airport Operations	Everett	TG: 3056	Snohomish
OPS 2 Airport Operations	Everett	TG: 3088	Snohomish
Tower to Fire for Crash Alerts	Everett	TG: 3120	Snohomish
Snohomish County Air Reserve	Snohomish	034.1000	Simplex
	Snohomish	244.4000	AM

Miscellaneous

Name		Location	Frequency	
Alderwood Mall				
Management / Security	[DPL 311]	Lynnwood	451.8625	456.8625
GAP Store		Lynnwood	467.8500	Simplex
467.8750 467.9000		Lynnwood	467.9250	Simplex
JC Penney Store		Lynnwood	461.6125	Simplex
Macy's Department Store		Lynnwood	452.9750	457.9750
Nordstrom Store	[PL 192.8]	Lynnwood	460.9125	Simplex
Custodial & Maintenance	[PL 127.3]	Lynnwood	464.7750	469.7750
Angel of the Winds Casino	451.3625	Arlington	451.1875	Simplex
464.9875 451.5125	451.6625	Arlington	451.3125	Simplex
BNSF Railway Company		Everett	160.5000	161.5350
		Snohomish	160.7100	Simplex
Boy Scouts of America		Edmonds	151.6250	Simplex
Daily Herald Newspaper	[PL 110.9]	Everett	173.2750	Simplex
		Everett	461.2625	Simplex
461.2875 461.8125	464.8125			
Everett Golf & Country Club		Everett	464.7750	469.7750
		Everett	462.2375	Simplex
Everett Mall				
Custodial & Maintenance	[PL 103.5]	Everett	461.2875	466.2875
Security & Operations	[PL 103.5]	Everett	464.3125	469.3125
Everett Naval Station				
TRUNKED RADIO SYSTEM (Motorola): TrunkTracker Setting: Type II UHF				
Base Frequency: 406.000; Offset 25 KHz				
407.1500* 407.9500* 408.7500* 409.5500				
POLICE Dispatch		Everett	TG: 48	
POLICE Tac 3		Everett	TG: 80	
Everett Silvertips Hockey Team		Everett	452.7500	Simplex
461.4875 463.3375	452.2375	Everett	461.3875	Simplex
Evergreen Speedway		Monroe	154.5700	Simplex
154.6000 463.9625	461.8375	461.5875		
Imagine Children's Museum		Everett	463.4875	Simplex
463.9875 461.4625		Everett	461.8625	Simplex
Kayak Point Golf Course		Stanwood	461.2000	Simplex
Lynnwood Convention Center		Lynnwood	465.3375	Simplex
465.4000 465.5000		Lynnwood	465.3125	Simplex
Operations Repeater		Lynnwood	460.2125	465.2125
Nike Factory Store		Tulalip	466.0250	Simplex
		Tulalip	467.2000	Simplex
		Tulalip	467.3750	Simplex
Nordstrom Stores				
Alderwood Mall Store	[PL 192.8]	Lynnwood	460.9125	Simplex
Northgate Shopping Center		Lynnwood	464.7750	469.7750
Port of Edmonds		Edmonds	155.9850	Simplex
Port of Everett		Everett	155.8800	Simplex
		Everett	458.0875	
Red Dragon Casino		Mountlake Ter	452.6875	Simplex
451.3875 452.0875	452.1375	Mountlake Ter	452.1875	Simplex
Residence Inn Hotel		Lynnwood	464.7125	Simplex
		Lynnwood	464.8125	Simplex
Seattle Premium Outlets				

Name		Location	Frequency	Input/Mode
Mall Security		Tulalip	463.4000	468.4000
Smokey Point Motor Inn		Arlington	154.6000	Simplex
Stevens Pass Alpine Club		Edmonds	154.5700	Simplex
		Edmonds	154.6000	Simplex
Tulalip Casino				
Security	[PL 162.2]	Marysville	458.8875	Simplex
Casino Operations	[DPL 023]	Marysville	462.6125	Simplex

Medical & Hospitals

Name		Location	Frequency	Input/Mode
Everett Clinic		Everett	462.1000	467.1000
Cascade Valley Hospital		Arlington	464.7750	469.7750
LifeMed		Mill Creek	151.9550	Simplex
		Mill Creek	151.9250	Simplex
Providence General Medical Center		Everett	461.6500	466.6500
461.7500 463.5500 464.5250				
Security		Everett	157.4500	Simplex
Stevens Memorial Hospital		Edmonds	TG: 61488	King
Hospital Security	[PL 218.1]	Edmonds	463.4625	468.4625
		Edmonds	464.8250	469.8250
		Edmonds	466.5125	Simplex
		Edmonds	466.5625	Simplex

Schools

Name		Location	Frequency	Input/Mode
Arlington School District		Arlington	151.9550	Simplex
464.3250 464.4250 464.4750		Arlington	452.2875	Simplex
School Bus Operations		Arlington	155.2950	Simplex
451.4625 452.0375 452.1625		Arlington	451.3375	Simplex
Edmonds Community College		Lynnwood	463.8000	468.8000
		Lynnwood	464.3500	469.3500
Edmonds School District		Edmonds	461.3875	Simplex
464.9625 469.9625 461.9375 463.2125 467.8750				
		Edmonds	155.2050	Simplex
Eisenhower Middle School		Everett	469.5375	Simplex
469.5875 469.9125				
Emerson Elementary School		Snohomish	151.8350	Simplex
Everett Community College Security		Everett	462.4250	Simplex
Everett School District		Everett	154.6000	Simplex
154.5400 159.7950 461.6750				
Evergreen Middle School		Everett	469.3125	Simplex
		Everett	469.5875	Simplex
Lake Stevens School District		Lk. Stevens	043.0000	Simplex
461.9625 461.8375				
Lakewood School District		Lakewood	155.2650	Simplex
Marysville School District				
Allen Creek school		Marysville	151.8950	Simplex
School Bus Operations	[PL 162.2]	Marysville	151.6850	Simplex
Marysville Junior High School		Marysville	151.6550	Simplex
Monroe School District		Monroe	155.2350	Simplex

151.6850 464.5750 464.6250

Mountlake Terrace High School	Mountlake Ter.	463.2125	Simplex
	Mountlake Ter.	461.9375	Simplex
Mukilteo School District	Mukilteo	154.5400	Simplex
School Bus Operations [PL 146.2]	Mukilteo	154.5150	Simplex
Northshore School District	Bothell	155.2650	Simplex
Operations Simplex	Bothell	151.6550	Simplex

151.7150 151.8350 151.7450 151.8950 463.2625

Puget Sound Christian College	Mountlake Ter	451.0375	Simplex
	Mountlake Ter	451.1375	Simplex
Riverview School District	Monroe	461.8750	466.8750
School Bus Operations	Monroe	864.7875	819.7875
Snohomish School District			
School Bus Operations	Snohomish	155.1750	Simplex
Snohomish High School	Snohomish	461.0500	Simplex
Stanwood School District	Stanwood	854.6375	809.6375

462.1500 463.8000 462.0000 463.5000

Sultan School District	Sultan	159.7950	Simplex

Public Utilities

Alderwood Water District	Lynnwood	158.2050	Simplex
	Lynnwood	158.1900	Simplex
Everett City Utilities	Everett	451.2000	456.2000

451.2375 451.4375 456.2375 456.4375 453.7875

Highland Water Association	Monroe	151.9250	Simplex
Mukilteo Water District	Mukilteo	047.8200	Simplex
Olympic View Water District	Edmonds	045.4000	Simplex
Silverlake Water District	Everett	153.4700	Simplex
	Mill Creek	157.2650	Simplex
Stanwood City Utilities	Stanwood	153.5150	Repeater

Weather Information

National Weather Service – NOAA Weather Radio

Conditions & Forecasts – Seattle	162.5500	Simplex
Conditions & Forecasts – Puget Sound Marine	162.4250	Simplex

Snohomish County SAME Alert Code: 053061

THURSTON COUNTY

Government & Public Safety

American Medical Response		Countywide	855.1625	810.1625
Intercity Transit Authority		Countywide	453.8250	458.8250
Primary	[PL 110.9]	Countywide	453.5250	458.5250
Alternate		Countywide	460.2000	465.2000
Alternate		Countywide	460.5500	465.5500
Lacey City Government				
Public Works – Primary		Lacey	453.1625	458.1625
Public Works – Alternate		Lacey	453.4625	Simplex
458.4625 458.7125		Lacey	453.7125	Simplex
Wastewater Treatment Department		Lacey	159.9600	Simplex
Wastewater Treatment Department		Lacey	160.1250	Simplex

Lacey Fire District #3 – See Thurston County / Olympia FDs Radio

Note: Serves City of Lacey and surrounding unincorporated areas

Station Locations & Apparatus

31	1231 Franz St SE, Lacey (HQ)	E31, T31, WT31, M3, Aid 31, BC3, E312, M9, M2
32	10910 Yelm Hwy SE, Olympia	E32
33	6500 Mullen Road SE, Olympia	E33, BR33
34	8400 Steilacoom Rd SE, Olympia	E34, Aid 34, E342, M6, BR34
35	3701 Williamette Dr, Lacey	E35, BR35

Nisqually Indian Tribe	Olympia	158.7300	Simplex
Olympia City Government	Olympia	151.0025	159.1875
Olympia Search & Rescue Team	Countywide	155.2200	Simplex

Olympia Fire Department – See Thurston County / Olympia FDs Radio

Station Locations & Apparatus

1	Headquarters	100 Eastside Street NE	E1, BC10, E4, T1, M10
2	Westside	330 Kenyon Street NW	E2, M4

3 Eastside 2525 22nd Avenue SE E3

Olympic Ambulance Olympia 462.9500 467.9500

Rainier City Government Rainier 153.9800 Simplex

Tenino City Government

Thurston County / Olympia Fire Departments

Ch1	Countywide Dispatch	[PL 88.5]	Countywide	154.4300	Simplex
Ch2	Tactical – Olympia	[PL 88.5]	Olympia	154.3700	
Ch3	Tactical – Tumwater	[PL 88.5]	Tumwater	153.8000	
Ch4	RedNet – Mutual Aid	[No PL]	Countywide	153.8300	Simplex
Ch5	Tactical – Northeast Cnty	[PL 88.5]	Countywide	154.0100	150.7750
Ch6	Tactical – Southwest Cnty	[PL 88.5]	Countywide	154.1750	150.8050
Note:	County areas East of I-5 on Ch5 County areas West of I-5 on Ch6				
	Future Use – Olympia		Olympia	158.8575	155.9175
	Patient Info to St. Peter		Countywide	155.3400	Simplex

Thurston County Fire District Numbers & Area Served

1 Rochester & Grand Mound
2 Yelm
3 Lacey
4 Rainier
5 Black Lake
6 East Olympia
7 North Olympia
8 South Bay
9 McLane
10 Olympia (City)
11 Littlerock
12 Tenino
13 Griffin
14 Tumwater (City)
15 Munn Lake (contract to Tumwater FD)
16 Gibson Valley
17 Bald Hills

Thurston County / Olympia Law Enforcement 453.2750 458.2750

Ch1	Dispatch – Sheriff/Towns	[PL 88.5]	Countywide	155.5500	154.8300
Ch2	Dispatch – Cities	[PL 88.5]	Countywide	156.1500	155.0700
Ch3	Car to Car – Cities	[PL 88.5]	Countywide	155.0100	Simplex
Ch4	LERN – Interagency	[No PL]	Countywide	155.3700	Simplex
Ch5	Car to Car – Sheriff/Towns	[PL 88.5]	Countywide	154.8000	Simplex
Ch6	Records Access	[PL 88.5]	Countywide	154.8750	155.9850
Note:	Dispatch for small towns (other than Olympia, Lacey & Tumwater) is on Ch1 Dispatch for Olympia, Lacey and Tumwater is on Ch2				
	Sheriff Jeep Patrol		Countywide	155.2800	Simplex
	Correction Facility		Olympia	155.2500	153.7400
	Correction Facility		Olympia	155.0400	155.9550
	Unknown Use		Olympia	155.1450	155.8050
	Search & Rescue		Countywide	155.1600	Simplex

Thurston County Government Countywide 453.3750 458.3750

Public Works	[PL 110.9]	Countywide	453.9750	458.9750
151.1150 156.0750 158.7750		Countywide	159.0750	Simplex
Unknown Use		Countywide	155.3100	Simplex
453.2125 453.6375 458.2125		Countywide	458.6375	Simplex
Road Maintenance Department		Countywide	453.0500	Simplex
Road Maintenance Department		Countywide	458.0500	Simplex

Thurston County Explorer Search & Rescue 047.5000 Simplex
Countywide 047.5800 Simplex

Washington State Capital

Trunked Radio System

Frequencies: 856.7375 857.7375 858.7375

[#]	Auxiliary	Olympia	TG: 16
[1]	Operations – Primary	Olympia	TG: 16400
[2]	Operations – Secondary	Olympia	TG: 16432
[3]	Emergency Incidents	Olympia	TG: 16464
[4]	Parking Control	Olympia	TG: 16496

Washington State Department of Labor & Industries
Olympia 453.2625 458.2625

		Location	Freq	Freq 2
		Olympia	453.5625	458.5625
Washington State Patrol				
District 1: Pierce & Thurston Counties				
Ch4 West Pierce County Disp	[PL 151.4]	Tacoma	155.5200	Simplex
Ch11 Districtwide Car to Car	[PL 151.4]	Countywide	154.9050	Simplex
Ch19 Thurston County Dispatch	[PL 151.4]	Olympia	154.9200	Simplex
Ch31 East Pierce County Disp	[PL 151.4]	East Pierce	154.7400	Simplex
Thurston County Car-Car	[PL 151.4]	Olympia	155.4600	Simplex
Yelm Police Department		Yelm	155.4150	Simplex

Airports

	Location	Freq	Freq 2
Hoskins Field Airport	Olympia	122.9000	AM
Olympia Airport			
ATIS – Airport Information	Olympia	135.7250	AM
Ground Control	Olympia	121.6000	AM
Seattle Approach / Departure	Olympia	121.1000	AM
Tower / CTAF – Common Traffic	Olympia	124.4000	254.2500
UNICOM	Olympia	122.9500	AM
Port of Olympia	Olympia	453.2000	Simplex
453.3125 453.7625	Olympia	453.9375	Simplex
458.7625 453.8875	Olympia	458.9375	Simplex
R & K Skyranch Airport	Rochester	123.0000	AM

Miscellaneous

	Location	Freq	Freq 2
Capital Medical Center	Olympia	463.4750	468.4750
	Olympia	462.1250	Simplex
	Olympia	463.3500	Simplex
Clearwood Community Association	Yelm	464.4750	469.4750
Deschutes Animal Clinic	Olympia	047.5800	Simplex
Hawks Prairie Casino	Lacey	451.3375	Simplex
451.3625 451.4125 451.4375	Lacey	451.4625	Simplex
KIRO Television	Olympia	863.0375	818.0375
LOTT Alliance – Metro Olympia's Wastewater Treatment System			
458.3250 458.5750 458.9000	Olympia	453.1000	458.1000
	Olympia	453.6000	458.6000
Lost Lake Resort	Olympia	151.5500	Simplex
159.7950 159.8700	Olympia	160.1100	Simplex
Lucky Eagle Casino (Chehalis Indian Tribe)	Rochester	462.2000	467.2000
464.3750 464.8750	Rochester	464.5750	469.5750
468.6000 469.2000	Rochester	466.3250	Simplex
Red Wind Casino (Nisqually Indian Tribe)	Olympia	451.3125	456.3125
	Olympia	451.9625	456.9625
	Olympia	452.7125	457.7125
	Olympia	461.5000	Simplex
	Olympia	464.3000	Simplex
Providence Saint Peter Hospital	Olympia	452.8250	457.8250
	Olympia	461.3500	466.3500
	Olympia	463.2500	Simplex
	Olympia	464.1500	Simplex
Tri-City & Olympia Railroad	Olympia	160.3350	161.0500

Tumwater Municipal Golf Course	Tumwater	462.8875	Simplex
	Tumwater	467.8875	Simplex
Westfield Capital Mall	Olympia	464.4750	469.4750
461.7125 462.0375 464.3250	Olympia	464.6750	Simplex
464.7750 464.9750	Olympia	467.0375	Simplex

Schools

A G West Black Hills High School	Tumwater	151.9550	Simplex
Evergreen State College	Olympia	155.6850	Simplex
Griffin School District 324	Olympia	155.2650	Simplex
North Thurston Public Schools	Lacey	152.9000	Repeater
School Bus Operations	Lacey	155.2950	Simplex
School Bus Operations	Lacey	155.1750	Simplex
151.6850 159.6900 152.4050	Lacey	159.7200	Simplex
Olympia Public Schools	Olympia	152.3900	157.6500
	Olympia	152.9600	159.9900
151.7750 159.5250	Olympia	157.6800	Simplex
462.1375 467.1375 151.5950	Olympia	152.3900	Simplex
Rochester School District 401	Rochester	159.9750	Simplex
160.0500 160.0200 160.1400	Rochester	160.1850	Simplex
Saint Martin's University	Lacey	159.8925	Simplex
	Lacey	159.9825	Simplex
South Puget Sound Community College		155.8200	154.9950
452.3000 452.6250	Olympia	153.0200	159.7500
451.4000 452.0500	Olympia	151.8650	Simplex
Tumwater School District 22	Tumwater	159.6150	Simplex
School Bus Operations	Tumwater	157.5600	Simplex
Tumwater Middle School	Tumwater	151.7150	Simplex
Tumwater Elementary School	Tumwater	151.8950	Simplex
Tumwater High School	Tumwater	151.8050	Simplex
Yelm Community Schools	Yelm	464.6750	469.6750
	Yelm	464.7750	469.7750
	Yelm	154.5400	Simplex

Weather Information

National Weather Service – NOAA Weather Radio

Conditions & Forecasts – Capitol Peak	162.4750	Simplex
Thurston County SAME Alert Code: 053067		

WAHKIAKUM COUNTY

Cathlamet City Government		Cathlamet	453.9000	
Saint James Family Center		Cathlamet	151.8650	Simplex
Wahkiakum County Government		Skamokawa	155.9850	155.1150
		Chinook	154.8000	158.9400
Wahkiakum Fire & EMS	[PL 123.0]	Chinook	155.0700	158.8500
Wahkiakum County Sheriff		Cathlamet	460.4750	
Wahkiakum Port District 2		Cathlamet	463.8500	

Washington State Patrol

Note: Not all WSP units or dispatchers transmit PL tone at all times

District 5: Lewis, Klickitat, Cowlitz, Clark, Skamania, Wahkiakum Counties

Ch2	Cowlitz County Dispatch		Countywide	154.7700	Simplex
Ch7	Clark / Skamania County	[PL 136.5]	Countywide	154.6800	155.5800
Ch7	Klickitat / Okanogan Counties		Countywide	154.6800	Simplex
Ch14	Lewis County Dispatch		Countywide	155.8500	Simplex

Wahkiakum County Fire Department Numbering

1 Puget Island 2 Skamokawa 3 Grays River 4,5 Cathlamet

Weather Information

National Weather Service – NOAA Weather Radio

Conditions & Forecasts – Astoria	162.4000	Simplex
Wahkiakum County SAME Alert Code: 053069		

WHATCOM COUNTY

Government & Public Safety

Bellingham City Fire: See Whatcom County Fire Departments

Bellingham City Police	[PL 114.8]	Bellingham	154.0250	Simplex
Police Dispatch	[PL 91.5]	Bellingham	453.2250	458.2250
Police Tactical	[PL 114.8]	Bellingham	155.0700	

453.1125 453.3875 453.4125
453.4375 453.4875 453.7375 453.9125

Bellingham City Government		Bellingham	155.0700	

453.8000 453.6000 852.7625 852.8125

Public Works		Bellingham	854.3125	809.3125
Public Works (repeat 854.3125)	[PL 173.8]	Bellingham	453.7500	458.7500
Whatcom Transit Authority		Bellingham	453.4250	

TRUNKED RADIO SYSTEM (Unknown): TrunkTracker Setting: Unk

Not yet in use

851.1625 851.2625 851.8125 851.9125 851.9625
852.1125 852.6625 852.7125 853.4125 854.1625

Bellingham Port Police	[PL 123.0]	Bellingham	453.8750	
		Bellingham	464.4625	Simplex
		Bellingham	464.9375	Simplex

Blaine City Fire Department: See Whatcom County Fire Departments

Blaine City Government	Blaine	154.0850	

Everson Police: See Whatcom County Sheriff

Ferndale Police: See Whatcom County Sheriff

Glacier Fire & Rescue: See Whatcom County Fire Departments

Lynden City Streets	Lynden	154.0400	
Nooksack City Government	Nooksack	154.0850	155.8200

Sumas City Government			Sumas	155.8200	
United States Border Patrol - Blaine Sector					
Ch1	Galbreath Mountain	[PL 100.0]	Blaine	163.6250	Repeater
Ch2	Sumas Mountain		Blaine	163.6750	Repeater

Note: Blaine, Everson, Lynden and Sumas Police also use USBP Ch1

Radio Unit Identifiers on USBP Radio

- 1 ## Admin
- 2 ## Blaine Units
- 3 ## Bellingham Units
- 4 ## Lynden Units (Guide Meridian)
- 6 ## Sumas Police
- 8 ## Blaine Police
- 9 ## Air/Marine Units

Washington State Patrol

Note: Not all WSP units or dispatchers transmit PL tone at all times

District 7: Snohomish, Skagit, Whatcom, San Juan, Island Counties

Ch17	Districtwide Car to Car	[PL 141.3]	Countywide	154.9350	Simplex
Ch20	Whatcom County	[PL 141.3]	Bellingham	154.7550	155.2500
Western WA Univ. Police		[PL 100.0]	Bellingham	453.5500	458.5500
			Bellingham	153.9200	Simplex
Whatcom County Fire Departments			Bellingham	046.1400	
F1	Dispatch – County & City	[PL 114.8]	Countywide	154.4300	Simplex
F2	Responses – County	[PL 114.8]	Countywide	154.2200	Simplex
F3	Responses – Bellingham	[PL 114.8]	Bellingham	154.3400	Simplex
F4	Alternate	[PL 114.8]	Countywide	154.3250	Simplex
F5	Responses – Repeater	[PL 114.8]	Countywide	153.6050	160.2000
	Alternate		Countywide	154.3850	Simplex
	Fire District 3			158.8800	

TRUNKED RADIO SYSTEM (Planned, at press time): TrunkTracker Setting: Unk

Various Sites:	851.1625	851.2625	851.8125	851.9125	852.1125
	852.6625	852.7125	853.4125	854.1625	
Blaine Tower:	851.2125	852.5375	853.2875	853.6625	854.0375
	857.2625	857.7375	858.7625	859.7625	859.9875

Fire Station Locations & Apparatus

1	Bellingham	1800 Broadway	E51, M1, M2, BC1, M51, Rehab
2	Bellingham	1590 Harris St	E52, M52, SERP (HazMat)
3	Bellingham	1111 Indian St	E53, M53, R91, L41
4	Bellingham	2306 Yew St	E54, M54, L42
5	Bellingham	3314 Northwest Ave	E55, M55, SERP (HazMat)
6	Bellingham	4060 Deemer Rd	E56, M56
M4	Bellingham	858 E Smith Rd	M4
#1	Everson/Deming/Nooksack		E151, E152, A111
#2	Geneva/Sudden Vly	Lake Whatcom	E251, A211, Car 210
1	North Whatcom	Lynden	Car 301, Car 302
2	North Whatcom	Laurel	E352, WT322, U332, U382, A312
3	North Whatcom	Delta	E353, WT323, U333, U383, A313
4	North Whatcom	Northwood	E354, WT324, U334, A314
5	North Whatcom	Wiser Lake	E355, WT325, U335, A315
	North Whatcom	Point Roberts Station 1	E551, A511
	North Whatcom	Point Roberts Station 2	E552, U532
	North Whatcom	Birch Bay Station	E1351, E1361, L1341, A1311
	North Whatcom	Custer Station	E1352, WT1322, A1312
	North Whatcom	Haynie Station	E1353, WT1323, A1313
	North Whatcom	Blaine Station	E1354, E1364, A1314
	North Whatcom	Semiahmoo Station	E1355, A1315
#4	Bellingham	Bakerview / VanWyck / Agate	E451, R492, Car 410
1	Ferndale FD	2020 Washington St	E751, A711
2	Ferndale FD	4047 Brown Rd	E752, R712, WT722
3	Ferndale FD	5368 NW Rd, N Bellingham	E753, A713, WT723
4	Ferndale FD	5491 Grandview Rd, Blaine	E754, R714

5	Ferndale FD	1886 Grandview Rd	E755, M3 (Bellingham FD)
1	Marietta FD	Bennett	E851, E861, R891, A811, U832
2	Marietta FD	Home Road	E852, A812, Quad
3	Marietta FD	Curtis Road	E853
4	Marietta FD	Airport	E854, E864
5	Marietta FD	Gooseberry Point	E855, A815, U831
#11	Lummi Island	Lummi Island	E1151, WT1121, R1113, A1111
1	Sumas FD	Sumas	E1451, E1461, WT1421, A1411
2	Sumas FD	Kendall	E1452, A1412, WT1422, U1432
3	Sumas FD	Welcome	E1453, U1453, WT1423, A1413

Whatcom County Government

	Parks & Recreation Department		Blaine	453.5000	458.5000
	Flood Control Operations		Bellingham	154.1000	
	Highway Maintenance	[PL 100.0]	Alger	150.9950	156.1800
	Water District		Bellingham	153.4850	
	Juvenile Detention Center		Bellingham	458.5375	Simplex

Whatcom County Sheriff

			Countywide	453.4000	
Ch1	Sheriff Dispatch	[PL 85.4]	Countywide	453.3250	458.3250
	Tactical Simplex	[PL 100.0]	Countywide	155.6100	
	Operations Simplex		Countywide	453.4000	
Ch2	Sheriff East County		Countywide	453.6750	458.6750
	Jail Operations	[DPL 023]	Countywide	453.9750	
			Countywide	453.2500	458.2500

Whatcom County Law Enforcement Callsigns

- A Sheriff – ID Numbers
- B Bellingham PD – Administration
- D Bellingham PD – Detectives
- E Everson Police Department (+V)
- F Ferndale Police Department – IDs
- G Ferndale Police Department – Units
- H Lummi Police Department
- I Point Roberts Police Department
- J Blaine Police Department
- K Bellingham PD – Special Detail
- K9 Bellingham PD – K-9 Units
- L Lynden Police Department
- M Bellingham PD – Traffic Units
- O Bellingham PD – Off Duty Officers
- P Bellingham PD – Patrol Units
- R Bellingham PD – Reserves & DARE
- Q Sumas Police Department
- S Sheriff – Units
- S Bellingham PD – Sergeants
- V Everson Police Department (+E)
- Z Bellingham PD – Special Investigations

Airports

Bellingham International Airport

RCO (Seattle FSS)	Bellingham	122.1500	AM
UNICOM	Bellingham	122.9500	AM
Tower	Bellingham	124.9000	385.6000
Ground	Bellingham	127.4000	385.6000
Vancouver BC Approach/Departure	Bellingham	132.7000	290.8000
ATIS	Bellingham	134.4500	AM

Blaine Municipal Airport	Blaine	122.8000	AM
Floathaven Seaplane Base	Bellingham	122.9000	AM
Lynden Airport	Lynden	122.9000	AM
Point Roberts Airpark	Point Roberts	122.8000	AM
Skywest Airlines	Bellingham	460.6500	465.6500

Hotels & Resorts

Baker Lake Resort	Concrete	461.6625	Simplex
Bellingham Golf & Country Club	Bellingham	461.5125	466.5125
Best Western Lakeway Inn	Bellingham	154.5700	Simplex
Birch Bay Resort Park	Bellingham	151.7750	Simplex
	Bellingham	151.8050	Simplex
Four Winds Westward Ho	Eastsound	461.3250	
Loomis Trail Golf & Ctry Club	Bellingham	150.8600	
	Bellingham	464.8375	Simplex
Mount Baker Ski Area	Glacier	151.6850	Simplex

151.7450 151.8050 151.8650 460.9250 461.3500

TRUNKED RADIO SYSTEM (LTR): TrunkTracker Setting: LTR

153.2900 153.3950 153.4400 153.5000

Nooksack River Casino	Deming	461.0375	Simplex

461.1625 451.1875 451.2375 461.0375

461.1625 467.8500 467.8750

Paradise Lakes Country Club	Sumas	151.7750	Simplex
	Sumas	464.4750	
Silver Reef Casino	Ferndale	159.7725	Simplex

159.8175 159.8625 159.9625 160.2075 152.2275

152.3875 152.4275 157.5375 157.6425 157.6875

Semiahmoo Hotel	Blaine	461.2500	
The Leopold	Countywide	463.6875	Simplex
	Countywide	464.8250	

Miscellaneous

American Red Cross	Bellingham	047.4200	
Bellingham Mountain Rescue	Bellingham	155.1600	Simplex
	Bellingham	155.2050	Simplex
Birch Bay Water & Sewer District	Blaine	153.6500	
Cascade Natural Gas	Bellingham	048.5000	
Life Rescue Inc	Countywide	464.0375	Simplex
	Countywide	464.8875	Simplex
Northwest Washington Fair	Lynden	464.5750	469.5750

451.5000 462.2250 462.3500 462.4000 462.5250

Puget Sound Energy Inc	Bellingham	153.4100	

153.5450 153.6200 158.1450

Saint Josephs Hospital	Bellingham	154.6000	Simplex

155.3400 453.0750 457.5500 457.5750 467.7750

467.8000 467.8250 467.8500 467.9000 467.9250

Silver Reef Casino	Ferndale	159.7725	Simplex

159.8175 159.8625 159.9675 160.2075

Sisters of Saint Joseph of Peace	Bellingham	155.2800	
	Bellingham	155.3400	
	Bellingham	464.6375	Simplex
Whatcom Museum of History & Art	Bellingham	464.5000	Simplex
464.5500 469.5000	Bellingham	469.5500	Simplex
Whatcom Security Agency	East Sound	463.7250	468.7250

Retail Stores

Bed Bath & Beyond Store		Bellingham	467.9000	Simplex
Bellis Fair Mall		Bellingham	461.3000	466.3000
		Bellingham	464.6375	Simplex
Costco Warehouse Store	[PL 67.0]	Bellingham	467.7625	Simplex
JC Penney Store		Bellingham	461.6125	Simplex
Kohl's Store	[DPL 612]	Bellingham	466.3375	Simplex
Macy's Store		Bellingham	469.6750	Simplex
Office Depot Store	[PL 67.0]	Bellingham	467.8500	Simplex
Office Max Store	[PL 67.0]	Bellingham	461.0875	Simplex
Old Navy Store		Bellingham	467.8500	Simplex
Sears Store	[PL 151.4]	Bellingham	461.1875	Simplex
Target Stores				
Channel Set 1 – Ops	[DPL 026]	Bellingham	461.0375	Simplex
Channel Set 1 – Security	[DPL 065]	Bellingham	462.9125	Simplex
Channel Set 2 – Ops	[DPL 032]	Bellingham	466.2875	Simplex
Channel Set 2 – Security	[DPL 047]	Bellingham	469.4875	Simplex
Walmart Store	[PL 67.0]	Bellingham	154.5400	Simplex

Schools

Bellingham School District		Bellingham	033.4000	
Ch1 Bus Transportation		Bellingham	047.6600	
Bus Transportation – Link to 47.66		Bellingham	460.5625	Simplex
Bellingham High School Operations		Bellingham	461.0375	Simplex
Bellingham High School Operations		Bellingham	451.2875	Simplex
Bellingham High School Operations		Bellingham	451.3375	Simplex
		Bellingham	461.1125	Simplex
Ferndale School District		Ferndale	151.8050	Simplex
		Ferndale	461.3500	
Meridian Middle School		Bellingham	154.5700	Simplex
Western Washington University		Bellingham	151.6550	Simplex
Police Dispatch	[PL 100.0]	Bellingham	453.5500	458.5500
Parking Operations	[PL 118.8]	Bellingham	453.0500	
Campus Maintenance	[PL 146.2]	Bellingham	155.0400	155.9100
Operations Simplex		Bellingham	152.9300	
461.5500 464.8250 467.8375				
Whatcom Community College		Bellingham	467.3250	Simplex
463.3000 463.5750 468.5750		Bellingham	463.2000	Simplex

Weather Information

National Weather Service – NOAA Weather Radio

Conditions & Forecasts – Puget Sound Marine	162.4250	Simplex
Conditions & Forecasts – Blaine	162.5250	Simplex
Whatcom County SAME Alert Code: 053073		

OTHER FREQUENCIES

Frequencies that don't fit in any other section.

Itinerants (Color Codes & Names)

These Star & Dot frequencies are used nation-wide for virtually every low-power simplex application imaginable. If you're looking for a store, school, club, apartment, hospital, recreation facility, or other group that uses simplex handheld radios for limited-area operations, chances are you will find them on these frequencies or those list in Common Frequencies below.

Channel 2	049.8450	Black Dot	462.6250
Channel 3	049.8600	Orange Dot	462.6750
Channel 5	049.8750	Brown Dot	464.5000
Channel 6	049.8300	Yellow Dot	464.5500
Channel 7	049.8900	Channel J	467.7625
Red Dot	151.6250	Channel K	467.8125
Purple Dot	151.9550	Silver Star	467.8500
Blue Dot	154.5700	Gold Star	467.8750
Green Dot	154.6000	Red Star	467.9000
White Dot	462.5750	Blue Star	467.9250

Common Low-Power Simplex Business Frequencies

The following frequencies are also commonly used in low-power business band radios, but do not have specific titles like the frequencies above. Information in parenthesis after the frequencies in the list below indicate preset "channel" numbers for Motorola XTN, Spirit, GT, and S series radios (programmed via dip switches inside the radio). Note one set of 27 channels for VHF radios and 56 channels for UHF radios.

154.6550 (25)	151.8800 (08)	467.7875 (48)	462.8875 (28)	461.0375 (09)
154.6000 (24)	151.8200 (07)	466.3375 (46)	462.8625 (27)	467.9250 (08)
158.4075 (27)	151.7600 (06)	466.3125 (45)	462.8375 (26)	467.9000 (07)
154.5400 (23)	151.7000 (05)	466.2875 (44)	462.8125 (25)	467.8750 (06)
158.4000 (26)	151.9550 (02)	466.2625 (43)	462.7875 (24)	464.5500 (02)
154.5275 (22)	154.6000 (04)	466.2375 (42)	461.3625 (22)	462.5750
154.5150 (21)	151.6250 (01)	466.2125 (41)	462.7625 (23)	462.6250
152.7000 (19)	154.5700 (03)	466.1875 (40)	461.3375 (21)	462.6750
151.9250 (18)	151.8050	466.1625 (39)	461.3125 (20)	467.8500 (05)
154.4900 (20)	151.8350	466.1375 (38)	461.2875 (19)	464.5000 (01)
151.8950 (17)	469.5625 (56)	466.1125 (37)	461.2625 (18)	467.8125 (04)
151.8650 (16)	469.5375 (55)	466.0875 (36)	461.2375 (17)	464.3250
151.7750 (15)	469.5125 (54)	466.0625 (35)	461.2125 (16)	467.7625 (03)
151.7450 (14)	467.9125 (52)	464.5625 (33)	461.1875 (15)	464.8250
151.7150 (13)	469.4875 (53)	466.0375 (34)	461.1625 (14)	469.2625
151.6850 (12)	467.8875 (51)	464.5375 (32)	461.1375 (13)	469.5000
151.6550 (11)	467.8625 (50)	464.5125 (31)	461.1125 (12)	469.5500
151.5125 (10)	467.8375 (49)	462.9125 (29)	461.0875 (11)	
151.9400 (09)	466.3625 (47)	464.4875 (30)	461.0625 (10)	

Family Radio Service (FRS)

Ch1	462.5625	Ch2	462.5875	Ch3	462.6125	Ch4	462.6375

Ch5	462.6625	Ch8	467.5625	Ch11	467.6375	Ch14	467.7125
Ch6	462.6875	Ch9	467.5875	Ch12	467.6625		
Ch7	462.7125	Ch10	467.6125	Ch13	467.6875		

Amateur Radio

6 Meter Band	050.000 - 054.000
Simplex:	052.5250
2 Meter Band	144.000 - 148.000
Simplex:	146.460 146.480 146.500 146.520 146.540 146.560 146.580
222 MHz Band	222.000 - 225.000
Simplex:	223.5000
70 CM Band	420.000 - 450.000
Simplex:	441.000 441.025 446.000 446.025

Medical Paging

The following frequencies are generally used at hospitals large enough to require their own area paging system, although they can be acquired by any medical facility which chooses to set-up this type of system. Although intended to cover only a *very* small area (usually the facilities building or campus), they can usually be heard clearly throughout the city the hospital is in (at least a good 15 to 20 miles).

152.0075 157.4500 163.2500 453.0250 453.0750 453.1250 453.1750

Weather Radio (NOAA / National Weather Service)

These frequencies are used nationwide, and at least one can be heard in virtually every major city in the United States. Many weather radio transmitters can also be heard from smaller cities and rural areas due to mountain top locations and reasonably high power. Channel 1 (162.550) and Channel 2 (162.400) are the most common and widely used. See the County sections in NWFD for local weather radio in Oregon and Washington.

Ch1	162.550	Ch3	162.475	Ch5	162.450	Ch7	162.525
Ch2	162.400	Ch4	162.425	Ch6	162.500		

Wireless Mics

169.4450	171.1050	177.6000	184.6000	192.6000	199.6000
169.5050	171.8450	180.0000	186.0000	194.6000	202.4000
170.2450	171.9050	180.6000	186.6000	195.6000	203.4000
170.3050	174.0000	181.6000	190.6000	196.6000	204.0000
171.0450	174.6000	183.6000	192.0000	198.0000	210.0000

Retail Stores

Retail stores general have only a handful of frequencies likely to be used at any one of their stores. The following list was compiled from radio usage at retail stores throughout the country. When attempting to locate the frequency the personnel at a particular store operate on, the Common Low Power Business Frequencies and Itinerates lists above should be checked. More specifically, the following list identifies a smaller set of frequencies likely to be in use at a particular store. Common PL tones are shown, but the frequency shown is more likely to be in use than the precise PL shown, so start with the frequencies listed and no PL tone when searching.

Ace Hardware: 461.6625; 462.1375; 463.2875; 469.8625

Banana Republic: 151.835; 151.955; 154.600

Best Buy: 460.8875; 461.8375; 461.9125

Costco: 151.6250; 154.5700; 154.6000; 464.500; 451.4375; 452.1375; 452.6375; 452.8375

Crate & Barrel 467.7625; 467.875

Express 154.540; 464.550; 467.7625; 467.875; 467.900

Foot Locker: 154.570

GAP: 151.625; 151.805; 151.955; 154.515; 154.540; 154.570; 154.600

Home Depot: 461.2125; 463.2875; 464.6375

JC Penny 154.600; 463.6375; 463.7375; 466.2625; 466.9625; 467.8125

Limited: 464.550; 462.625

Lord & Taylor: 467.1375

Macys: 468.6625; 469.8125; 451.3625; 461.0375; 461.0625 461.0875; 461.1125; 461.1875 461.2375; 461.2625; 461.2875 461.3875; 461.6625; 461.7625 461.7875; 461.8125; 462.7875 463.2625; 463.2875; 463.4875 463.6375; 463.6625; 463.7625 463.9875; 464.1375; 464.2125 464.5500; 464.7125; 464.7375 464.8625; 464.8250; 466.0625 466.2875; 466.6625; 466.7875 466.8125; 467.9250; 468.2875

Mervyns:
464.5000 Ch1 Operations
464.5500 Ch2 Alternate

Nordstrom: 451.3125; 451.6625 451.7375; 452.2875; 456.3125 456.6625; 456.7375; 457.2875 460.9125; 461.5125; 463.5625 463.8375; 464.0875; 464.2625 464.6125; 468.4375; 469.0875

OfficeMax: 467.9250; 467.8500 467.8750; 467.9000

Old Navy: 151.6250; 151.7750 151.9550; 154.5700; 154.6000 464.5000; 464.5500; 467.7625

Pottery Barn: 151.6250; 151.7600; 151.9550; 154.5700 154.6000; 464.5500; 467.7625

Robinsons-May: 464.8125

Sears: 151.8350; 461.1625; 461.1875; 466.1125; 467.9000 467.9250; 469.8125

Structure: 467.7625; 467.8125

Target:
151.6250 Ch1 Operations
151.9250 Ch2 Security
151.9550 Ch2 Security
461.0375 Ch1 Operations
466.2875 Ch2 Operations
469.4875 Ch3 Security
462.9125 Ch4 Security

Toys R Us: 151.6250; 151.9550; 154.6000 457.5375; 457.5875; 468.7625; 469.6625

Walmart: 154.5700; 154.6000; 464.5000

Warner Bros Studio Store: 154.5150; 154.6000

Williams-Sonoma: 151.6250; 151.7600; 151.9550; 154.5700 464.5500; 467.8125; 467.9250

General Mobile Radio Service

The General Mobile Radio Service is much like the new Family Radio Service in that it is to be used for family and personal activities (non-business). The GMRS system allows higher power output than FRS (and therefor more expensive radios, usually $200 and up) and requires an FCC license (usually included with purchased radios).

462.5500	462.5875[1]	462.6250	462.6625[1]	462.7000
462.5625[1]	462.6000	462.6375[1]	462.6750	462.7125[1]
462.5750	462.6125[1]	462.6500	462.6875[1]	462.7250

[1] These frequencies are limited to 2 watts simplex

Toy Walkie Talkies

The following are the most common frequencies used by toy walkie talkies sold at Radio Shack and most toy stores.

049.8300	049.8450	049.8600	049.8750	049.8900

Police, Fire, & Medical Interagency Frequencies

The following are frequencies used by various police, fire, and medical agencies throughout the United States to communication with one another.

Name	Freq.	Name	Freq.
VHF (V-CALL)	155.7525	800 MHz (ITAC-5)	868.5125
VHF (VTAC-1 Simplex)	151.1375	Fire Mutual Aid Radio	154.2800
VHF (VTAC-2 Simplex)	154.4525	EMSCOM Ch1 Base	463.0000[3]
VHF (VTAC-3 Simplex)	158.7375	EMSCOM Ch1 Mobiles	468.0000[3]
VHF (VTAC-4 Simplex)	159.4725	EMSCOM Ch2 Base	463.0250[3]
VHF (RTAC-1 Repeater)	161.8500	EMSCOM Ch2 Mobiles	468.0250[3]
VHF (RTAC-1 Input)	157.2500	EMSCOM Ch3 Base	463.0500[3]
VHF (RTAC-2 Repeater)	161.8250	EMSCOM Ch3 Mobiles	468.0500[3]
VHF (RTAC-2 Input)	157.2250	EMSCOM Ch4 Base	463.0750[3]
VHF (RTAC-3 Repeater)	161.8750	EMSCOM Ch4 Mobiles	468.0750[3]
VHF (RTAC-3 Input)	157.2750	EMSCOM Ch5 Repeater	463.1000[3]
UHF (U-CALL)	453.2125	EMSCOM Ch6 Base	463.1250[3]
UHF (UTAC-1)	453.4625	EMSCOM Ch6 Mobiles	468.1250[3]
UHF (UTAC-2)	453.7125	EMSCOM Ch7 Base	463.1500[3]
UHF (UTAC-3)	453.8625	EMSCOM Ch7 Mobiles	468.1500[3]
800 MHz (I-CALL)	866.0125	EMSCOM Ch8 Base	463.1750[3]
800 MHz (ITAC-1)	866.5125	EMSCOM Ch8 Mobiles	468.1750[3]
800 MHz (ITAC-2)	867.0125	EMSCOM Ch9 Repeater	462.9500[3]
800 MHz (ITAC-3)	867.5125	EMSCOM Ch10 Repeater	462.9750[3]
800 MHz (ITAC-4)	868.0125		

[2] Designated for agencies using only 800 MHz conventional & trunked radio systems.
[3] Used statewide for ambulance to hospital communications.

Phonetic Alphabet

APCO Phonetic Alphabet

A	Adam	G	George	M	Mary	S	Sam	Y	Young
B	Boy	H	Henry	N	Nora	T	Tom	Z	Zebra
C	Charles	I	Ida	O	Ocean	U	Union		
D	David	J	John	P	Paul	V	Victor		
E	Edward	K	King	Q	Queen	W	William		
F	Frank	L	Lincoln	R	Robert	X	X-Ray		

International Phonetic Alphabet

A	Alpha	G	Golf	M	Mike	S	Sierra	Y	Yankee
B	Bravo	H	Hotel	N	November	T	Tango	Z	Zulu
C	Charlie	I	India	O	Oscar	U	Uniform		
D	Delta	J	Juliet	P	Papa	V	Victor		
E	Echo	K	Kilo	Q	Quebec	W	Whiskey		
F	Foxtrot	L	Lima	R	Romeo	X	X-Ray		

CTCSS ("PL Tone" by Motorola) Frequencies

Name Freq.										
	XZ	67.0	ZA	71.9	WA	74.4	XB	77.0	SP	79.7

		Name	Freq.	Name	Freq.	Name	Freq.	Name	Freq.	Name	Freq.
YZ	82.5	YA	85.4	1A	103.5	3A	127.3	5A	156.7	7A	192.8
		YB	88.5	1B	107.2	3B	131.8	5B	162.2	M1	203.5
		ZZ	91.5	2Z	110.9	4Z	136.5	6Z	167.9		
		ZA	94.8	2A	114.8	4A	141.3	6A	173.8		
		ZB	97.4	2B	118.8	4B	146.2	6B	179.9		
		1Z	100.0	3Z	123.0	5Z	151.4	7Z	186.2		

Civil Air Patrol

Air to Air	123.4500	Repeater – Input	143.9000
Air to Air	123.4750	Repeater – Output	143.9250
Air to Air	123.1000	Repeater – Input	143.7500
Air to Air	122.9000	Operations Repeater	143.7500
Repeater – Output	148.1500	Operations Repeater	143.9000

Various Air Show Frequencies

The following is a brief list of a portion of the frequencies that may be in use at various air shows state-wide.

Air Force Thunderbirds

Air to Air (Primary)	143.850	Ground Support	413.075
Air to Air Solos (Planes 5 & 6)	236.550	Ground Support	413.100
Air to Air Solos (Planes 5 & 6)	241.400	Ground Support	413.150
Air to Air Solos (Planes 5 & 6)	273.500	Ground Support	413.275
Air to Air Solos (Planes 5 & 6)	283.500	Ground Support	413.350
Air to Air Solos (Planes 5 & 6)	322.950	Ground Support	413.400
Air to Air	140.000	Maintenance	148.550
Air to Air	140.400	Miscellaneous Use	140.000
Air to Air	141.850	Miscellaneous Use	148.175
Air to Air	236.600	Miscellaneous Use	283.500
Air to Air	294.700	Miscellaneous Use	319.700
Air to Air	322.300	Miscellaneous Use	392.800
Air to Air	322.600	Miscellaneous Use	394.000
Air to Air	382.900	Show Operations	120.450
Air to Air	383.900	Team Leader Command (Pri.)	250.850
Air to Air	394.000	Team Leader Command	235.250
Ground Operations (F-1)	413.100	Victor (F1)	141.850
Ground Operations (F-2)	413.025	Victor (F2)	143.850
Ground Support	413.025		

Air Show Common Control Frequency	123.400
Air Show Common Control Frequency	123.450
Air Show Stunt Coordination	120.450
Air Show Time Control	126.400

Canadian Snowbirds

Ch1		275.800	Ch8		266.300	Ch15	236.600
Ch2		295.600	Ch9		294.500	Ch16	283.900
Ch3		310.800	Ch10		322.800	Ch17	363.800
Ch4		227.600	Ch11	Primary	245.700	Ch18	289.400
Ch5	Primary	243.400	Ch12		316.500	Ch19	245.000
Ch6	Primary	240.500	Ch13		344.500	Ch20	239.800
Ch7		378.500	Ch14		356.600		

U.S. Navy - Blue Angels

Air to Air (Primary)	275.350
Air to Air (Primary)	251.600
Air to Air	241.400
Air to Air	250.800
Air to Air	395.900
Air to Air	360.400
Air to Air	384.400
Air to Air	391.900
Solos (Ch8 Planes 5 & 6)	251.600
Air to Air (Ch16 All 6 Aircraft)	307.700
Air to Air (Ch17 All 6 Aircraft)	302.150
Air to Air (Diamond Formation)	275.350
Air to Air Solos	246.400
Air to Air (Ch19 Planes 5 & 6)	251.800
Ground Support	142.025
Maintenance (Ch10)	143.600
Maintenance (Alpha)	142.000
Maintenance (Bravo)	143.000
Maintenance (Charlie)	141.560
Maintenance (Delta)	142.025
Maintenance (Delta)	142.625
Flightline Common Maint.	263.350
Refueling	121.900
Team Operations	142.000
Team Operations	142.265
Team Operations	143.000

U.S. Army Golden Knights

Air to Ground	123.000
Air to Ground	123.400
Air to Ground	123.450
Air to Ground	123.475
Air to Ground	123.500
Operations	032.300
Operations (Primary)	042.350

Marine Radio (United States)

Ch#	Transmit	Receive	Use
01	156.050	160.650	Ship to Shore - Telephone
02	156.100	160.700	Ship to Shore - Telephone
03	156.150	160.750	Ship to Shore - Telephone
04	156.200	160.800	Ship to Shore - Telephone
05	156.250	160.850	Ship to Shore - Telephone
06	156.300		Intership - Safety
07	156.350		Intership - Ship to Shore - Commercial (WA: Foss Tuggs)
08	156.400		Intership - Commercial
09	156.450		Intership - Ship to Shore (WA: Pleasure Boats)
10	156.500		Intership - Ship to Shore - Commercial (WA: Red Stack Tugs)
11	156.550		Vessel Traffic Management
12	156.600		Vessel Traffic Management (WA: Navy Bangor Operations)
13	156.650		Bridge to Bridge - 1 watt
14	156.700		Vessel Traffic Management (WA: Puget Sound, South of Whidbey)
15	156.750		EPIRB Buoy
16	156.800		Distress / Safety / Calling
17	156.850		Pilotage - Vessel Docking / Manoeuvers
18	156.900		Intership - Ship to Shore - Commercial (WA: Vessel Assistance)
19	156.950		Port Operation (WA: Argosy Tour Boats)
20	157.000		Port Operation
21	157.050		US Coast Guard - Authorized Stations (WA: Air Operations)
22	157.100		US / Canadian Coast Guard - Public Working (WA: Surface Ops)
23	157.150		Port Operation (USCG)
24	157.200	161.800	Ship to Shore - Telephone
25	157.250	161.850	Ship to Shore - Telephone
26	157.300	161.900	Ship to Shore - Telephone
27	157.350	161.950	Ship to Shore - Telephone
28	157.400	162.000	Ship to Shore - Telephone
60	156.025	160.625	Ship to Shore - Telephone
61	156.075		Intership - Ship to Shore
62	156.125		Intership - Ship to Shore
63	156.175	160.775	Ship to Shore - Telephone
64	156.225	160.825	Ship to Shore - Telephone
65	156.275	160.875	Ship to Shore - Telephone
66	156.325	160.925	Ship to Shore - Telephone / Port Operation
67	156.375		Intership - Ship to Shore
68	156.425		Intership - Ship to Shore - Non-Commercial (WA: Pleasure Boats)
69	156.475		Intership - Ship to Shore (WA: Pleasure Boats)
70	156.525		Digital Selective Calling - Distress and Safety
71	156.575		Vessel Traffic Management
72	156.625		Intership (WA: Pleasure Boats)
73	156.675		Intership - Ship to Shore (WA: Harbor Tug Operations)
74	156.725		Vessel Traffic Management (WA: Harbor Tug Operations)
77	156.875		Pilotage - Vessel Docking / Maneuvers (WA: Harbor Tug Ops)
78	156.925		Intership - Ship to Shore - Commercial
79	156.975		Intership - Ship to Shore - Commercial (WA: State Ferries)
80	157.025		Intership - Ship to Shore - Commercial
81	157.075		Port Operation (USCG) (WA: Special Operations, i.e. SeaFair)
82	157.125		Port Operation (USCG)
83	157.225		Intership / Port Operation (WA: Special Operations, i.e. SeaFair)
84	157.225	161.825	Ship to Shore - Telephone
85	157.275	161.875	Ship to Shore - Telephone
86	157.325	161.925	Ship to Shore - Telephone
87	157.375	161.975	Ship to Shore - Telephone
88	157.425		Intership

INDEX